EXTERIOR HOME IMPROVEMENT COSTS
Updated
Eighth Edition

The practical pricing guide
for Homeowners
& Contractors

RSMeans

EXTERIOR HOME IMPROVEMENT COSTS

Updated
Eighth Edition

The practical pricing guide
for Homeowners
& Contractors

RS**Means**

Copyright 2002
Construction Publishers & Consultants
63 Smiths Lane
Kingston, MA 02364-0800
(781) 422-5000

The editors for this book were Barbara Balboni and Andrea Keenan. The managing editor was Mary Greene. The production manager was Michael Kokernak. The production coordinator was Marion Schofield. The electronic publishing specialist was Paula Reale-Camelio. The proofreader was Robin Richardson. The book and cover were designed by Norman R. Forgit. Cover photographs by Norman R. Forgit. Illustrations by Carl Linde, Richard W. Lowrey, R.A, Robert Megerdichian & Associates, and Barbara Balboni.

Printed in the United States of America

10   9   8   7   6   5   4   3

Library of Congress Catalog Number 2002511398

ISBN 0-87629-657-6

Reed Construction Data

TABLE OF CONTENTS

TABLE OF CONTENTS *Continued*

Acknowledgments

The editors would like to express appreciation to Scot Simpson for allowing us to reprint selected terms from the glossary of his book, *Builder's Essentials: Framing and Rough Carpentry*.

The "Questions for the Building Department" and "Contractor Evaluation Sheet" forms in this book were provided by Richard Connolly, President of Cornerstone Consulting, Inc., Weymouth, Massachusetts. Cornerstone Consulting assists consumers in identifying and hiring designers, contractors and subcontractors. The company also helps homeowners develop specifications and negotiate bids, and in many cases, coordinates the overall project, whether new construction or remodeling.

INTRODUCTION

This book is designed to take some of the mystery out of exterior projects, both by helping you understand what is involved – in terms of work, materials, and expertise – and by giving you an idea of what it all costs.

This revised eighth edition includes 64 projects – with current cost information and building practices.

Project-by-project, material and labor costs are listed, along with the hours required for installation. The discussion that accompanies each project will help you determine which parts of the job are appropriate for your level of expertise, and which are better left to a contractor. If you are thinking about doing the work yourself in order to save money, consider how much you will really save. If your work has to be corrected later by a contractor, you may actually spend much more in the long run.

In each project in this book, material quantities are those required in a standard renovation. The contractor's charge includes an estimate of overhead and profit. Overhead includes both direct and indirect costs of doing the work. This includes the contractor's site visits, planning, estimating time, and the miscellaneous costs associated with doing business.

A section called "Adjusting Project Costs to Your Location," found at the back of the book, provides cost factors for over 900 cities and towns, organized alphabetically and by zip code. Multiply the project cost by the factor for your location to get the most accurate estimate.

Exterior projects share certain common features and requirements, such as preparation of the site. Shrubbery, ground cover, and lawns may need to be temporarily removed or protected before starting a project.

Doing your own work can save you money and give you the satisfaction that comes from creating something of value. But not everyone can or should undertake every task. Experienced do-it-yourselfers can handle many basic tasks, but beginners may have to spend some extra money for professional help. Correction of structural problems, particularly, should be left to qualified tradespeople and experts. Also, remember that most renovation projects require strength and a good back. If you don't have these attributes, it's probably best to leave major projects to a professional.

Before undertaking a project that will require structural changes (that is, the removal or alteration of a supporting wall that will affect the strength of the structure), consult an architect/engineer or local building official. The proposed changes may require a permit and plans approved or certified by a licensed engineer or architect.

A permit is essentially a license to do the work. Depending on the scope of the project, a permit application may require a description of the property, a drawing of the proposed changes, and a site plan drawing. Most jobs that require the assistance of licensed professionals also require a permit.

A permit generally has a fee attached. Regulatory agencies are established to enforce minimum code requirements and to promote public safety, and permit fees are a small investment compared with the penalties and problems that can arise if this part of the construction process is ignored.

When you obtain a permit, ask about the inspection schedule. An inspection is generally required to ensure that the work was done in accordance with code requirements. Contractors generally take care of the permit and arrange for the inspection.

Exterior remodeling projects can be a bit tricky, especially in older houses where the need to match existing trim and architectural details may require purchasing special materials or architectural components. The projects in this book include the cost of basic carpentry, plumbing, and electrical work. Architectural details and milling costs have not been included. Difficult access or conditions have not been included, and additional costs should be considered before completing your budget. Refer to the "Details" section of this book for help in determining these charges.

Scheduling is a major consideration, especially in landscaping and exterior home improvement projects. If you intend to do the work yourself, determine whether you really have the time to focus on the project and get it done quickly. If the project will take a total of eight days, and you have only weekends to devote to the work, it will really be four weeks before it is complete. Keep in mind that ideal conditions for seeding, exterior painting, reroofing, and so forth are directly related to the time of year, which may be a relatively short period. Again, while hiring a contractor increases your project cost, it may be worth it when compared to the problems that may arise because of poor timing or the inconvenience of a long-term project.

Another consideration is delivery and storage of materials. Plan how you will transport the materials to your house. Most manufacturers and suppliers will recommend proper storage and care of the product prior to installation. In some cases, these recommendations affect the warranty of the product. It may be worth planning to arrange deliveries for a time when you are actually ready to install the materials. Alternatively, you might purchase tarps or other weather protection for stored items.

Finally, consider how the completed work will affect the resale of your home. If you think you'll sell your home within the next few years, weigh any remodeling investments carefully. Avoid spending more than you can get back at sale time. Also, remember that most lending institutions expect homes they finance to meet code requirements. In your desire for a beautiful enhancement to your home, don't neglect environmental and safety issues.

To ensure that your improvements will appeal to future buyers, use fairly neutral colors, particularly on items that are costly to change, such as bathroom fixtures, tile, walls and floors, and countertops. Use quality materials that wear well and look good; and plan the improvements to blend in with the style and character of your home.

The following suggestions are from professional contractors who have learned by experience the most efficient ways to approach remodeling projects. Those who take the time to read this section improve the odds for starting their projects in a more organized manner, enjoying the experience, and winding up with a result they can be proud of.

Before getting started, establish a time frame for the project. Clearly, the season of year and your own needs and convenience have a great deal to do with the time period chosen for certain projects. Major projects can often be done in phases, which can help to break up the period of disruption; but this approach may also require some rearranging of your living space or interference with the use of your yard.

When working on home improvements, try to plan your day so as to be finished at a reasonable hour. Most outdoor tasks can be accomplished safely and effectively only when natural light is available, and you may have to allow some time to clean up, restack materials, and so forth to eliminate hazards or obstacles.

If your work will involve disrupting water, gas, or electric service, make sure it takes place at a time when you can arrange for emergency repair or professional assistance. In other words, do not choose a holiday weekend or a Sunday afternoon. In addition, if water supplies will be disconnected, be sure to fill several pails for emergency use.

Do not depend on supply houses (for plumbing, electrical, rental equipment, or some landscape products) to be open on Saturday afternoons and Sunday mornings. Call in advance to avoid frustration.

Check your local trash pickup or dump rules and schedule. Make sure you understand the policy on what refuse materials are acceptable for the local dumping station or landfill.

The motivation for most of us tackling a home improvement project on our own is to save money by investing our time. Be careful about trying to save time as well. In the rush to get started, many homeowners have broken windows, scratched finishes, and even damaged their new materials with improper handling, storage, or installation methods.

Assess your collection of equipment and tools, and make a list of what you will need. Research the best brands and deals. Call around for the best price, and keep your eyes open for a tool or material on sale, or start purchasing materials in advance to save money or to get an early start on the project. (A convenient worksheet is included at the back of this book to help you list and price individual materials.) If you are looking for bargains, make sure the product is complete and of acceptable quality, and that it can still be serviced. Keep in mind that the better the quality or grade of material, the easier it is, generally, to work with. Also, prior to purchasing an item, check on any delivery charges or additional fees for special treatments that may increase the actual cost of the item.

Read and follow the instructions accompanying tools, materials, appliances, and fixtures. If the can of paint says, "let the first coat dry for eight hours before applying the second coat," wait eight hours. Sometimes we are in such a rush to step back and enjoy the fruits of our labor that we eliminate steps such as the recommended sanding, final coat, or buffing. Overcome the urge, and you will be thankful later. The work may take longer, but it will last and look better.

An addition may add to your home's square footage. Keep in mind that only areas with a ceiling height of six feet or more in a 1-1/2 story residence are considered living area.

There is no greater satisfaction than having accomplished a major project ourselves, especially when it comes to a home improvement. But, if we all could do it ourselves, the professionals would be out of business. Recognize your limitations; plan your project and evaluate the impact of the construction on your family's comfort; read the safety section that follows; seek expert advice; and most importantly, enjoy the experience.

Keep a journal of your project, and track the costs and time involved. Photograph the steps along the way, especially for major, time-consuming projects. This will give you a record of the "event," as well as a document that may be useful for future work on your home or for resale.

Remodeling the exterior of your home presents some definite challenges, but it also promises rewards that are realized

daily in the improved appearance, function, and value of your house. Being aware of what is involved in a project and knowing the costs in advance will help you to plan the best renovation.

All of the projects covered in this book deal with work that will affect the exterior appearance of your home or enhance your yard. In some cases, the purpose of the exterior improvement is to create new interior space for your home, and these projects are complete in terms of both exterior and interior finishes. If you are interested in refinishing an existing room or building in creative storage space, consult *Interior Home Improvement Costs*. It contains 66 projects, ranging from bathrooms, kitchens, and home offices to a variety of closets, painting, and flooring.

Note: Whether you are hiring a contractor or doing the work yourself, it is important to have a quality standard for workmanship in mind. We recommend *Residential & Light Commercial Construction Standards*, a publication that defines quality in construction, to help avoid disputes with your contractor and to answer installation questions with authority. The book is based on building code requirements and guidance from leading professional associations, product manufacturer institutes, and other recognized experts. The book is available at the contractor desk of many major home improvement centers and bookstores, or can be ordered from R.S. Means Co. at 1-800-334-3509 or www.rsmeans.com

COST VERSUS VALUE

When you decide to undertake any home remodeling project, it's important to consider how that improvement will add to the value of your property. How much of the cost of remodeling will be recouped if you sell your house?

Traditionally, updating a kitchen and adding a bathroom have been the projects that have the highest resale value. Improvements such as these can boost a home's resale value by more than they cost as soon as one year after the job is completed. Real estate agents say that a minor kitchen remodel (average cost: $9,000) will generally recoup 98% of the cost. A major kitchen remodel (average cost: $22,000) will recoup about 85% of the cost. A bathroom addition (about $13,000) can regain 89% of its cost at sale time. The percentage varies depending on where you live. Other improvements that are almost as valuable as a kitchen remodeling or bathroom addition include adding a family room or master bedroom suite. As the number of people who work at home continues to grow, the home office will become an increasingly valuable addition as well.

Remodeled homes sell better and faster in today's real estate market. Older homes can be remodeled to include the same features that are standard in new homes. However, don't "overimprove" your house – by making it considerably more elaborate than the other homes in your immediate neighborhood. When you add to your home to make it conform to the neighborhood, the cost of the improvement is easier to regain at sale time.

Quality is important. Use good quality products and make sure the job is well done. If you are not adept at carpentry, hire a professional. A poorly executed remodeling job will not add value to your home.

Don't base all remodeling decisions on resale value, however. You need to live in your house, and it may be several years before you even think of selling it. Your project should be primarily for your own convenience and enjoyment. Also, if you're trying to decide whether to move or improve, weigh the cost and value against the cost of selling and moving into another house.

Because we spend so much time in our homes, we tend to overlook some of the very basic cosmetics that make a home appealing: paint, wallpaper, and other finishes that are especially subject to everyday wear and tear.

Upgrading these items can make a tremendous difference in the saleability of a house, with a modest investment by the do-it-yourselfer.

Prior to undertaking any project, you should contact your local building department to clarify the need for permits or licensed contractors to do the work. Later on when you wish to sell the property, this could be a very important issue.

You might also want to talk to real estate professionals in your area or check newspapers and web sites for current market trends. This may help you determine whether a proposed investment in home improvements can be recovered.

SAFETY TIPS

Each project description in this text features a "What To Watch Out For" section, which often includes safety tips relevant to that project. You must be prepared to do your work safely and meet local building codes. If you don't think you can attain these standards, you're better off hiring a licensed professional. Following are some additional recommendations.

- Make a point of keeping the work area neat and organized. Eliminate tripping hazards and clutter, especially in the areas used for access. Take the time to clean up and reorganize as you go along. This means not only keeping neat piles of materials, but also remembering not to leave nails sticking out; pull them or bend them over. You may be called away or run into difficulties that delay the project, and you should not leave a hazard.

- Wear the proper clothing and gear to protect yourself from possible hazards.

- Avoid loose or torn clothing if you are working with power tools. Serious injury can be caused if clothing is caught in moving parts.

- Wear heavy shoes or boots to protect your feet, and a hardhat in any situation where materials or tools could fall on your head.

- Wear safety glasses when working with power tools or in any other circumstance where there is the potential for injury to the eyes.

- Use hearing protection when operating loud machinery or when hammering in a small, enclosed space.

- Wear a dust mask to protect yourself from inhaling sawdust, insulation fibers, or other airborne particles.

- Wear suitable gloves whenever possible to minimize hand injuries.

- When lifting, always try to let your leg muscles do the work, not your back. Keep your back straight, your chin tucked in, and your stomach pulled in. Maintain the same posture when setting an item down. Seek assistance when moving heavy or awkward objects and, remember, if an object is on wheels, it is easier to push than to pull it.

- When working from a ladder, scaffold, or any temporary platform, make sure it is stable and well braced. When walking on joists, trusses, or rafters, always watch each step to see that what you are stepping on is secure.

- When working with adhesives, protective coatings, or other volatile products, be sure to follow manufacturers' recommendations on ventilation. Pay particular attention to drying times and fire hazards associated with the product. If possible, obtain from your supplier a Material Safety Data Sheet, which will clearly describe any associated hazards.

- Do not use or store flammable liquids or use gasoline-fueled tools or equipment inside of a building or enclosed area.

- When working with electricity or gas, be sure you know how to shut off the supply; then make sure it is off in a safe way by testing the outlet fixture or equipment. It may be wise to invest in a simple current-testing device to determine when electric current is present. Be sure you know how to use it properly. If you don't already have one, purchase a fire extinguisher, learn how to use it, and keep it handy. Have emergency telephone numbers and utility telephone numbers at hand.

- When using power tools, never pin back safety guards. Choose the correct cutting blade for the material you are using. Keep children or bystanders away from the work area, and never interrupt someone using a power tool or actively performing an operation. Always unplug tools when leaving them unattended or when servicing or changing blades.

A few tips on hand tools:

- Do not use any tool for a purpose other than the one for which it was designed. In other words, do not use a screwdriver as a pry bar, pliers as a hammer, etc. Not only can the tools be easily ruined, but the impact that causes the damage may also injure the user.

- Do not use any striking tool (such as a hammer or sledgehammer) that has dents, cracks, or chips; shows excessive wear; or has a damaged or loose handle. Also, do not strike a hammer with another hammer in an attempt to remove a stubborn nail, get at an awkward spot, etc. Do not strike hard objects like concrete or steel, which could chip the hammer, causing personal injury.

- Do not hold an item in one hand while using a screwdriver on it. One slip and your hand is wounded. Do not use a screwdriver for electrical testing or near a live wire.

- If you rent or borrow tools and equipment, take time to read the instructions or have an experienced person demonstrate proper usage.

Seek further advice on proper tool selection and use from your local building supply dealer, or from the Hand Tools Institute at 25 N. Broadway, Tarrytown, NY 10591 or http://www.hti.org. Always review manufacturers' instructions and warnings.

Home remodeling can be a satisfying and rewarding experience. Proper planning, common sense, and good safety practices go a long way to ensure a positive, money-saving experience. Take your time, know your limitations, get some good advice, and have fun.

SPECIAL CONSIDERATIONS FOR REMODELING AN OLDER HOME

There is something special about working on an old house. In some regions of the country, it may seem like you don't really own your house until you move away from it: "The Smiths lived in the old Haywood house until they sold it to the Joneses, who now live in the old Taylor house." You can get the sense that any work you do should meet with the approval of the original owners. And what about the craftsmen who first created the home? When you work on an older home, you cannot help but feel a kinship with them. Often, you will find a signature or mark that identifies the craftsperson. In some instances, you may recognize a detail that is associated with a particular builder. True craftsmen respect the work of those who came before them, and good carpenters, masons, plumbers, and electricians are remembered for generations.

People who own old homes usually discover, at some point, that upkeep and routine maintenance generally cost more than work on newer homes. The satisfaction of preserving the home's personality – shaped by its designer, builder, and occupants – needs to outweigh the costs. Many levels of commitment are necessary to maintain the old structure exactly as it was. The same dedication is needed when you are introducing modern conveniences in a way that maintains the original character while providing an easier-to-maintain structure. You will need to decide how "pure" you want to be and to what level you wish to preserve your home's original features.

Houses constructed prior to 1940 have unique characteristics not seen in homes built today. From foundation to finishes, the differences can be significant. The casual observer might not recognize these features, but knowledgeable homeowners and building professionals can make a

long list. Consider the following conditions as you contemplate a remodeling project for your old house.

Old houses are often built with true dimensional lumber. This means that a 2 x 4 is actually 2" x 4". Whereas, the actual dimensions of a 2 x 4 purchased today are 1-1/2" x 3-1/2". This difference in dimensions is consistent for all framing lumber produced today. In old houses, the boards used for both rough and finished carpentry may be a full 1" thick, compared to stock used today that is actually 3/4" thick. If you are pursuing an exact restoration, you will need to find a mill producing the exact dimensional lumber you require. If exact reproduction is not important, then you must be prepared to introduce shims to build up narrower new members to match what is already in place. In cases where new members (both cosmetic and structural) will be exposed, true dimensional lumber will need to be used, and it can be costly.

Some older homes are balloon-framed. With this style of wood framing, the vertical structural members (the posts and studs) are continuous pieces from sill to roof plate. The intermediate floor joists are supported by ledger boards spiked to, or let into, the studs. Temporary bracing and floor support can be critical to working on both interior and exterior walls that have been balloon framed. Adding new wires and utilities may actually be easier in a balloon-framed wall, because the studs run uninterrupted from sill to roof plate, with the exception of firestops.

Old houses usually have higher ceilings that require longer lengths of lumber for wall framing. If the walls are over 8' high, you may incur the cost of greater waste when purchasing wall finish materials. You may also find that the rooms are smaller, and that the partitions are structural (load-bearing), requiring

more material when reframing. In some cases, the original builder (or a subsequent remodeler) may have placed additional blocking or bracing within the wall structure. This may obstruct the placement of new pipes or wiring, which could lead to additional demolition and reframing.

Doors and windows are often unique shapes and sizes. Replacement may involve the services of a custom millwork company. Reworking old doors and sash can be extremely time consuming. There may also be a long lead time for ordering custom units. The different dimensions of framing members may result in varying wall thicknesses, which means stock door frames. Window frames may require extension jambs, adding to the cost and time of construction.

Door and window frames may be part of the structural components that make up a typical wall framing system and may require greater care during removal. Similarly, cabinets and bookcases are often built in place. Removing them may necessitate a fair amount of patching and matching of finishes. Lath and plaster in older homes is generally thicker than drywall used today. It can be difficult to patch this wall covering so that it matches existing surfaces, if you want to achieve a perfect finish. Often, it is necessary to reconstruct and finish an entire wall rather than patch a small area.

To match interior and exterior mouldings, you may need to have the material custom-milled. When removing existing mouldings and trim, try to determine the nailing pattern and fastening method that was used originally. Take care and try to save as much of the material as possible. Even small pieces may come in handy for patching and replacing small sections of trim, siding, mouldings or other details.

When you are trying to match existing materials, be sure to remove all paint, to establish true dimensions.

Exterior trim replacement or matching of new to existing trim can sometimes be almost impossible. Many details become obscured or disappear under numerous coats of paint. Hand-planed mouldings, and those of the same shape but different size, may be costly to reproduce. Before you consider eliminating mouldings or details, take a good look at what the finished product will look like without them. Many stock mouldings can be used in combination to produce a detail that is very similar to an original design.

When rebuilding cornices and overhangs, it is a good idea to dismantle them carefully to understand how they were put together, as well as to make patterns from the original pieces. Carefully document the lengths of overhangs and flashing details so that you can reproduce the look that is so pleasing and common to older homes. Taking things apart carefully offers another advantage, too. You will be able to fabricate parts and do some of the assembly on the ground, rather than on a scaffold. We recommend priming and applying first coats of paint prior to attaching any exterior trim or finish materials.

If your renovation involves replacing siding, it is a good idea to mark the spacing of the old siding on the cornerboards. Then, you can apply the new material to match the exact layout of the existing siding.

Many companies specialize in reproducing building components, lighting fixtures, bath and kitchen fixtures, and cabinetry. Often, they can supply almost anything else you think of that would complete your renovation in keeping with the period. Using reproductions can help you satisfy current building codes more easily, while ensuring the safety of the occupants.

There is no reason you should avoid tackling any of the projects outlined in this guide. However, owners of older homes should consider working with professionals to address structural concerns, fireplace and chimney rebuilding, electrical, plumbing, and abatement of hazardous materials (such as lead paint or asbestos pipe covering). It may also be worth your while to visit your library or to seek the assistance of the local historical society prior to planning your projects. You may discover some interesting history regarding your house and its original design and construction. This knowledge may help you decide how many of the original details you want to preserve.

Working on an old house is special. You may do it for sentimental reasons, as a reminder of a simpler time, to preserve its aesthetic "character," or out of respect for the craftsmanship. Whatever the reason, be patient and enjoy the satisfaction that comes from knowing you have added your signature to something that will endure because you cared.

Part One
WORKING WITH A CONTRACTOR

DECIDING TO WORK WITH A CONTRACTOR

The "What to Watch Out For" section in each project description points out some of the difficulties that can emerge during the course of a home renovation. In addition, some projects indicate that the time involved for a do-it-yourselfer can be 100-150% or more than the time estimated for a professional to complete the work. This is not to say that you cannot complete the work; sometimes, however, it makes more sense to hire a professional to do all or some of the work.

You may choose to act as general contractor for the project. It is important to understand what the term "general contractor" implies: he or she has a general knowledge of the construction process and manages the tradespeople who participate in the actual construction of the project. The responsibility for scheduling, planning, quality control, and overall job performance belongs to the general contractor.

Because of the close relationship you have with your own home, you need to be completely satisfied with your contractor's commitment to the work to be done. Many people put more time and energy into choosing a contractor for their home than they do into choosing a physician for themselves! The following lists are suggestions for selecting and working with a contractor. They are by no means exhaustive, but can be helpful in beginning to think about the relationship you will be establishing if you choose to hire someone to work on your home improvement project. The following discussion assumes that you are acting as a general contractor, hiring subcontractors to work with you. Should you hire a general contractor, he or she will undoubtedly use the same principles.

SELECTING A CONTRACTOR

Hiring a contractor who is licensed guarantees that he or she has satisfied the state's requirements to perform a certain type of work—it does not, however, guarantee that the work will be high quality. You want a contractor who has a good track record and will be pleasant to work with.

- Ask friends and colleagues for recommendations when looking for a carpenter, plumber, electrician, painter, or other contractor. You may have heard about your friend's experiences (both good and bad); you may have seen the completed project; you may have even met the contractor.
- As you walk or drive around your area, pay attention to the construction that is taking place. You may be surprised both at the amount of work being done and at the number of contractors doing the work. Look for signs and names on trucks, or ask one of the workers for a business card.
- Make inquiries at your local lumber yard, building supply house, or building official's office. While they will typically not make specific recommendations, they may give you some general suggestions. Often very specialized tradespeople – stair builders, for instance – can be discovered this way.
- Look for a contractor who works with a qualified team of subcontractors and suppliers. Contractors who do repeat business with subcontractors and suppliers receive discounts on materials and labor – cost savings that will be passed on to you.
- Ask a contractor whom you trust from one trade about another. For example, ask your carpenter about plumbers; ask your plumber for the name of a good electrician. You may begin to see that, if you ask enough people, the same names keep coming up.
- Check with your local builder's association or trade associations.
- While asking for names, also ask about the contractor's reputation for follow-up and warranties.
- When choosing a contractor, age should not be a factor. Sometimes younger tradespeople have more knowledge about newer materials and methods than do professionals who have been in the business for years.
- Ask to see pictures or a list of addresses where you can see the contractor's previous work. No one should be hesitant to provide references.

ESTABLISHING A RELATIONSHIP WITH A CONTRACTOR

There is no reason to believe that hiring a contractor will be anything but a positive experience. As long as you both agree to the terms of the work and follow through on your obligations, the relationship should be productive and mutually profitable.

- Don't be shy about asking a lot of questions. Make sure your questions are answered to your satisfaction and that you and the contractor both understand the terms of the work to be done. Be sure to communicate your needs clearly, especially if you intend to request materials made by specific manufacturers or other special items.

- Your contractor should be able to gauge the approximate length of time the project will take, identify the nature of disruptions to your daily life, and estimate the impact of his or her work on existing plumbing, heating, and electrical systems and on your landscaping.
- Request a written proposal that clearly defines the scope of the work and the time in which it is to be performed.
- A professional contractor will provide you with a schedule of the job so you can track the progress of the project. The schedule should include deadlines for selections you must make, such as paint colors or flooring materials.
- Agree on a payment schedule and adhere to it.
- Be reasonable in your requests. Establish the terms of the work to be done before it begins; try not to change the terms of the agreement after the job is under way. If you do request a change, realize that it will probably involve additional time and expense.
- When establishing a time frame in which the work will be completed, realize that the forces of nature (rain, snow, earthquakes) and other extenuating circumstances (illness, availability of materials) can delay the project. Be firm but fair in your expectations.

CONTRACTUAL ARRANGEMENTS

Even a small job requires a legally binding document that spells out the terms and conditions of your agreement. A professional contractor will probably insist on a detailed contract with all aspects of the job in writing. A fair contract will address the needs of both parties. There are several types of contractual

arrangements that can be used for construction work. The following is a partial list providing basic information; you should consult a legal advisor if you need further advice.

- A *lump sum contract* stipulates a specific amount as the total payment due to the contractor for performance of the contract.
- A *cost-plus-fee contract* provides for payment of all costs associated with completion of the work, including direct and indirect costs as well as a fee for services, which may be a fixed amount or a percentage of costs.
- A *cost-plus-fee contract with not-to-exceed clause* is similar to the cost-plus-fee contract, but the profit is a set amount. The contractor guarantees that the project cost will not exceed a set amount.
- In a *labor-only contract*, the contractor is paid for labor only. The materials are furnished by the owner or others.
- In a *construction-management-fee-only contract*, the owner pays all subcontractors on the project, and pays a supervision fee to the general contractor in a lump sum or as a percentage of the work involved.

WHAT TO WATCH OUT FOR

- Don't pay any amount that does not seem justified or that was not agreed to beforehand without asking for an explanation. You should not be required to pay for a job in full before the work is complete. On the other hand, most contractors will require at least one partial payment before and/or during the course of the work.
- Ensure that the contractor you intend to work with is fully insured and licensed for the work he or she will

perform. Be sure you know the licensing requirements in your jurisdiction. (You can contact your state's department of business regulations or licensing for information.) Contractors should carry Workers' Compensation and general liability insurance. Ask him or her for a current certificate of insurance.
- Be sure to ask for all warranties, manuals, and other literature related to new equipment or materials being installed. Request documentation of brands being specified.
- If you intend to do some of the work yourself, you and your contractor should both understand where (and when) your work ends and the contractor's work begins.
- Maintain a businesslike relationship with the contractor. If you become too friendly or casual, the terms of the agreement may become blurred, time may be wasted, and disagreements may ensue.
- Make sure you understand *who* will be doing the actual work. The person you make arrangements with may not be the same person who will do the physical work. Again, ask plenty of questions.

ETHICAL CONSIDERATIONS

- When soliciting bids for the job, provide equal information to all parties. Provide the same opportunity for site visits to all bidders as well.
- Once the decision has been made to hire a certain contractor, inform unsuccessful bidders (keeping in mind that you may wish to solicit them in the future).

- As much as possible, stay out of the workers' way while they are doing their jobs. Be available to answer questions or make last-minute decisions, but try not to create distractions or engage in needless conversation. You do not want to be responsible for accidents or delays in the completion of the project.
- If you have agreed to do certain prep work yourself, make sure it is complete before the workers arrive to do their jobs.
- Respect safety issues and job site conditions. If a contractor asks you to move all breakable items and keep small children out of the area, there is good reason to do so.
- Do not withhold information that could be valuable to the contractor. If you know of any condition or situation that could pose a problem during the course of the work, let the contractor know.
- Realize that the contractor is there to perform the agreed-upon work; don't ask for "freebies." For example, if the contractor is there to install a new window, don't expect him or her to also repair other windows in the room for no additional cost.

CLOSING OUT THE PROJECT

When you decide to complete a home improvement project on your own, you really have no obligation to finish the project. Many of us occupy a new space before it is complete. If you hire an outside contractor, however, you assume that the work will be completed. The following items should not be overlooked.

- Ensure that any necessary inspections are completed.
- Obtain occupancy permits if necessary.
- Have in your possession all warranties, manuals, and other literature associated with new equipment or materials that have been installed.
- Ask the contractor about materials that may be left over from the job; e.g., if you had a ceramic tile floor installed, you will want to have some extra tiles on hand for repairs. If you have no need for additional materials and you have paid for them in advance, ask about credit for materials not used.
- Ensure that items remaining on the punch list have been completed. A punch list is comprised of items that remain to be replaced or completed at the time of substantial completion, in accordance with the requirements of the contract.

PROJECT FORMS

The following pages contain forms that can be helpful as you begin making decisions about your home improvement project. The first form, "Questions for the Building Department," might be completed over the telephone for a small project. For larger projects, you will probably have to make an appointment with your local building department. Many homeowners wrongly assume that permits and inspections are not required for small projects, or that the permitting process will cause more aggravation than it is worth. This is a false assumption. The purpose of establishing and enforcing codes is to promote public safety and adherence to minimum building standards. When in doubt, do not hesitate to inquire with your local building officials. Obtaining answers to the questions on the form will enhance your understanding of the building process.

The second set of forms, the "Contractor Evaluation Sheet," reflects important information that you should have for every contractor you consider hiring. You may need to make several copies of this document.

NOTES

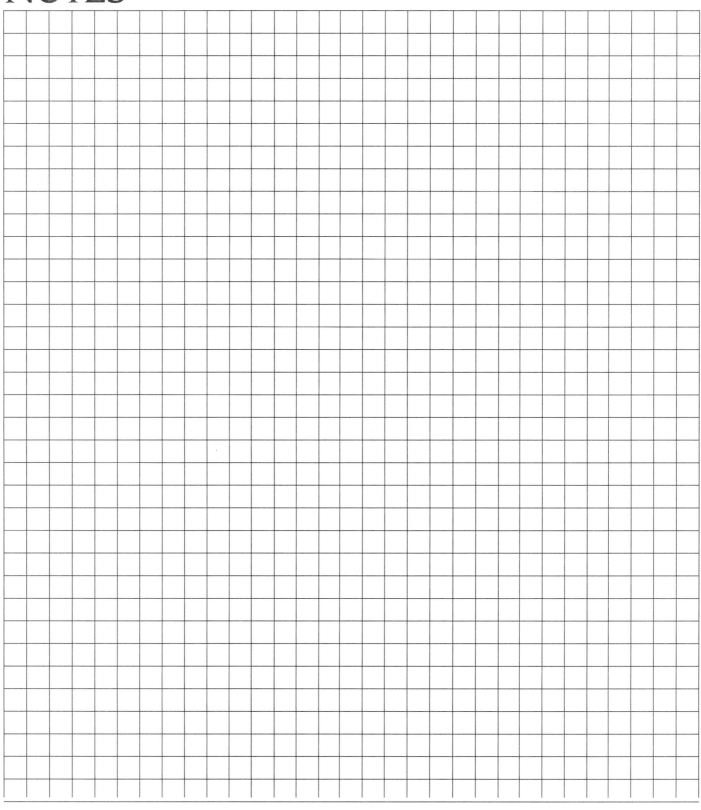

QUESTIONS FOR THE
BUILDING DEPARTMENT

1. What general information do you require on permit forms?

2. Do I need to submit plans? □ yes □ no

   If yes, how many copies do I need to file for permit?

   Do the plans need to be prepared and/or stamped by an architect or engineer? □ yes □ no

3. Do you require the following on the blueprints?

   1/4″ scale floor plan? □ yes □ no
   Exterior elevations? □ yes □ no
   Floor framing? □ yes □ no
   Roof framing? □ yes □ no
   Interior door schedule? □ yes □ no
   Exterior door schedule? □ yes □ no
   Window schedule? □ yes □ no
   Interior elevation? □ yes □ no
   Wiring schematic? □ yes □ no
   Lighting plan? □ yes □ no
   Heating or air conditioning plan? □ yes □ no
   Labor specifications? □ yes □ no
   Material specifications? □ yes □ no

4. Do I need a certified plot plan by a registered engineer? □ yes □ no

   If yes, how many copies?

   Is the plot plan required to include septic design? □ yes □ no

   Site utility locations? □ yes □ no

   What elements need to be shown?

5. Is a septic plan and evaluation/ inspection required? □ yes □ no

   Is one or the other required for the building permit? □ yes □ no

6. Are there fire alarm requirements? □ yes □ no

   If yes, do I obtain that information from the fire department? □ yes □ no

7. How much advance notice is required to schedule an inspection?

8. Does the town require soil compression tests? □ yes □ no

   If yes, by whom is the testing done?

9. Does the town require an "as built" drawing upon completion of the foundation? □ yes □ no

10. Does the town require that the foundation be sited by a registered engineer before pouring of the concrete? □ yes □ no

11. Does the town require the foundation to be a specified height above the crown of the street? ☐ yes ☐ no

    If yes, what is the height?

12. Does the town require a footing for the foundation? ☐ yes ☐ no

    If yes, does the town require a footing inspection? ☐ yes ☐ no

13. Does the town require a foundation inspection before backfilling? ☐ yes ☐ no

14. Does the town require a post-footing inspection? ☐ yes ☐ no

15. Does the town require that the electrical and plumbing work be completed before framing inspection? ☐ yes ☐ no

16. Does the town require an insulation inspection? ☐ yes ☐ no

17. Does the town require fireproofing? ☐ yes ☐ no

    If yes, solid blocking? ☐ yes ☐ no

    If yes, mineral wool? ☐ yes ☐ no

If yes, fireproof caulking for electrical and plumbing holes? ☐ yes ☐ no

If an attached garage is being installed are there fire-related material requirements?

    For sheetrock on wall between garage and house? ☐ yes ☐ no

    For sheetrock on ceiling of garage? ☐ yes ☐ no

    Fire-rated garage door? ☐ yes ☐ no

If a new heating system is being installed are there fire-rated material requirements?

    For sheetrock on walls in system area? ☐ yes ☐ no

    For sheetrock on ceiling in system area? ☐ yes ☐ no

18. What are the requirements for a Certificate of Occupancy?

19. What is the fee for the building permit?

20. How long does it take to get the building permit?

21. How long does the building permit remain in effect?

Courtesy of Cornerstone Consulting, Inc., Weymouth, MA

CONTRACTOR EVALUATION SHEET
Part 1: Telephone Interview

Company: _____

Address: _____

Town: _____

State: _____ Zip: _____

Telephone: _____

Beeper: _____

Main contact: _____

If a post office box is given for the address above, ask for a residential address.

Address: _____

Town: _____

State: _____ Zip: _____

POINTS TO COVER WITH A SUBCONTRACTOR:

1. Are you required to have a license?　☐ yes　☐ no
   If yes, what is your license number?

   If yes, what type of license?

2. Are you required by the state to be registered?　☐ yes　☐ no

3. Do you carry Workers' Compensation?　☐ yes　☐ no

4. Do you carry Liability Insurance?　☐ yes　☐ no

5. Can your insurance company send me a Certificate of Insurance?　☐ yes　☐ no

6. Is this business your sole means of support?　☐ yes　☐ no
   If no, what percentage of your time is dedicated to your other interests?

   If no, of the remaining time, how much will be dedicated to my project?

7. How long have you operated your business?

8. How many years of experience in the industry do you have?

9. Do you belong to any professional organizations?　☐ yes　☐ no
   If yes, what organizations?

10. Do you subscribe to and read any professional publications?　☐ yes　☐ no
    If yes, what publications?

Courtesy of Cornerstone Consulting, Inc., Weymouth, MA

11. Are you involved personally with the work on a physical basis?   ☐ yes ☐ no
    If yes, what portions?

19. Do you or your employees work on holidays?   ☐ yes ☐ no

12. Will you be involved in other projects while this one is underway?   ☐ yes ☐ no
    If yes, what percentage of the time?

20. In the event you cannot be at the work site as scheduled, how will I know?

13. Will you be able to start and complete the work before you undertake another project?   ☐ yes ☐ no

21. What is your policy on changes I request that may add or subtract from the work?

14. Does your work require you to respond to emergencies?   ☐ yes ☐ no
    If yes, what percentage of the time?

22. What is your policy on changes initiated by you that may add or subtract from the work?

15. What are your daily starting and ending times?

23. What is your policy on my giving instructions to your employees?

16. What are your operating and office hours?

24. What is your policy on cleaning up after yourself for the work that you or your workers do?

17. Will you or your employees take vacations during the project?   ☐ yes ☐ no

18. Do you or your employees work on weekends?   ☐ yes ☐ no

Courtesy of Cornerstone Consulting, Inc., Weymouth, MA

25. What is your policy on being a resource to me for work outside your expertise?

26. Do you provide warranties for your workmanship? ☐ yes ☐ no
    If so, please explain your warranty coverage.

27. Do you provide warranties from manufacturers? ☐ yes ☐ no
    If yes, please explain how these warranties are implemented.

28. Are you or your employees required by any manufacturer to have special training in order to install its product? ☐ yes ☐ no
    If yes, what products?

    If yes, is the warranty affected without this training?

29. Do you provide me with a "Release from Payment" from your suppliers? ☐ yes ☐ no

30. Do you allow the homeowner to provide his or her own materials? ☐ yes ☐ no

31. Do you subcontract any of the work? ☐ yes ☐ no
    If yes, what parts?

    If yes, do you provide me with a "Release from Payment" from your subcontractor(s)? ☐ yes ☐ no

32. What is your policy regarding decisions I make that have a negative effect on you, such as a delay in making a decision?

33. Are your workers made aware of your policies and procedures? ☐ yes ☐ no

34. When are payments made?

35. What form of payment do you accept?

36. Are all checks for payment made out to your company? ☐ yes ☐ no
    If no, what is your Social Security number?

37. Are you willing to be paid according to a bank's schedule? ☐ yes ☐ no

Courtesy of Cornerstone Consulting, Inc., Weymouth, MA

38. Have you previously done work similar to this project?   ☐ yes ☐ no

 If no, how are you qualified to work on this project?

39. Do we agree on who is supplying which materials under what circumstances?   ☐ yes ☐ no

40. Do you have any other policies or procedures I need to be aware of?   ☐ yes ☐ no

 If yes, what policies?

 If yes, what procedures?

41. As references, please provide me with the names and telephone numbers of three of your former customers.

 Name _____

 Telephone _____

 Time to Call _____

 Name _____

 Telephone _____

 Time to Call _____

 Name _____

 Telephone _____

 Time to Call _____

42. As references, please provide me with the names and telephone numbers of three of your professional contacts.

 Name _____

 Telephone _____

 Time to Call _____

 Name _____

 Telephone _____

 Time to Call _____

 Name _____

 Telephone _____

 Time to Call _____

43. As references, please provide me with the names and telephone numbers of three of your suppliers.

 Name _____

 Telephone _____

 Time to Call _____

 Name _____

 Telephone _____

 Time to Call _____

 Name _____

 Telephone _____

 Time to Call _____

Courtesy of Cornerstone Consulting, Inc., Weymouth, MA

CONTRACTOR EVALUATION SHEET
Part 2: Points to Cover Before Signing a Contract

44. Do we agree on the basic costs of the work? ☐ yes ☐ no
    If no, what costs need to be resolved?

45. Do we agree on the optional costs
    of the work? ☐ yes ☐ no
    If no, what optional costs need to be
    resolved?

46. Do we agree on those items of work ☐ yes ☐ no
    that are not included?
    If yes, what are those items of work that
    are not included?

47. Do you have any miscellaneous charges ☐ yes ☐ no
    of which I need to be aware?
    If yes, what charges?

48. Please indicate your payment schedule:
    Payment 1:
    Payment 2:
    Payment 3:
    Payment 4:
    Payment 5:
    Payment 6:

49. Please indicate what sections of the work
    each payment covers:
    Payment 1:
    Payment 2:
    Payment 3:
    Payment 4:
    Payment 5:
    Payment 6:

50. Are there any unresolved issues? ☐ yes ☐ no
    If yes, what issues are unresolved?

    If yes, whose responsibility is it
    to resolve each issue?

51. Do you have two copies of ☐ yes ☐ no
    the specifications?

52. Have you attached to your contract ☐ yes ☐ no
    a copy of the specifications?

Courtesy of Cornerstone Consulting, Inc., Weymouth, MA

53. Does your contract read: "Refer to Attached Plans and Specifications"? ☐ yes ☐ no

54. Have you signed both copies of the contract? ☐ yes ☐ no

55. Do you have my signature on both copies of the contract? ☐ yes ☐ no

56. Do I have a copy of the signed contract? ☐ yes ☐ no

FOR THE HOMEOWNER:

The above points were discussed with the contractor as a basis of understanding and do not represent contractual commitments by either party.

Signed _____

Signed _____

Date _____

FOR THE SUBCONTRACTOR:

The above points were discussed with the client as a basis of understanding and do not represent contractual commitments by either party.

Signed _____

Title _____

Date _____

Courtesy of Cornerstone Consulting, Inc., Weymouth, MA

WORKING WITH AN ARCHITECT

Many of us have great ideas or have seen pictures, drawings, or actual homes that have features we admire. Often, when we see that certain something that we think will be perfect in our home, we fail to realize that the reason we liked it in the first place was because it was part of a total picture. The proportion and relationships of size, shape, and structure are all part of the appeal.

By working with an architect, you can see your home remodeling project on paper – the "total picture" – before construction begins. The architect can approve contractors' work and maintain quality control. An architect can bring a new perspective to the task and identify short- and long-term solutions to your design needs. An architect can help you define your project in terms that provide meaningful guidance for design. Architects can also do site studies, help secure planning and zoning approvals, and perform a variety of other predesign tasks.

In selecting an architect, you will want to look for an appropriate balance among design ability, technical competence, professional service, and cost. As you search for a suitable architect, consult other homeowners who have undertaken similar projects and ask who they selected. Ask who designed buildings and projects that you've admired. Many local chapters of The American Institute of Architects maintain referral lists and can assist you in identifying architects in your area who specialize in certain types of projects.

When interviewing an architect, ask him or her to show you projects that are similar to yours or that have addressed similar issues. Also, ask how the architect will approach your project and who will be working on it (including consultants). Ask for references you may contact. Ask how the architect will gather information, establish priorities, and make decisions. Evaluate the architect's interest in your project: Will your needs be a major or minor concern? Also, don't be afraid to ask up front about fees and terms of the agreement.

Before interviewing an architect, ask yourself the following questions:

* What do you like about your home? What's missing? What don't you like?
* Why do you want to add to or renovate your house?
* What is your lifestyle? Are you home a great deal? Do you work at home? Do you entertain often?
* What functions/activities will take place in the new space?
* What do you think the addition or renovation should look like?
* How much can you realistically afford to spend?
* How soon would you like to be settled in your new space? Are there rigid time constraints?
* Do you have strong ideas about design styles?
* Who will be the primary contact with the architect, contractor, and others involved in designing and building the project?
* What qualities are you looking for in an architect?
* Do you plan to do any of the work yourself?
* How much disruption can you tolerate as you renovate or add on to your home?

The answers to these questions will help to focus your discussion with prospective architects.

This information is based on the recommendations of The American Institute of Architects (AIA). AIA offers helpful information, in addition to referrals, for consumers who are considering working with an architect, including 20 *Questions to Ask Yourself Before You Get Started*, *The Steps Involved in Design and Construction – What to Expect*, and *You and Your Architect*. Contact AIA at 1-800-AIA-3837, or 1735 New York Avenue, N.W., Washington, D.C. 20006. Visit AIA's Web site at **http://www.aia.org.**

NOTES

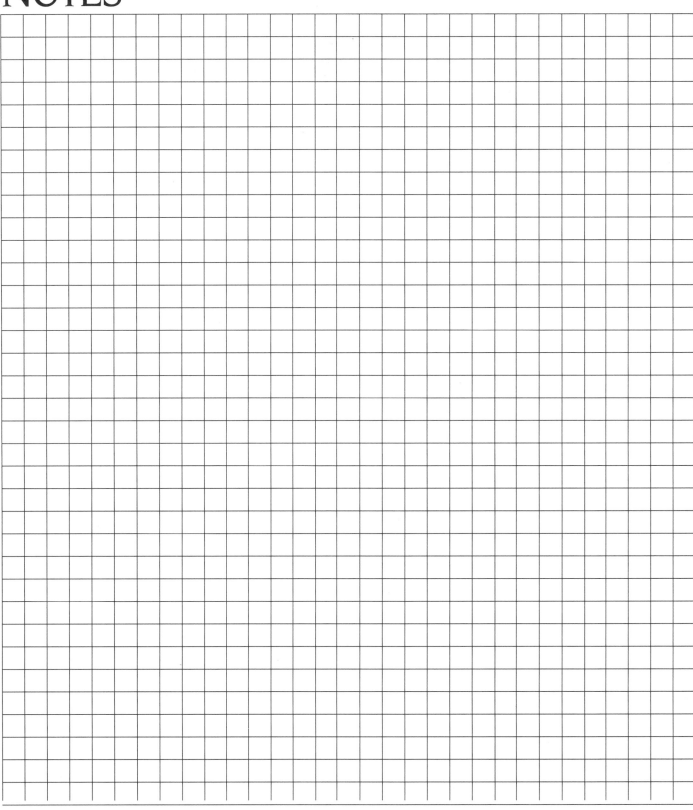

Part Two
EXTERIOR PROJECTS

HOW TO USE PART TWO

Part Two, "Exterior Projects," provides descriptions and cost estimates for complete home improvement projects. Part Three, "Details," provides cost information for individual construction items. Tips for using "Details" can be found at the beginning of that part.

Each project plan in Part Two contains three types of information to help

you organize the job. First, a detailed illustration shows the finished project and the relationship of its components. Second, a general description of the plan includes a review of the materials and the installation process. It also evaluates the level of difficulty of the project. The "What To Watch Out For" section highlights ways to enhance the project

or to cope with particularly difficult installation procedures. Third, a detailed project estimate chart lists the required materials, the professional installation times, and their corresponding costs. The components of this chart are described in detail below.

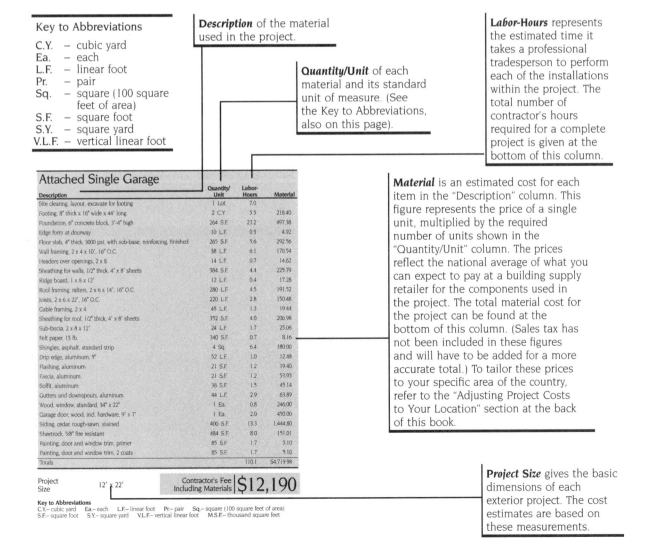

Key to Abbreviations

C.Y. – cubic yard
Ea. – each
L.F. – linear foot
Pr. – pair
Sq. – square (100 square feet of area)
S.F. – square foot
S.Y. – square yard
V.L.F. – vertical linear foot

Description of the material used in the project.

Quantity/Unit of each material and its standard unit of measure. (See the Key to Abbreviations, also on this page).

Labor-Hours represents the estimated time it takes a professional tradesperson to perform each of the installations within the project. The total number of contractor's hours required for a complete project is given at the bottom of this column.

Material is an estimated cost for each item in the "Description" column. This figure represents the price of a single unit, multiplied by the required number of units shown in the "Quantity/Unit" column. The prices reflect the national average of what you can expect to pay at a building supply retailer for the components used in the project. The total material cost for the project can be found at the bottom of this column. (Sales tax has not been included in these figures and will have to be added for a more accurate total.) To tailor these prices to your specific area of the country, refer to the "Adjusting Project Costs to Your Location" section at the back of this book.

Attached Single Garage

Description	Quantity/Unit	Labor-Hours	Material
Site clearing, layout, excavate for footing	1 Lot	7.0	
Footing, 8" thick x 16" wide x 44' long	2 C.Y.	5.5	218.40
Foundation, 6" concrete block, 3'-4" high	264 S.F.	23.2	497.38
Edge form at doorway	10 L.F.	0.5	4.92
Floor slab, 4" thick, 3000 psi, with sub-base, reinforcing, finished	265 S.F.	5.6	292.56
Wall framing, 2 x 4 x 10', 16" O.C.	38 L.F.	6.1	170.54
Headers over openings, 2 x 8	14 L.F.	0.7	14.62
Sheathing for walls, 1/2" thick, 4' x 8' sheets	384 S.F.	4.4	225.79
Ridge board, 1 x 8 x 12'	12 L.F.	0.4	17.28
Roof framing, rafters, 2 x 6 x 14', 16" O.C.	280 L.F.	4.5	191.52
Joists, 2 x 6 x 22', 16" O.C.	220 L.F.	2.8	150.48
Gable framing, 2 x 4	45 L.F.	1.3	19.44
Sheathing for roof, 1/2" thick, 4' x 8' sheets	352 S.F.	4.0	206.98
Sub-fascia, 2 x 8 x 12'	24 L.F.	1.7	25.06
Felt paper, 15 lb	340 S.F.	0.7	8.16
Shingles, asphalt, standard strip	4 Sq.	6.4	180.00
Drip edge, aluminum, 5"	52 L.F.	1.0	12.48
Flashing, aluminum	21 S.F.	1.2	19.40
Fascia, aluminum	21 S.F.	1.2	53.93
Soffit, aluminum	38 S.F.	1.5	45.14
Gutters and downspouts, aluminum	44 L.F.	2.9	63.89
Wood, window, standard, 34" x 22"	1 Ea.	0.8	246.00
Garage door, wood, incl. hardware, 9' x 7'	1 Ea.	2.0	450.00
Siding, cedar, rough-sawn, stained	400 S.F.	13.3	1,444.80
Sheetrock, 5/8" fire resistant	484 S.F.	8.0	151.01
Painting, door and window trim, primer	85 S.F.	1.7	5.10
Painting, door and window trim, 2 coats	85 S.F.	1.7	5.10
Totals		110.1	$4,719.98

Project Size	12' x 22'		Contractor's Fee Including Materials	**$12,190**

Project Size gives the basic dimensions of each exterior project. The cost estimates are based on these measurements.

Key to Abbreviations
C.Y.– cubic yard   Ea.– each   L.F.– linear foot   Pr.– pair   Sq.– square (100 square feet of area)
S.F.– square foot   S.Y.– square yard   V.L.F.– vertical linear foot   M.S.F.– thousand square feet

Section One
DECKS

A deck's size and design should be determined by its intended use, the type and size of the house to which it will be attached, and your own budget limitations. Following are some tips for designing and purchasing deck materials.

- Be sure to integrate the new deck with existing structures and landscaping. For example, posts and railings for a new deck should be the same color as the house trim.

- When deciding on a size for the deck, add a few more feet than you think you'll need. Most people underestimate the space they'll need for furniture and activity.

- When shopping for deck materials, try to stay away from boards with excessive knots, voids, twisting, warping, and cupping.

- The factory coloring of pressure-treated deck material will gradually change to a natural wood color over several years, but if you want to finish your deck, transparent stain looks good and requires little upkeep. If you use a heavy-bodied stain, regular maintenance will be needed to preserve its appearance. Finishing your deck will generally ensure a consistent aging process; in addition, finishing allows you to use different materials. If you are adding to an existing deck and wish to match an existing deck, painting or staining is probably necessary.

- Even if premium length stock is more expensive, the fewer the joints and the longer the uninterrupted member, the better.

- If the deck will be 2'-6" or more above the ground, the railing must be strong enough to prevent a 200-pound person from falling, and its rails must be spaced close enough to prevent a 4" ball from passing through it.

- Most retail lumber stores or home improvement centers should have a selection of framing anchors and hangers to assist you in securing deck members. Some of these elements are required by local building codes.

- In damp areas, decks can develop a mossy patina. This can be removed with a mild bleach solution.

- Sturdy railings are essential if your deck is elevated more than one step above the ground. You may want to consult your local building department for spacing of balusters or decorative rail components. Many communities have definite requirements for rail height and spacing of members.

- Exposure to extreme weather conditions dries wood out, making it susceptible to water retention and rot unless it is treated annually with sealer-preservative.

Depending on your location, your material selection may be cedar, pressure-treated pine or even recyclable wood substitutes. Check your dealers' prices and evaluate the benefits when opting for this type of material. Do not be afraid to mix composite recycled material with traditional framing lumber, such as using pressure-treated framing material with composite material decking.

ELEVATED DECK

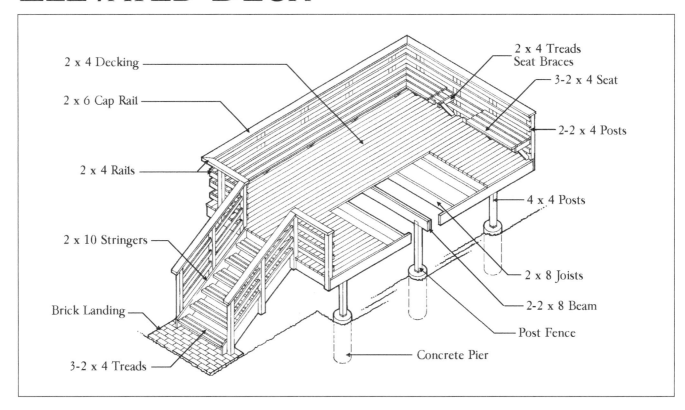

2 x 4 Decking

2 x 6 Cap Rail

2 x 4 Rails

2 x 10 Stringers

Brick Landing

3-2 x 4 Treads

2 x 4 Treads
Seat Braces

3-2 x 4 Seat

2-2 x 4 Posts

4 x 4 Posts

2 x 8 Joists

2-2 x 8 Beam

Post Fence

Concrete Pier

MATERIALS

This plan presents a 10′ x 16′ deck, complete with stairway and two-sided bench railing with a cost adjustment for a deluxe 12′ x 12′ redwood deck with a built-in perimeter bench and bannistered safety railing.

The components of the 10′ x 16′ deck are all constructed from pressure-treated products available at most building supply outlets. The 12′ x 12′ deck uses redwood, which is more costly than standard lumber and may be more difficult to get in some areas of the country. Be sure to arrange delivery, as lumber for either deck requires trucking. The purchase will probably warrant free curb delivery if the site is a reasonable distance from the supplier.

The support system and frame for the pressure-treated deck plan consist of six 4 x 4 posts set in 3′ concrete footings. The 2 x 8 joists are spaced 2′ on center, with doubled support joists at 6′ intervals, and doubled 16′ headers. It is most important that you lay out the footing and post locations accurately, since the rest of the structure depends on their

alignment. Double-check all measurements as you go and be sure to use 8″ diameter tube forms for the foundations. Before you begin construction of the frame, review the layout template to be sure that the doubled support joists will be correctly aligned when they are placed.

Square the four corners before locating the post positions, and make sure that you follow the plan. Include six, not four, supports for the redwood deck. Economizing by eliminating structural components will cost you more money and aggravation in the long run because of sagging, loosening of the structure, and other conditions that require repair.

As with any deck, but especially for the redwood deck that is located 8′ above ground, it is important that the deck frame be plumb, level, and square. Before fastening the joists, review the accuracy of the layout to be sure that the doubled joists are precisely aligned with the 4 x 4 support posts. This alignment must be exact to ensure safe bearing of the load of the structure. Before laying the decking, double-check the frame for level and square. Then tie the outside corners temporarily at right angles while you place the decking.

The decking in this plan provides an attractive and durable surface, but boards of other widths can also be used. Changing the pattern of the decking will affect the cost in either plan. Diagonal decking creates attractive surfaces, but wastes material and demands more installation time. For conventional decking, be sure to measure accurately, and make precise saw cuts; the cut-off pieces will be used later to fill out the courses.

As you place the decking, place the cupped side of the boards down, and use enough nails. To keep the rows of nail heads straight, lay the entire deck with minimum fastenings, and then complete the nailing by following chalk lines. Use a spacer to keep the opening between courses at about 1/8″ or slightly wider. This space is an important feature, as it allows for drainage of rainwater, helps to ventilate the structure, and prevents the build-up of moisture-laden matter between the deck boards.

There are many options for the railing, bench, and stairway design. Standard 4 x 4 posts and 2 x 4 rails are used for the most economical railing system, but 2 x 2 balusters can also be installed at more expense to enhance the railing design or to make it child-safe. The built-in railing bench included in the redwood deck plan can be altered to suit personal taste or practical needs. The railing consists of 2 x 4 rails, and 4 x 4 fitted and lagged posts. The rails add considerably to the materials cost, but they are an important safety feature in any deck set this high off the ground.

Costs can be cut by eliminating the bench altogether, or by reducing its length and limiting it to one side of the deck. Seek assistance from a qualified person if you have not installed a stairway before. Safety is a primary consideration for any set of stairs, and consistent vertical spacing from tread to tread is one way of ensuring it. Proper support and correct placement of stringers and stairway railings are also critical safety issues.

LEVEL OF DIFFICULTY

The pressure-treated deck can be built by most people with a reasonable level of skill in home improvement projects. Its 10' x 16' size makes it a major project, but the work time can be spread out because of the convenience of the location. A basic knowledge of carpentry is needed for the installation of all of the deck's components. Beginners can tackle this deck if they are given instruction before they start and guidance during construction. They should plan on tripling the estimated times, even with professional assistance. Intermediates should add 40% to the professional time for the framing and decking tasks and 75% for the footing, post, railing, bench, and stairway installations. Experts should add 10% to the estimated labor-hours for all parts of this project.

The redwood deck project requires considerable carpentry skill. Several features make it a more demanding do-it-yourself project than the other deck plan. One is the expense of the redwood material, which may be twice the cost of conventional deck lumber. Another is the deck's height off the ground, which requires working from a stepladder and causes general inconvenience in some operations. A third challenging feature is the finish work required for the bench, railing, and stairway installations. Beginners and less skilled intermediates should not attempt to complete this project on their own. They should hire a contractor for the framing and finish work and add 100% to the estimated labor-hours for the tasks that they intend to take on. Intermediates and experts should add 50% and 20%, respectively, for the support, framing and decking jobs, and 100% and 40% for the bench, railing, and stairway installations.

WHAT TO WATCH OUT FOR

If a deck is 4' or less above grade level, the installation of the supports is accomplished fairly easily; but if a deck is higher than 4' or situated over sloping terrain, the job becomes more difficult. The redwood deck, while moderate in size, is challenging because of its placement at 8' above ground level. Allow extra time for this inconvenience. Plan to use a stepladder for some of the railing work, as much of the layout and fastening can be approached most efficiently from outside the structure.

SUMMARY

Although these are large deck projects, the pressure-treated model can be built by most intermediates willing to invest the time, get help in starting out, and seek guidance as they go. The redwood deck is more challenging, but it is an attractive structure that will add comfort and value to your home.

For other options or further details regarding options shown, see

Ground-level deck

Elevated L-shaped deck

Patio & sliding glass doors

Elevated Deck*

Description	Quantity/Unit	Labor-Hours	Material
Layout, excavate post holes	0.50 C.Y.	0.5	
Concrete, field mix, 1 C.F. per bag, for posts	6 Bags		44.64
Forms, round fiber tube, 1 use, 8" diameter	18 L.F.	3.7	39.10
Deck material, pressure-treated posts, 4 x 4 x 4'	24 L.F.	0.9	41.76
Post base, 4 x 4	6 Ea.	0.4	34.20
Post cap, 4 x 4	6 Ea.	0.4	16.78
Headers, 2 x 8 x 16'	64 L.F.	2.3	111.36
Joists, 2 x 8 x 10'	120 L.F.	4.3	208.80
Decking, 2 x 4 x 12'	576 L.F.	16.8	345.60
Stair material, pressured-treated stringers, 2 x 10 x 10'	20 L.F.	0.8	44.40
Treads, 2 x 4 x 3'-6", 3 per tread	70 L.F.	2.0	42.00
Landing, brick on sand, laid flat, no mortar, 4.5 brick per S.F.	16 S.F.	2.6	41.28
Railing material, pressured-treated posts, 4 x 4	48 L.F.	1.7	83.52
Railings, 2 x 4 stock	240 L.F.	7.0	144.00
Cap rail, 2 x 6 stock	48 L.F.	1.5	64.51
Bench material, seat braces, 2 x 4 stock	108 L.F.	3.1	64.80
Joist and beam hangers, 18 ga. galvanized	12 Ea.	0.6	7.34
Nails, #10d galvanized	15 lbs.		21.24
Bolts, 1/2" lag bolts, 4" long, square head, with nut and washer	28 Ea.	1.6	25.20
Totals		50.2	$1,380.53

Project Size   10' x 16'

Contractor's Fee Including Materials | **$4,456**

Key to Abbreviations
C.Y.– cubic yard   Ea.– each   L.F.– linear foot   Pr.– pair   Sq.– square (100 square feet of area)
S.F.– square foot   S.Y.– square yard   V.L.F.– vertical linear foot   M.S.F.– thousand square feet
*For a redwood deck add $1302 to the contractor's fee.

ELEVATED L-SHAPED DECK

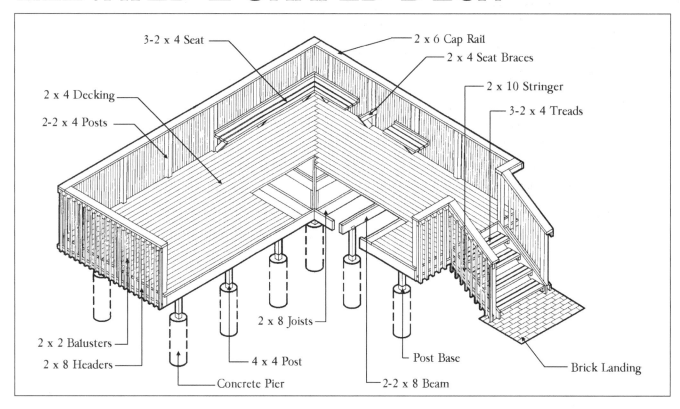

- 3-2 x 4 Seat
- 2 x 6 Cap Rail
- 2 x 4 Seat Braces
- 2 x 10 Stringer
- 3-2 x 4 Treads
- 2 x 4 Decking
- 2-2 x 4 Posts
- 2 x 2 Balusters
- 2 x 8 Headers
- 4 x 4 Post
- 2 x 8 Joists
- Post Base
- Brick Landing
- Concrete Pier
- 2-2 x 8 Beam

One of the most exciting features of deck construction is that an unlimited variety of configurations can be used to style the facility to the house and surrounding terrain. Imaginative decks may be custom-planned to complement the design features of the home. This L-shaped plan demonstrates how a deck installation can be modularized to create an attractive and functional facility. Basically, it consists of two separate decks that have been tied into one L-shaped structure, bordering two sides of the house. The same modular construction method can also be used for U-shaped, angled, and multi-level deck designs. With proper instruction before they start, and guidance along the way, many do-it-yourselfers can tackle this deck with professional results.

MATERIALS

The materials used in this deck plan are the standard pressure-treated products employed for other deck projects. The legs of the "L" are formed from two separate, but attached, deck frames, each with its own 4 x 4 post-and-footing support system. More time is required to lay out the footings for this deck because there

are more of them, but the process is the same as with other decks of simpler design. The most important aspect of the footing and post placement is the establishment of a square and level support system for the rest of the structure. If you are inexperienced in this procedure, get some help from a knowledgeable person and take the job in steps, building the frame for the large 8' x 16' section first before laying out the footing and post support for the 8' x 12' section. By proceeding in this way, you can adjust for small errors in alignment of the section and prevent potential materials waste resulting from potential inaccurate layout.

The frame consists of 2 x 8 joists and headers that are assembled as two separate units. As in the layout for the footings and support posts, be sure to double-check all measurements and make sure that the corners of each section are properly squared. Since this L-shaped facility is built only 3' off the ground, most of the frame for the 8' x 16' section should be built in place. If the adjacent terrain allows, it may be laid out and started on the ground and then positioned

on the 4 x 4s to be finished in place. The 8' x 12' frame can be built in the same way.

The decking used in this L-shaped format consists of 2 x 4s, but 2 x 6s may also be installed on frames with 24" joist spacing. If thinner deck boards are to be installed, the joists will have to be placed at closer intervals and their number will be increased, so plan accordingly. Measure and cut the boards accurately, so that the shorter, cut-off pieces can be used to fill out the courses as the decking is placed. When laying the boards, place the cup side down and leave a space between courses. Also, keep a check on the gain at both ends and the middle as you proceed, to be sure that the decking is square to the frame. After all of the boards have been placed and fastened, snap a chalk line and saw the ragged ends to an even finished length.

The railing, bench, and stairway design offer the opportunity to add an imaginative touch to the deck. This plan, for example, suggests using a corner bench design, but various other configurations are possible. The railing system, too, can be modified to suit the desired function and personal taste. In both cases, the cost of the project will be affected by alterations. A balustered railing design is attractive and provides the required safety, but the expense of the material used for the balusters, as well as installation time, is considerable. Built-in benches also use a substantial amount of material, so plan to spend more money and allow more installation time if you are adding to the length or the number of benches. The ship's ladder stairway design enhances the open motif of the deck, but conventional stringers and treads can also be placed. The materials cost is the same for both stairway designs, but the cutting of the stringers takes a little longer. If you are not experienced in stairway building, seek instruction before you start. Accurate layout of tread height is a must to ensure a usable set of stairs.

LEVEL OF DIFFICULTY

As noted in the introduction to this deck plan, the carpentry skills required to complete this project are the same as for standard straight, square, or rectangular decks. A more intricate layout is involved in the alignment of the two sections; but if the project is taken in steps, the layout is simply a matter of more time, not more skill. Beginners, if they are given ample instruction and guidance, can do much of the work on this project, but they will need time to proceed slowly and must be willing to seek advice as they go. Intermediates and experts should also allow additional time for the layout and placement of the support system for this deck. Otherwise, the project should pose no challenges greater than those encountered in other deck projects.

Beginners should triple the professional time estimates for most tasks; intermediates should add about 60% to the estimates; and experts should add about 20%. If the facility is to be built morethan 6′ off the ground, more time will be required for do-it-yourselfers of all ability levels.

WHAT TO WATCH OUT FOR

Modular decks of different shapes and levels are more expensive to build because they usually demand more material and installation time. But if your budget allows for a deluxe deck, the resulting structure can add considerably to the appearance and value of your home. Take some time to draw up the plans and do some reading on the subject before you decide on a deck design. Multi-level decks are not that difficult to build, particularly if they are close to ground level. U-shaped, angled configurations, and even hexagonal or octagonal designs, where appropriate, can be built to professional standards by do-it-yourselfers who are willing to learn first and then work slowly. In addition to imaginative basic designs, other amenities like built-in planters, eating and serving facilities, and sitting areas can be included to enhance the deck's appearance and function.

Elevated L-Shaped Deck

Description	Quantity/ Unit	Labor- Hours	Material
Layout, excavate post holes	1 C.Y.	1.0	
Concrete, field mix, 1 C.F. per bag, for posts	12 Bags		89.28
Forms, round fiber tube, 1 use, 8″ diameter	42 L.F.	8.7	91.22
Post base, 4 x 4	14 Ea.	0.9	79.80
Post cap, 4 x 4	14 Ea.	0.9	39.14
Deck material, pressure-treated posts, 4 x 4 x 3′	42 L.F.	1.5	73.08
Headers, 2 x 8 x 16′	80 L.F.	2.8	139.20
Joists, 2 x 8 x 8′	96 L.F.	3.4	167.04
Joists, 2 x 8 x 12′	72 L.F.	2.6	125.28
Decking, 2 x 4 x 12′	720 L.F.	20.9	432.00
Stair material, pressure-treated stringers, 2 x 10 x 6′	12 L.F.	0.5	26.64
Treads, 2 x 4 x 3′-6″, 3 per tread	60 L.F.	1.8	36.00
Landing, brick on sand, laid flat, no mortar, 4.5 brick/S.F.	16 S.F.	2.6	41.28
Railing material, pressure-treated posts, 4 x 4	56 L.F.	2.0	97.44
Balusters, 2 x 2 stock	540 L.F.	14.4	265.68
Cap rail, 2 x 6 stock	72 L.F.	2.3	96.77
Bench material, pressure-treated seats & braces, 2 x 4 stock	96 L.F.	2.8	57.60
Joist and beam hangers, 18 ga. galvanized	18 Ea.	0.8	11.02
Nails, #10d, galvanized	20 lbs.		28.32
Bolts, 1/2″ lag bolts, 4″ long, square head, with nut and washer	40 Ea.	2.3	36.00
Totals		**72.2**	**$1,932.79**

Project Size	8′ x 12′ + 8′ x 16′		Contractor's Fee Including Materials	**$6,317**

DECKING OPTIONS
Cost per Square Foot, Installed

Description	
2 × 6 Composite Material	$ 4.45
2 × 4 Treated Pine	$ 3.01
2 x 6 Treated Pine	$ 2.61
1 × 4 Redwood	$ 5.60
2 × 6 Cedar	$10.04

SUMMARY

This L-shaped plan demonstrates how deck construction can be modularized to create an attractive and practical exterior recreation and sitting area. Many do-it-yourselfers can complete this plan with professional results and considerable savings.

For other options or further details regarding options shown, see

> *Elevated deck*
> *Ground-level deck*
> *Patio & sliding glass doors*

Key to Abbreviations
C.Y.– cubic yard   Ea.– each   L.F.– linear foot   Pr.– pair   Sq.– square (100 square feet of area)
S.F.– square foot   S.Y.– square yard   V.L.F.– vertical linear foot   M.S.F.– thousand square feet

MULTI-LEVEL DECK

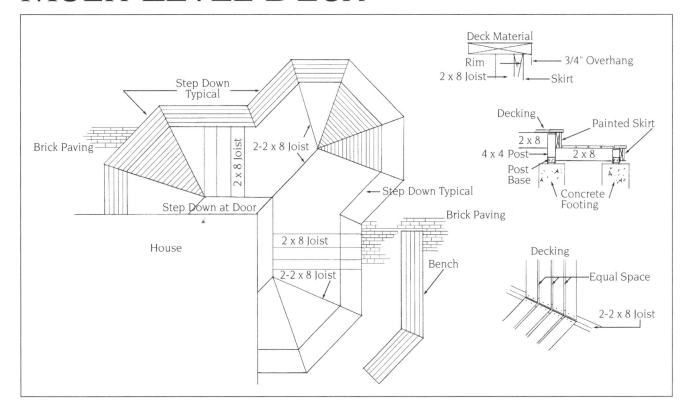

Decks have always been high on the do-it-yourselfer's list of projects. There are many "how to" books available on deck construction, and home centers and building supply dealers are providing ideas, add-on products, and training courses. What they are saying is, "You don't have to be satisfied with a simple rectangular design; be creative."

Most decks are one-level, accessed from the home through a slider or atrium door, with stairs leading to the yard below. This project goes a bit further, with multiple levels and interesting angles. Multi-level decks appeal to the eye and eliminate that "raised" look. The design is easily adapted to sloped yards and allows you to eliminate stairs by using step-downs and platforms. While this approach increases the size of the deck, it also makes a much more pleasing area that lends itself to creative furniture arrangements and planting schemes. This layout can also nicely conceal the amount of exposed foundation.

MATERIALS

Synthetic/recycled decking made from plastics and/or wood products has gained popularity with builders and homeowners

for its durability, wood-like appearance, and low maintenance. Wood-polymer decking and lumber is highly resistant to moisture and insects and never requires preservative treatments, staining, or sealing. It can be sawed, sanded, nailed, screwed, and drilled. An experienced do-it-yourselfer should have no problem working with this material. Because all synthetic/recycled decking and lumber is not alike, it is important to read the specific manufacturer's recommendations on where and how to use the different-sized members, recommended joist spans and space between boards, and material performance in extreme weather conditions.

This deck wraps around the corner of the home. The step-down consists of a second platform that extends around the perimeter of the upper level. The framing is accomplished with pressure-treated lumber per building code requirements. The skirt that covers the riser at the step-downs could be synthetic lumber, or as in this case, painted wood trim.

The design features an octagonal extension at the corner of the deck, and 45° angles at the other two outer corners. The center of the octagonal section is located off of the corner of the house by

bisecting the right angle created by extending the exterior wall lines from the corner (this should be a 90° angle). The pattern of the decking is designed so that no ends are showing. Each piece of decking requires careful measuring and cutting. The key to this design is to perfect the layout so that when the decking material is applied, the angle cuts at the joints are equal, or 22.5°. Any error in the layout will cause the lengths of the angle cut at the end of the decking to be unequal, which would detract from the appearance of the pattern.

Wood-polymer lumber does not noticeably swell or shrink in wet or dry conditions. Changes in temperature will cause some expansion and contraction. The boards should be spaced appropriately—both the spaces between the decking boards and where end joints meet. Equal spacing between the boards is especially important on this project and can be accomplished with removable spacers.

Because synthetic decking does not chip, split or crack, you should be able to minimize waste by reversing the cut-off board pieces as you go around the deck. Consider purchasing decking in lengths that work well for using up the entire piece.

You may want to wait until you have installed most of the decking before purchasing just enough additional material to complete the job. Because of the deck pattern, we recommend beginning at the outside edges of the deck and working towards the center to ensure a proper and consistent overhang. (The outside piece of decking should overhang the skirt.) Synthetic/recycled plastic decking used for a deck requires joist spacing similar to that used for traditional lumber. When using 5/4" x 6" or 2" x 6" decking boards, the joist spacing should be 16" on center.

Most deck fasteners work well with wood-polymer lumber. We recommend stainless steel and hot-dipped galvanized fasteners. Because of the longevity of synthetic decking and lumber, we also recommend upgrading the fasteners and anchors used for the deck framing and anchoring systems. This plan requires more framing anchors and joist hangers than typical wood rectangular or square decks because of the greater weight of the material and the more demanding design. Check with your material supplier for custom hangers that may not be on

the shelf, but can be special-ordered. Planning ahead to use the most appropriate materials can save a lot of aggravation for the inexperienced builder, and ensures a high quality job for the more advanced do-it-yourselfer.

LEVEL OF DIFFICULTY

This deck project requires an ability to visualize the end result while planning the job, laying it out, and building it. Working from a multi-level design is different than working from one elevation. Care must be taken to maintain proper riser heights (7" is recommended). It is also necessary to understand the basic geometry of bisecting angles and to be able to maintain straight and true lines. While it is not strictly necessary, it might be easier to use a builder's level rather than a standard level, if you have access to one, because it allows you to set multiple elevations from one point. The elevation of the top of the footings in relation to the top of the finished deck and step-downs is critical for safety. Cutting the decking and minimizing waste is also somewhat difficult for a beginner.

Constructing this type of deck requires two people—for pulling and snapping chalk lines, and for holding the decking pieces at the correct angle for nailing. Beginners should add 100% and intermediates and experts 50% and 20% respectively, to the estimated labor-hours for layout and construction.

WHAT TO WATCH OUT FOR

Although this deck is large, you should not assume that it can carry greater loads. It must be built properly for structural stability. Consult with construction professionals or the building inspector when heavier than normal loads, such as large groups, hot tubs, and landscape planters are anticipated. Make sure the structure complies with local zoning and building ordinances for setbacks from property lines. Do not underestimate the amount of fastening required to safely attach the deck to the house. Take time to locate the fasteners so that they will not interfere with the joist layout. Before undertaking this project, we suggest that you read the other deck projects in this section. The landscape projects may provide ideas on projects to consider in conjunction with this deck. Also think about and plan for access to outlets or wiring for lighting, speakers or appliances. As mentioned earlier, different types of synthetic decking have different installation, performance and maintenance expectations. Research the proper cleaning procedures for each type, and what you can expect in terms of fading and aging.

SUMMARY

The introduction of wood-polymer lumber has opened up new possibilities in deck design. What was once thought of as a pile of wood that will need to be maintained, cleaned, stained, and repaired annually has, with this material, become a long-lasting, low-maintenance structure that can add significant value to your home. This project is an advanced design that requires good carpentry skills. Investing adequate time in planning and layout is critical.

For other options or further details regarding options shown, see

> *Elevated deck*
>
> *Elevated L-shaped deck*
>
> *Entryway steps Redwood hot tub*

Multi-Level Deck

Description	Quantity/ Unit	Labor- Hours	Material
Layout, excavate post holes	3 C.Y.	3.0	
Concrete, field mix, 1 C.F. per bag, for posts	26 Bags		193.44
Forms, round fiber tube, 1 use, 8" diameter	78 L.F.	16.1	169.42
Post base, 4 x 4	26 Ea.	1.6	148.20
Deck material, pressure-treated posts, 4 x 4 x 3'	61 L.F.	2.2	106.14
Headers, 2 x 8 x 12'	396 L.F.	14.1	689.04
Joists, 2 x 8 x 10',12',14',16'	640 L.F.	22.8	1,113.60
Decking, wood-polymer lumber, 5/4 x 6 x 16'	1,760 L.F.	38.5	3,104.64
Wood skirt board, 1 x 8 x 10'	240 L.F.	8.5	391.68
Drilling anchors, 3/4" diameter, 4" deep	20 Ea.	3.6	3.36
Expansion anchors, 3/4" diameter, long	20 Ea.	2.5	78.00
Lag screws, 3" long, 1/2" diameter	20 Ea.	1.2	6.00
Bolts & hex nuts, 1/2 x 4	40 Ea.		16.80
Joist hangers, heavy duty 12 ga., galvanized, 2 x 8	96 Ea.	4.7	49.54
Joist hangers, heavy duty 12 ga., galvanized, two 2 x 8	21 Ea.	1.1	55.44
Connector plates	6 Ea.	1.9	172.80
Paint skirt	240 L.F.	3.0	17.28
Stainless steel nails, plain	25 lbs.		146.10
Totals		124.8	$6,461.48

Project Size	800 S.F.		Contractor's Fee Including Materials	$15,361

Key to Abbreviations
C.Y.– cubic yard Ea.– each L.F.– linear foot Pr.– pair Sq.– square (100 square feet of area)
S.F.– square foot S.Y.– square yard V.L.F.– vertical linear foot M.S.F.– thousand square feet

GROUND-LEVEL DECK

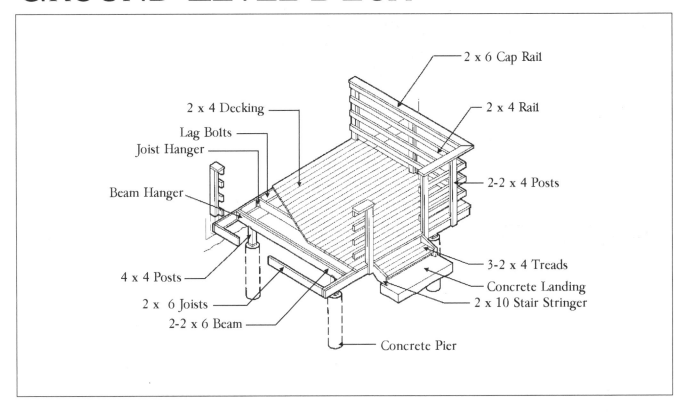

2 x 6 Cap Rail

2 x 4 Decking

Lag Bolts

Joist Hanger

2 x 4 Rail

Beam Hanger

2-2 x 4 Posts

4 x 4 Posts

3-2 x 4 Treads

2 x 6 Joists

Concrete Landing

2-2 x 6 Beam

2 x 10 Stair Stringer

Concrete Pier

The addition of a deck is one of the easiest and most economical ways to improve the exterior of your home. Basic single-level decks are usually uncomplicated in their construction and are, therefore, manageable undertakings for homeowners who have limited building experience. This deck plan demonstrates how an attractive and functional facility can be constructed for a modest investment of both time and money.

MATERIALS

Most decks built at foundation or ground level are comprised of the same basic components. These vary in size and number, according to the method of installation, the dimensions of the structure, its height above grade, and the complexity of its design. The materials used are basic decking products available at most building supply outlets, but they vary in quality and price, so do some comparison shopping before you buy. When you purchase the materials, select them yourself, if possible, looking for straight deck boards and sound support members. You should also clarify the conditions of delivery at the time of the

purchase, since most suppliers will make curb deliveries free of charge for a complete deck order.

All of the wood products for this deck are specially designed for use in exterior structures that are exposed to the elements. Pressure-treated deck materials are more costly than conventional, untreated components, but they are a much better choice. The additional expense for pressure-treated wood is well worth the investment to guarantee the deck's longevity.

The support system for this 8' x 10' ground-level deck consists of footings, 4 x 4 posts, and the frame, which is comprised of 2 x 6 joists and headers. Before you begin the excavation for the footings, lay out their locations accurately to ensure a square and level support system for the rest of the structure. Several methods can be employed to establish the location of the support posts and their footings, but one of the most accurate is to approximate the layout with batter boards at the four corners, then establish, by the use of stringlines, equal diagonals. Failure to place the footings and posts precisely at square with each other and with the side of your house will adversely affect the rest of the project. Double-

check the post locations before you set them in concrete and be sure to use cylindrical "tube" forms for the foundations. These are available at most building supply yards and are well worth the price. Review the layout of the 2 x 6 frame as well, to be sure that the doubled joists are correctly aligned with the 4 x 4 supports.

The decking suggested in this plan consists of 2 x 4s laid across the frame and fastened with galvanized nails. However, decking materials of another thickness, width, or edge type can also be installed. Remember that the decking itself is the most visible component of the structure, so additional expense for aesthetic reasons may be a worthwhile investment. This deck plan includes a basic post-and-rail design for the railing system. It is one of the simplest railing designs to install, and it creates an attractive border with a 2 x 6 cap, which can double as a convenience shelf or a support for planters. More elaborate railings are also possible, but most of them require more materials and installation time. A more extensive railing, with spaces no wider than 4" between rails or balusters, would be required if the deck is elevated at 2'-6" or more above the ground. The system in this plan is very economical and basic in its design, but it is not child-proof. Safer, more expensive railings should be installed on decks that create a safety hazard for small children. Stairway railings are not included in this plan because of its one-step access, but they should be installed at extra cost on deck stairways of more than one step. The posts for the railings should be constructed from 4 x 4s that are mortised into the frame and then fastened with lag bolts. Although the fitting of the posts takes a little longer and the lag bolts are more expensive than nails, the finished installation looks better, is more secure, and will last longer.

LEVEL OF DIFFICULTY

This basic deck plan is an approachable project for beginners and less-skilled intermediates. A few basic carpentry skills, along with some experience in measuring, plumbing, and leveling, are required. Beginners should seek advice before they start and as the project progresses, and should double the professional time for all tasks involved in this project. Intermediates and experts should add 30% and 10%, respectively, to the estimated labor-hours for the installation process.

WHAT TO WATCH OUT FOR

Because this deck is built at ground level with limited access to the terrain underneath, precautions should be taken to restrict unwanted vegetation from growing in this area. Some weeds and hardy grasses can establish themselves under the deck, eventually restricting ventilation and creating an eyesore. Decks that are built in sunny locations are particularly susceptible to this problem. To prevent this unwanted growth, remove about 3" to 4" of topsoil from the area to be covered by the deck and then cover it with a layer of polyethylene. Place enough gravel over the plastic to bring the entire area to grade level. This procedure will increase the cost of the job, but it will prevent a problem that is difficult to correct after the deck is in place.

Ground Level Deck

Description	Quantity/ Unit	Labor- Hours	Material
Layout, excavate post holes	0.50 C.Y.	0.5	
Concrete, field mix, 1 C.F. per bag, for posts	4 Bags		29.76
Forms, round fiber tube, 1 use, 8" diameter	12 L.F.	2.5	26.06
Post base, 4 x 4	4 Ea.	0.3	22.80
Post cap, 4 x 4	4 Ea.	0.3	11.18
Deck material, pressure-treated posts, 4 x 4 x 4'	16 L.F.	0.6	27.84
Headers, 2 x 6 x 10'	40 L.F.	1.3	53.76
Joists, 2 x 6 x 8'	72 L.F.	2.3	96.77
Decking, 2 x 4 x 10'	280 L.F.	8.2	168.00
Stair material, pressure-treated stringers, 2 x 10 x 3'	6 L.F.	0.2	13.32
Treads, 2 x 4 x 3'-6", 3 per tread	12 L.F.	0.4	7.20
Landing, precast concrete, 14" wide	4 L.F.	0.1	16.32
Railing material, pressure-treated posts, 2 x 4 x 3'	30 L.F.	0.9	18.00
Railings, 2 x 4 stock	52 L.F.	1.5	31.20
Cap rail, 2 x 6 stock	26 L.F.	0.8	34.94
Joist and beam hangers, 18 ga. galvanized	7 Ea.	0.3	3.61
Nails, #10d galvanized	8 lbs.		11.33
Bolts, 1/2" lag bolts, 4" long, square head, with nut and washer	18 Ea.	1.0	16.20
Totals		21.2	$588.29

Project Size	8' x 10'	Contractor's Fee Including Materials	$1,884

Key to Abbreviations
C.Y.– cubic yard Ea.– each L.F.– linear foot Pr.– pair Sq.– square (100 square feet of area)
S.F.– square foot S.Y.– square yard V.L.F.– vertical linear foot M.S.F.– thousand square feet

SUMMARY

This project provides an excellent opportunity for inexperienced do-it-yourselfers to improve their construction skills. With some advice and instruction, they can complete this deck for the cost of materials, while adding an attractive, functional, and valuable addition to their homes.

For other options or further details regarding options shown, see

Elevated deck

Elevated L-shaped deck

Entryway steps

Redwood hot tub

ROOF DECK

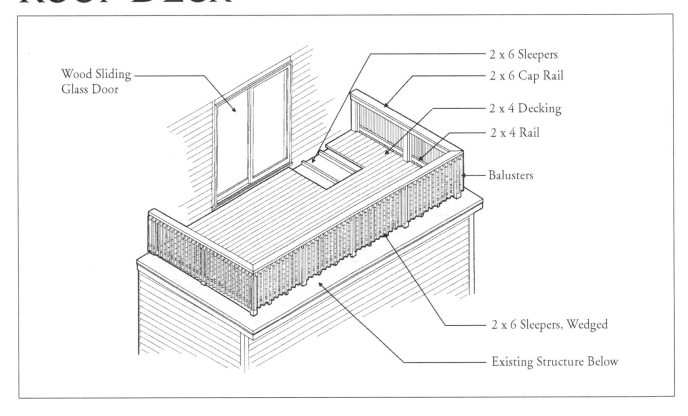

Wood Sliding Glass Door

2 x 6 Sleepers
2 x 6 Cap Rail
2 x 4 Decking
2 x 4 Rail
Balusters
2 x 6 Sleepers, Wedged
Existing Structure Below

Not all houses can accommodate a roof deck, but those that can have the potential for a unique outdoor living area. This plan demonstrates how normally unused roof area can be turned into deck space. It is a challenging project for the do-it-yourselfer because of the general inconvenience involved in all roof work and because of the many carpentry skills that are required. Generally, a flat or gently sloping roof is best suited to a deck. A steeper roof can also accommodate a deck, but a different support system must be employed. The critical factor is the dormer or exterior wall, which must be able to accommodate a door for access to the proposed deck. Without this basic structural condition, the project cannot be accomplished without costly preliminary work.

MATERIALS

Conventional deck materials and design are used in this roof deck plan, with the addition of sleepers that function as the deck frame and support system. Pressure-treated lumber is employed for the basic components, including the sleepers, decking, and railing. If the roof is flat, the 2 x 6 sleepers can be laid on edge as in the example; but if the roof is slanted, they have to be sawn at an angle to compensate for the slope. Be sure to make the angled cut precise and consistent for all of the sleepers to ensure stability and equal load distribution. Some shimming will also be needed. Before you place the sleeper system, carefully check the condition of the roof and its rafters. If there are questionable areas, take steps to correct the situation, seeking professional advice and assistance if needed. Take care not to damage or puncture the roofing material during the deck installation process. Be sure to use proper caulking at points where the deck is attached to the roof.

After the sleeper frame has been placed, fastened, and squared, the decking and railing are installed as in any other deck. The 2 x 4 boards are laid with staggered end joints and with a space between courses to allow for drainage. The railing consists of doubled 2 x 4 posts, fitted and lagged into the sleeper frame, 2 x 4 rails, 1 x 1 balusters, and a 2 x 6 cap. Great care should be taken in the design of the railing system for the roof deck; safety should be a primary feature. Check local building codes and regulations before construction begins to determine the minimum requirements for safety railings on off-ground decks.

The doorway to the roof deck should be installed before the deck is constructed to provide a convenient means of access to the roof site during construction. Many options are available in the choice of doors, as long as they are compatible with the structure and size of the wall. This plan includes the cost of materials for framing a 6' sliding glass door. The preparations will vary for other types and sizes of doors. Work carefully when cutting the opening to prevent excessive damage to the interior wall and exterior siding. Whenever possible, it pays to use an existing window space as part of the new door opening. You will have to enlarge and modify the rough window opening, but you will still be a step ahead in your door installation by using this approach.

LEVEL OF DIFFICULTY

Beginners should not attempt to build this deck on their own unless it is to be located on a flat roof. Even then, they should seek professional advice before they begin the job. They should also leave the entire door installation to a professional, as the cutting of the opening and placement of the new unit involves considerable carpentry know-how. Intermediates should be able to complete the deck construction in about twice the professional labor-hours, but they should add more time if the roof area is not easily accessible. Plan to work slowly and seek guidance. Hole-cutting and door installation are not tasks for the intermediate who is inexperienced in these areas. Experts should add about 30% to the professional time for all tasks. They should allow a bit more time for the hole-cutting and door installation, as hidden problems tend to arise in these operations, especially in older homes. If electrical wiring is located in the wall, shut off the supply before cutting the hole. Hire an electrician to rewire the affected area if you are unfamiliar with electrical work. All do-it-yourselfers should consider the risks and inconvenience of working on a roof before they start this project, and make plans to acquire the appropriate equipment for safe and efficient roof work.

Roof Deck

Description	Quantity/ Unit	Labor- Hours	Material
Deck material, pressure-treated sleepers, 2 x 6 x 8'	112 L.F.	3.6	150.53
Header, 2 x 6 x 16'	16 L.F.	0.5	21.50
Decking, 2 x 4 x 12'	408 L.F.	11.9	244.80
Railing material, pressure-treated posts, 2 x 4 x 3', doubled	72 L.F.	2.6	125.28
Railings, 2 x 4 stock	72 L.F.	2.1	43.20
Balusters, 2 x 2 treated pine	288 L.F.	7.7	141.70
Cap rail, 2 x 6 stock	36 L.F.	1.2	48.38
Cut opng. for sliding glass door, flr. sheathing & flooring, to 5 S.F.	1 Ea.	1.6	
Cut wall, sheathing to 1" thick, not including siding	1 Ea.	1.1	
Cut wall, drywall to 5/8" thick	1 Ea.	0.3	
Frame for door, 2 x 4 to support header	48 L.F.	0.8	20.74
Header, 2 x 8, doubled	14 L.F.	0.7	14.62
Door, sliding glass, stock, wood frame, 6' wide, economy	1 Ea.	4.0	828.00
Nails, #10d galvanized	10 lbs		14.16
Bolts, 1/2" lag bolts, 4" long, square head, with nut and washer	24 Ea.	1.4	21.60
Totals		39.5	$1,674.51

Project Size   8' x 16'

Contractor's Fee Including Materials **$4,364**

Key to Abbreviations
C.Y.– cubic yard   Ea.– each   L.F.– linear foot   Pr.– pair   Sq.– square (100 square feet of area)
S.F.– square foot   S.Y.– square yard   V.L.F.– vertical linear foot   M.S.F.– thousand square feet

WHAT TO WATCH OUT FOR

Several important factors must be considered before installing a roof deck to avoid overloading the existing roof. Remember that the weight of the 8' x 16' deck will increase when in use. Be sure that the existing rafters and roofboards are in good condition and are sturdy enough to support *more* than the anticipated load. If additional support members are needed, they will have to be installed before you begin construction of the new deck. If you are in doubt as to how to reinforce the roof support, consult a contractor and hire him or her to do the work. The additional expense will be well worth the investment to ensure proper and safe support for the new deck.

SUMMARY

If the conditions of the house are right, a roof deck addition can provide a unique dimension in exterior living space. For do-it-yourselfers who can handle some or all of the work, the project can be completed at a surprisingly reasonable cost.

For other options or further details regarding options shown, see

Patio & sliding glass doors

Section Two
DORMERS

The addition of a dormer adds window area and living space to second or third story rooms and enhances the exterior appearance of your house. Consider the following tips when planning a dormer addition to your home.

- A dormer is a major visual element in a home's overall appearance. Its design, size and spacing relative to other features are crucial, and should be given careful consideration.

- Be sure to assemble all the tools and helpers you'll need before you begin the work in order to minimize the length of time the house is left open to the elements.

- For a large or complex dormer, be sure to have appropriate protective materials available to cover the opening overnight while the project is in progress.

- Wait for good weather and have a fairly accurate idea of how much you can complete in one day.

- The two most commonly installed dormer types are the gable dormer and the shed dormer. Both require adequate area and ridge height.

- The pitch of the dormer's roof must meet building code requirements, just as a full-size roof must do.

- The cutting of the roof opening is one of the most critical operations in a dormer project. Accurate measurement, both inside and out, is required to lay out the cuts before they are made.

- Don't assume that your existing ceiling joists will serve as adequate floor joists if your dormer project is part of a conversion of currently unoccupied space (e.g., attic). If the project requires that you increase the depth of your floor joists, don't overlook the effects on stairs, headroom, and window height.

- The thickness of existing materials must be matched, which can be tricky in older homes.

- Most roof work requires proper staging with adequate working space/platforms.

- If the new dormer is to be bedroom space, be sure that at least one window in the space meets building code requirements for egress (a clear opening space of 20″ × 24″ minimum).

- If the purpose of adding the dormer is to create new living space from attic space, be sure to check building code egress requirements for window size.

4′ WIDE GABLE DORMER

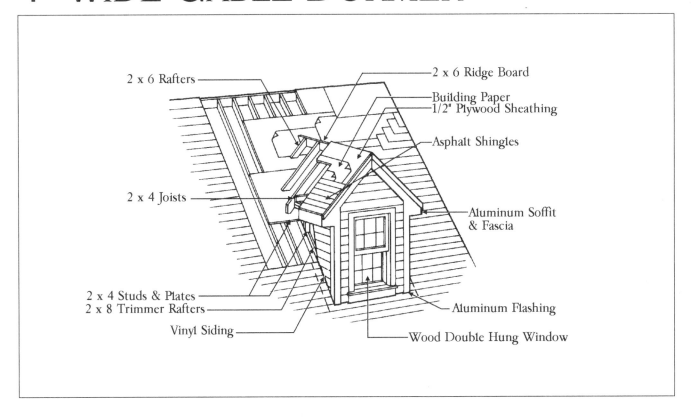

2 x 6 Rafters

2 x 4 Joists

2 x 4 Studs & Plates
2 x 8 Trimmer Rafters

Vinyl Siding

2 x 6 Ridge Board
Building Paper
1/2″ Plywood Sheathing

Asphalt Shingles

Aluminum Soffit
& Fascia

Aluminum Flashing

Wood Double Hung Window

The size and design of a dormer depends greatly on the style of the house and the amount of area available for the structure. While gable dormers are smaller than shed dormers, their installation still requires significant carpentry know-how and skill.

MATERIALS

Basically, dormers use the same building materials employed in the exterior wood construction of a complete house, just in smaller amounts. This dormer consists of a wood-frame support system, wall and roof materials, and other exterior construction products for the finish work. All of these materials are commonly used in home building and remodeling and are readily available from most lumber yards. Because of their abundant supply, it pays to shop for them and compare prices to get the best deal. The materials order for this project should be trucked, so make sure that delivery arrangements are made before the purchase is finalized.

Before the new dormer is constructed, extensive site preparation must be completed. The location of the dormer

on the roof must first be precisely determined, and then the opening cut from outside. Be sure to make the cut adjacent to existing roof rafters and strip the roofing materials carefully in the area of the opening before you saw the roof sheathing. After the rafters in the dormer opening have been cut and the sections removed, the cut ends have to be headed off, and trimmer rafters of the same cross section added as support to the full length of the existing rafters. These doubled support rafters are important in that they maintain the structural integrity of that part of the roof and help to support the load of the new dormer. After the opening has been prepared, the dormer can be framed in basically the same order as the framing of a house. The 2 x 4 wall studs are placed, and the 2 x 4 ceiling joists and 2 x 6 rafters are installed. The placing of the framing members for this gable dormer takes time because of the number of angle cuts and the extensive amount of fitting involved in the process. Remember, too, that the cutting and placement will probably be done under less than favorable conditions, with some traveling up and down a ladder or in and out of the opening as part of the job.

Once the frame has been completed, 1/2″ plywood sheathing is applied to the walls and roof and then covered with roofing felt. The roofing material suggested in this example plan is standard asphalt shingling, but your choice of materials should match as closely as possible the existing roofing. Cedar and some other materials will be more expensive than the standard roofing listed in the plan. The same applies to the selection of the siding, soffit, fascia, and window. The vinyl siding and the aluminum soffit and fascia, for example, should be used only if they are compatible with the materials that are already on the house. When selecting the window, purchase a quality unit that will be energy-efficient and require minimum maintenance. Before placing the siding, make sure that the seams between the old roof and the dormer walls have been correctly flashed. This process and the installation of the valley require time and roofing know-how, so seek the help of a professional if you are inexperienced in these tasks. After the dormer has been completed on the exterior, the insulation can be placed in preparation for the interior finish work. This cost estimate covers the work necessary to complete only the exterior of the dormer.

LEVEL OF DIFFICULTY

All of the dormer projects included in this section require moderate to advanced carpentry skills and a knowledge of roof structure and framing. There are, in addition, the challenges of awkwardness and height in any work done on the roof. Specialized equipment, such as roof jacks and staging, may need to be rented or borrowed. Steep roofs can be particularly difficult to work on without the right equipment. Hand-held power tools will have to be operated in awkward and inconvenient positions. For these reasons, beginners and do-it-yourselfers with limited skills are not encouraged to tackle dormer installations and should leave the entire project to a contractor. Experts and capable intermediates can complete most of the work, but they should proceed cautiously on the hole-cutting and all of the roofing and siding tasks. Even experts should consider hiring an experienced roofer to install the valley and to restore the disturbed sections of the old roofing. A poor roofing job can cause interior water damage that will exceed the roofer's fee many times over. Experts and intermediates should add 50% and 100%, respectively, to the professional time for most tasks, and more for the roofing and flashing installations if they are inexperienced in these areas.

WHAT TO WATCH OUT FOR

Because small gable dormers like this one are often installed in pairs, their symmetrical placement on the roof can further complicate the layout process. If the proposed opening is easily accessible from the inside, the job becomes much easier, especially if you have a helper. With two workers, one inside and one outside the roof, the rafters can be more easily located, the corners of the opening established, and the cutting efficiently accomplished. Remember that rapid completion of the dormer is desirable in order to restore the weathertight condition of the roof as soon as possible.

SUMMARY

Installing one or a pair of gable dormers can add space and window area to the upstairs of your house and improve its exterior appearance. With the right equipment and know-how, experts and capable intermediates can complete much of the dormer construction on their own to reduce the cost. Beginners, however, should leave all of this project to a professional contractor.

For other options or further details regarding options shown, see

7' wide gable dormer

Shed dormers – 12', 20', and 30' wide

Standard window installation

4' Wide Gable Dormer

Description	Quantity/Unit	Labor-Hours	Material
Removal of roofing shingles, asphalt strip	200 S.F.	2.3	
Roof cutout and demolition, sheathing to 1" thick, per 5 S.F.	5.20 Ea.	6.9	
Framing, rafters, 2 x 6 x 4', 16" O.C.	56 L.F.	1.5	38.30
Ridge board, 2 x 6 stock	8 L.F.	0.3	5.47
Trimmer rafters, 2 x 8 x 18' long	36 L.F.	1.1	37.58
Studs & plates, 2 x 4 stock	72 L.F.	1.3	31.10
Sub-fascia, 2 x 6 stock	12 L.F.	0.3	8.21
Headers, double, 2 x 6 stock	16 L.F.	0.7	10.94
Ceiling joists, 2 x 4, 16" O.C.	20 L.F.	0.3	8.64
Gable end studs, 2 x 4 stock	4 L.F.	0.1	1.73
Sheathing, roof, 1/2" thick plywood, 4' x 8' sheets	64 S.F.	0.7	37.63
Walls, 1/2" thick plywood, 4' x 8' sheets	64 S.F.	0.9	37.63
Building paper, 15 lb. felt	95 S.F.	0.2	2.28
Drip edge, aluminum, 0.016" thick, 5" girth	20 L.F.	0.4	4.80
Flashing, aluminum, 0.019" thick	35 S.F.	1.9	32.34
Shingles, asphalt strip, organic, class C, 235 lb. per square	1 Sq.	1.6	45.00
Soffit & fascia, 1 foot overhang, aluminum, vented	20 L.F.	2.7	35.52
Window, double-hung, 3' x 3'-6", vinyl clad	1 Ea.	0.9	366.00
Siding, vinyl, double 4" pattern	50 S.F.	1.6	33.00
Insulation, fiberglass, 3-1/2" thick, R-11, paper-backed	50 S.F.	0.3	13.80
Fiberglass, 6" thick, R-19	10 S.F.	0.1	4.08
Totals		26.1	$754.05

Project Size	4' x 6'-6"	Contractor's Fee Including Materials	$2,328

Key to Abbreviations
C.Y.– cubic yard   Ea.– each   L.F.– linear foot   Pr.– pair   Sq.– square (100 square feet of area)
S.F.– square foot   S.Y.– square yard   V.L.F.– vertical linear foot   M.S.F.– thousand square feet

7' WIDE GABLE DORMER

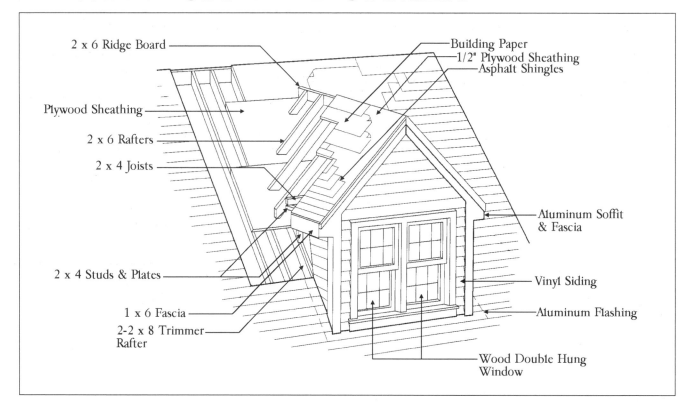

2 x 6 Ridge Board

Plywood Sheathing

2 x 6 Rafters

2 x 4 Joists

2 x 4 Studs & Plates

1 x 6 Fascia

2-2 x 8 Trimmer Rafter

Building Paper

1/2" Plywood Sheathing

Asphalt Shingles

Aluminum Soffit & Fascia

Vinyl Siding

Aluminum Flashing

Wood Double Hung Window

As noted in the small gable dormer plan, style, size, and location are crucial elements in dormer design. When placed on the roof of a house, they become a prominent feature, and the homeowner should ensure that they meet both aesthetic and practical needs. This double-window gable dormer plan demonstrates how a larger-size dormer can be installed for a reasonable cost.

MATERIALS

The exterior of a dormer is in some ways like a miniature wood-framed house in that it requires a wood support system, framed walls, and a roof. Because the materials, including 2 x 4s, 2 x 6s, siding, roofing materials, and the window unit, are usually in good supply, the do-it-yourselfer is encouraged to do some price comparing before buying. Once again, where standard building materials are concerned, it is to the homeowner's advantage to purchase everything from the same retailer. Consolidation of the order is less complicated and provides the buyer with the leverage to get the best prices. Further, an all-in-one order is probably large enough to warrant free local delivery from the supplier.

The first major procedure in this project is the layout and cutting of the opening in the roof to receive the new structure. Because gable dormers are often installed in twos, and sometimes threes, the location of the structures on the roof in relation to one another is paramount to a successful design. Whenever possible, the longitudinal cuts for the roof opening should be made next to existing rafters. The width of the dormer, therefore, may have to be adjusted to fit the rafter spacing on a given roof. The roofs of most houses have rafters spaced at 16" on center, but in some homes they may be spaced at 20", 24", or even larger intervals. In the latter case, plan on increasing the cost of this operation, as additional rafters and supports may have to be placed to restore the roof structure to safe standards. While the roof is open, check the condition of the other rafters and sheathing. If either of these components is questionable, have an inspection done by a professional. Any problems should be corrected before you proceed with the new installation. Placing a new dormer on a weak or deteriorated roof surface is a waste of time and money and can cause serious structural problems later on.

After the opening has been made, new support materials have to be placed, even on newer roofs. The most important of these supports are the trimmer rafters, which double the existing rafters next to the opening and span the entire length from the ridge to the wall plate. The ends of the cut rafters must be capped with doubled headers. In both cases, the lumber should match the dimensions of the existing rafter material.

The walls, ceiling, and roof of the dormer are framed and covered after the opening has been fully prepared. Because of the number of angle cuts and the individual fitting required for many of the framing members, this procedure is time consuming. If you are unfamiliar with the task, plan on making a few mistakes at first and allow some extra time for double-checking the measurements and angles as you work. Remember, too, that the cutting will usually have to be done in an area remote from the dormer site, on the ground or a suitable surface inside the opening; so you will use up more time going up and down a ladder or climbing in and out of the opening as the cuts are made. Plywood sheathing is placed on the dormer walls and roof after the framing has been completed. Be sure to scribe the angled side pieces accurately, as a tight fit is desirable. Remember, too, that the roofing, siding, and trim materials should match the products already in place on your house. Also, trim and structural features, such as the width of the fascia and the proportions of the soffit, should be duplicated as closely as possible. These matching materials and design features could change the cost estimates for both labor and materials, if the products in this plan have to be replaced to meet the needs of your particular installation. The most important decision in the selection of finish materials is the choice of the window

unit, the design and size of which should approximate or complement, if not match, the other windows in the house.

LEVEL OF DIFFICULTY

This project requires considerable carpentry skill and know-how. Beginners and intermediates whose remodeling experience is limited should not tackle it. The structure itself is a challenging undertaking, and its inconvenient roof location makes it a bit more complex. Do-it-yourselfers who undertake this project should anticipate some problems inherent to roof work. Some of these include height, limited access to the site, and the awkward position from which

power and hand tools must be operated. Two steps in the project are critical to the overall success of the installation. One is the adequate restoration of the roof support with trimmer rafters and headers after the opening has been cut. The other is correct roofing and flashing installation to restore weathertight conditions to the roof. The roofing work, particularly, may require the services of an experienced tradesperson, and even expert do-it-yourselfers are encouraged to consult a qualified roofer before attempting this task. Generally, experts should add 50% to the professional time for all procedures, and more for the finish roofing and flashing jobs. Intermediates should add 100% to the labor-hours estimates and should hire a professional for the roofing work if they are inexperienced in this area.

WHAT TO WATCH OUT FOR

Because of their inconvenient roof location, dormers are notoriously difficult to paint and maintain. As a result, you will want to select finish materials that will cut the time and difficulty of painting, window washing, and other maintenance chores. For example, window units with removable sashes may cost a little more initially, but the convenience of washing or painting from inside the house, rather than from the roof or an outside ladder, might make the extra expense worthwhile. In most cases, stained surfaces are easier to apply and maintain than painted ones. If you can match the trim color of your house with a stain, use it on the dormer instead of paint to reduce the scraping and application time of future paint jobs.

7' Wide Gable Dormer

Description	Quantity/ Unit	Labor- Hours	Material
Removal of roofing shingles, asphalt strip	200 S.F.	2.3	
Roof cutout and demolition, sheathing to 1" thick, per 5 S.F.	15 Ea.	20.0	
Framing, rafters, 2 x 6, 16" O.C.	108 L.F.	2.9	73.87
Ridge board, 2 x 8 stock	12 L.F.	0.4	8.21
Trimmer rafters, 2 x 8 x 18' long	36 L.F.	1.1	37.58
Studs & plates, 2 x 4 stock	120 L.F.	2.1	51.84
Sub-fascia, 2 x 6 stock	16 L.F.	0.4	10.94
Headers, double, 2 x 6 stock	24 L.F.	1.1	16.42
Ceiling joists, 2 x 6, 16" O.C.	36 L.F.	0.5	24.62
Gable end studs, 2 x 4 stock	10 L.F.	0.2	4.32
Sheathing, roof, 1/2" thick plywood, 4' x 8' sheets	105 S.F.	1.2	61.74
Walls, 1/2" thick plywood, 4' x 8' sheets	120 S.F.	1.7	70.56
Building paper, 15 lb. felt	210 S.F.	0.5	5.04
Drip edge, aluminum, 0.016" thick, 5" girth	28 L.F.	0.6	6.72
Flashing, aluminum, 0.019" thick	62 L.F.	3.4	57.29
Shingles, asphalt, multi-layered, 285 lb. per square	2 Sq.	4.0	117.60
Fascia, 1 x 6, rough-sawn cedar, pre-stained	28 L.F.	0.9	74.93
Soffit, 3/8" rough-sawn cedar, plywood, pre-stained	14 S.F.	0.3	20.16
Window, double-hung, 5' x 3', vinyl-clad, thermopane	1 Ea.	1.0	378.00
Siding, cedar, beveled, rough-sawn, pre-stained	120 S.F.	4.0	433.44
Insulation, fiberglass, 3-1/2" thick, R-11, paper-backed	120 S.F.	0.6	33.12
Fiberglass, 6" thick, R-19	120 S.F.	0.8	48.96
Totals		50.0	$1,535.36

Project Size	6' 8" x 10' 6"	Contractor's Fee Including Materials	$4,615

Key to Abbreviations
C.Y.– cubic yard   Ea.– each   L.F.– linear foot   Pr.– pair   Sq.– square (100 square feet of area)
S.F.– square foot   S.Y.– square yard   V.L.F.– vertical linear foot   M.S.F.– thousand square feet

SUMMARY

If you have good remodeling skills and are willing to work at a slow pace, you can complete this improvement at reasonable cost while adding to the value, living space, and appearance of your home.

For other options or further details regarding options shown, see

4' wide gable dormer

Shed dormers – 12', 20', and 30' wide

Standard window installation

12' WIDE SHED DORMER

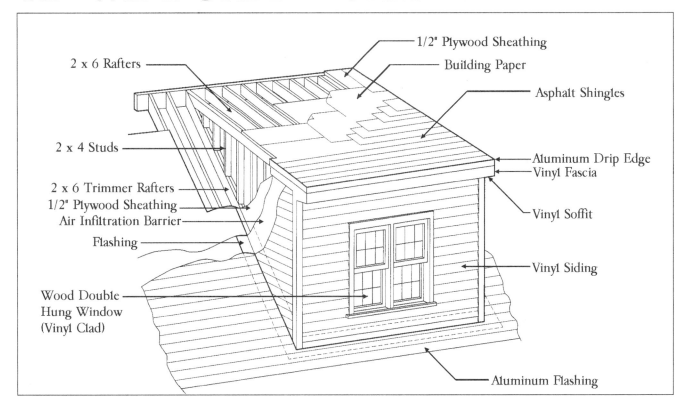

2 x 6 Rafters

2 x 4 Studs

2 x 6 Trimmer Rafters
1/2" Plywood Sheathing
Air Infiltration Barrier
Flashing

Wood Double
Hung Window
(Vinyl Clad)

1/2" Plywood Sheathing
Building Paper
Asphalt Shingles
Aluminum Drip Edge
Vinyl Fascia
Vinyl Soffit
Vinyl Siding
Aluminum Flashing

Although they are not as intricate as gable dormers, shed dormers are an efficient means of increasing the living space and window area of second- or third-floor rooms. Also, shed dormers are generally easier to install than those with gable designs. The framing and roofing operations can be accomplished by do-it-yourselfers with advanced skill levels. If you feel confident in your carpentry abilities, understand the structural composition of roofs, and are willing to cope with the inconveniences of roof work, you can complete this improvement on your own.

MATERIALS

The amount of construction material used for shed dormer projects is usually greater than for gable dormers because they tend to be larger structures. Yet the same basic materials are employed in essentially the same sequence of operations. A shed dormer roof often extends from the ridge of the existing roof to a point near or directly over the exterior wall. Because the dormer roof is flat, not gabled or peaked, the amount of usable open space enclosed by the structure is increased, and maximum

ceiling height is attained. As in most exterior house improvements, framing lumber of varying dimensions, plywood sheathings, roofing, siding, and various finish products are used in this plan. You can save by comparison shopping before buying. For an order this size, the difference can mean substantial savings.

The removal and reconditioning of the affected roof area is a critical operation in preparing the house for the new structure. The roofing, sheathing, and rafters must be torn out, the structural integrity of the old roof restored, and the rafter support for the dormer installed. In most cases, it is best to make the longitudinal cuts beside existing rafters, which are then reinforced with trimmers. These new supports must span from the ridge to the eaves. The existing rafters are usually spaced at 16" on center, but they may be located at odd intervals in some older houses. Be sure to match or exceed the dimensions of the existing rafters with the trimmer material. Any deficiencies in the existing roof or its support system should be corrected before the framing of the new dormer begins.

The frame for the shed dormer consists of 2 x 4 walls and 2 x 6 rafters and ceiling joists. After the front wall of the dormer

has been framed, placed, and temporarily braced, the joists and rafters are installed. As in other roof framing operations, the general rule is consistency of length, angling and notching, and spacing of the roof framing members. The framing of the side walls must be accomplished one piece at a time, with two different angles cut at each end of the studs to fit the roof lines. This process takes some time, so plan accordingly, especially if the task is new to you.

Once the framing has been completed, 1/2" plywood sheathing is applied and then covered with an air-infiltration barrier, either housewrap or 15-pound felt building paper. As in other dormer plans, the finish materials that are used on the dormer should match the products already in place on the rest of the house. This plan suggests low-maintenance vinyl siding and aluminum soffit and fascia coverings, but other materials, such as cedar shingles and clapboards and conventional wood trim, may be more appropriate for your house. A gutter and downspout for the shed roof have not been included in the plan, but it might be a good idea to add them if the dormer is located above a doorway. If the face of the dormer is a continuation of the house wall, a gutter and downspout will be necessary. The roofing should match the existing roof material, except where prohibited by the pitch of the roof. If the ridge is low and the pitch of the new shed roof nearly flat, rolled roofing of matching color might have to be used, as shingles are not recommended for roofs with shallow slopes. Add fiberglass batt insulation between the rafters. When selecting the window unit, try to use thermopane sash in the same design of your present windows. The extra expense for a high-quality window will be returned in energy savings and comfort over the years. These and other changes from the materials list in this plan will affect the cost of the finish work as well as the labor-hours estimate.

LEVEL OF DIFFICULTY

This dormer project is challenging, even for experts and accomplished intermediates, and is out of reach for beginners. Whenever you open a roof and tamper with major structural components, like rafters, you have to have the building skill to accomplish the work efficiently and correctly. All do-it-yourselfers should be prepared to cope with the inconvenience and difficulties associated with roof work. Skilled intermediates and experts should be able to complete all of the tasks involved in the project. They should add 50% and 20%, respectively, to the professional time for all of the procedures. The new roofing may pose some challenges and require some extra time, especially in the flashing and the restoration of disturbed sections of the old roof. Nevertheless, once the drip edge has been placed and the job correctly laid out, the roofing material should go on smoothly. Remember to allow extra time and to get some help for hauling bundles of roofing to the work area. As with most of the exterior wood structures outlined in this book, experience in using power tools is a prerequisite.

WHAT TO WATCH OUT FOR

All dormer installations should be completed or, at the least, closed in as rapidly as possible to protect the inside of the house from the elements. If you are doing the job on your own, arrange to have time for several successive days of work. Although sudden weather changes cannot be predicted, you can reduce the chance of getting caught off guard by scheduling the dormer project during the warm and dry time of year. Nevertheless, always have an emergency cover ready to use in the event of a sudden rainstorm, and then hold off on the resumption of construction until the roof area dries. Be aware that wet roofs can also be dangerous.

12′ Wide Shed Dormer

Description	Quantity/ Unit	Labor- Hours	Material
Removal of roofing shingles, asphalt strip	100 S.F.	1.1	
Roof cutout and demolition, sheathing to 1″ thick, per 5 S.F.	20 Ea.	26.7	
Framing, rafters, 2 x 6 x 12′, 16″ O.C.	144 L.F.	3.9	98.50
Trimmer rafters, 2 x 8 x 18′ long	36 L.F.	1.1	37.58
Studs and plates, 2 x 4 stock	192 L.F.	3.4	82.94
Sub-fascia, 2 x 6 stock	14 L.F.	0.4	9.58
Header, double, 2 x 8, 5′ long	10 L.F.	0.5	10.44
Header, double, 2 x 12, 20′ long	24 L.F.	1.3	48.96
Ceiling joists, 2 x 6, 12′ long	144 L.F.	1.8	98.50
Sheathing, roof, 1/2″ thick plywood, 4′ x 8′ sheets	198 S.F.	2.3	116.42
Wall, 1/2″ thick plywood, 4′ x 8′ sheets	90 S.F.	1.3	52.92
Housewrap, spun bonded polypropylene	90 S.F.	0.2	17.28
Building paper, 15 lb. felt	60 S.F.	0.1	1.44
Drip edge, aluminum, 0.016″ thick, 5″ girth	36 L.F.	0.7	8.64
Flashing, aluminum, 0.019″ thick	24 S.F.	1.3	22.18
Shingles, asphalt, strip, 235 lb. per square	6 Sq.	9.6	270.00
Siding, vinyl, double, 4″ pattern	90 S.F.	2.9	59.40
Soffit and fascia, aluminum, vented	36 L.F.	4.8	62.21
Window, 3′ x 4′, double-hung, vinyl clad	2 Ea.	1.8	592.80
Insulation, fiberglass, 3-1/2″ thick, R-11, paper-backed	132 S.F.	0.7	36.43
Insulation, fiberglass, 6″ thick, R-19	60 S.F.	0.4	24.48
Totals		66.3	$1,650.70

Project Size	12′ x 8′		Contractor's Fee Including Materials	**$5,600**

Key to Abbreviations
C.Y.– cubic yard   Ea.– each   L.F.– linear foot   Pr.– pair   Sq.– square (100 square feet of area)
S.F.– square foot   S.Y.– square yard   V.L.F.– vertical linear foot   M.S.F.– thousand square feet

WINDOWS OPTIONS
Cost Each, Installed

Description	
Wood Double Hung, 3′ × 4′	$243.00
Vinyl Clad Double Hung, 3′ × 4′	$387.00
Wood Casement, 2′ × 4′-6″	$352.00
Vinyl Clad Casement, 2′ × 4′-6″	$337.00
Metal Clad Casement, 2′ × 4′	$251.00

SUMMARY

This small shed dormer can increase bedroom area and other second- or third-floor space without the expense of putting on a full addition. Although it is a challenging undertaking, many accomplished do-it-yourselfers can complete the project on their own for the cost of the materials and the investment of several days' time.

For other options or further details regarding options shown, see

Gable dormers – 4′ and 7′ wide

Shed dormers – 20′ and 30′ wide

Standard window installation

20' WIDE SHED DORMER

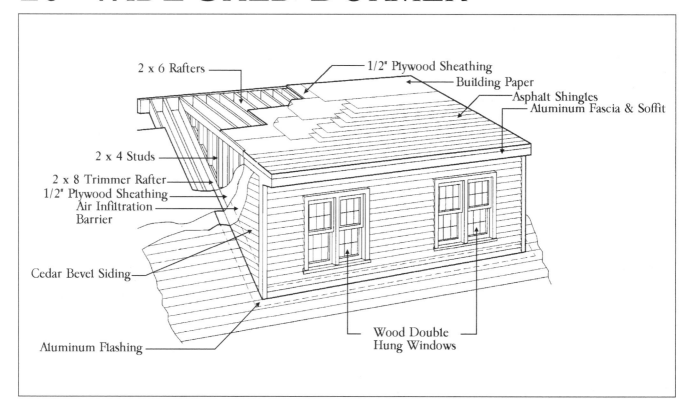

2 x 6 Rafters

1/2" Plywood Sheathing

Building Paper

Asphalt Shingles

Aluminum Fascia & Soffit

2 x 4 Studs

2 x 8 Trimmer Rafter

1/2" Plywood Sheathing

Air Infiltration Barrier

Cedar Bevel Siding

Wood Double Hung Windows

Aluminum Flashing

Wide shed dormers are usually placed on Cape-style houses and other structures with gable roof designs large enough to accommodate them. Before you begin the planning, measure your attic to see that you have enough ridge clearance to accept the dormer and enough roof and floor area to make the project worthwhile. A height of roughly 9' from the bottom of the ridge board to the floor of the room below or, in the case of an attic improvement, to the top of the ceiling joists, is a reasonable minimum measurement. Once the minimum height requirement has been met, the width and length of the dormer can be determined according to the intended use of the facility, the rafter spacing in the existing roof, and the location of interior partitions.

MATERIALS

This plan presents a basic design for the exterior shell of a 9' x 20' shed dormer with no interior materials included, except insulation. Although the structure is basic in its design, a substantial number of exterior materials is needed to construct it. If you are doing the building on your own, compare prices from

several suppliers before you place the order. Also, make sure that delivery is included in the final price, as the materials will have to be trucked. The process of constructing the dormer requires different materials for the various components as the structure is built, including 2 x 4s and 2 x 6s for the support system and frame, plywood sheathing, and various exterior coverings and finish materials.

The most critical step in the project is determining the location and laying out the section of roof to be opened for the new dormer. In most cases, the new structure will be located in an area that begins on the lower part of the roof and runs to the ridge board. This width measurement will vary with each case, depending on the size and pitch of the roof, aesthetic considerations, and the desired amount of new living space. The length of the roof area to be opened depends on such variables as the location of existing interior partitions, the design of the new living space and, wherever possible, the spacing of the existing roof rafters. The cost of this shed dormer project is based on a 9' x 20' structure placed on a roof with a steep pitch, but costs will vary somewhat if the dormer dimensions are modified. Once the

location has been determined and laid out, the roof section can be opened by removing the roofing, sheathing, and rafters.

After the opening has been made, the new work begins with the installation of trimmer rafters, the heading off of the cut rafter ends, and, if conditions warrant, the placing of additional support. Again, each case will vary, and additional costs and time may be required to bring the roof and dormer support system to safe standards. The frame of the new dormer should not be placed until its support system has been correctly installed and fastened. Professional consultation or subcontracting will save you money in the long run if extensive roof support reconditioning is required. The framing and closing in of the shed dormer proceeds in much the same order as that of other wood frame structures. The face of the dormer is framed with 2 x 4s and then braced temporarily while 2 x 6 ceiling joists and rafters are placed. The side walls are then framed and the entire enclosure sheathed with 1/2" plywood. An air-infiltration barrier is applied to the walls, and 15# felt to the roof. The roofing material, in this case asphalt shingles, is then applied to the shed roof before the window units are installed and the finish siding applied. Fiberglass batt insulation is installed between the rafters. The roofing, siding, and window units will vary with each installation, as the products that are selected for the dormer should match or approximate the materials already in place on the house.

If the materials differ from those listed in the estimate, the cost may be affected. For example, cedar roofing costs more than asphalt shingles, and quite a bit more than asphalt roll roofing. The windows included in this plan are quality thermopane double-hung units; but standard single-pane units can be installed for less expense. Remember, though, that the cost for storm windows, if you choose to install uninsulated windows, will have to be added. Generally, quality window units are worth the extra cost in the long run because of their return in energy savings and low

maintenance. Additional cost may also be incurred and more installation time required if a gutter and downspout are needed for the dormer. If the face of the dormer is far enough back from the edge of the existing roof, you probably won't need it; but if it is positioned close enough to the edge to cause spattering or icing problems from the roof run-off, then you should spend the extra money to install one.

LEVEL OF DIFFICULTY

This project is a major home improvement that demands a considerable amount of carpentry skill and know-how. Further, the roof location of the dormer increases the difficulty and adds significantly to the time and general inconvenience of the operations. In many instances, special equipment may have to be employed for roof and upper story work – staging, roof jacks, sturdy extension ladders, and

pump jacks, to name a few. Some of these devices may have to be rented, so allow for their rental cost and plan ahead to ensure their availability during the construction period. Be sure that you get some instruction in their operation before you use them if they are unfamiliar to you. Generally, beginners should not attempt this or other dormer projects, as the necessary skills and location of the work are too demanding. Experts and intermediates should add 25% and 75%, respectively, to the labor-hour estimates and be willing to seek professional advice when necessary.

WHAT TO WATCH OUT FOR

This dormer project plan presents a list of materials and estimated installation time for only the exterior of the structure. As the work progresses, though, take advantage of opportunities to make the subsequent interior finishing easier. Some of these odds and ends will take a little more time and money during the placement of the dormer, but they can provide substantial labor savings later on. For example, electrical or plumbing rough-in required for the interior of the new enclosure can be placed or relocated. Attic locations with hard-to-get-at places for insulation may be more accessible before the dormer is closed in. Plywood decking for a new attic room may be easier to stack or to place while the roof is open. Remember, however, that expediency is important for all dormer projects, so use discretion and take advantage of extra tasks that require only a minimum of time.

20' Wide Shed Dormer

Description	Quantity/ Unit	Labor- Hours	Material
Removal of roofing shingles, asphalt strip	300 S.F.	3.4	
Roof cutout and demolition, sheathing to 1" thick, per 5 S.F.	36 Ea.	48.0	
Framing, rafters, 2 x 6 x 12', 16" O.C.	192 L.F.	5.2	131.33
Trimmer rafters, 2 x 8 x 18'	36 L.F.	1.1	37.58
Studs and plates, 2 x 4 x 8', 16" O.C.	240 L.F.	4.2	103.68
Sub-fascia, 2 x 6	22 L.F.	0.6	15.05
Headers, double, 2 x 8 x 4' long	16 L.F.	0.8	16.70
Headers, double, 2 x 12 x 20' long	40 L.F.	2.1	81.60
Ceiling joists, 2 x 6 x 12'	192 L.F.	2.5	131.33
Sheathing, roof and walls, 1/2" thick, 4' x 8' sheets	384 S.F.	4.4	225.79
Housewrap, spun bonded polypropylene	130 S.F.	0.3	24.96
Building paper, 15 lb. felt	300 S.F.	0.7	7.20
Drip edge, aluminum, 0.016" thick, 5" girth	44 L.F.	0.9	10.56
Flashing, aluminum, 0.019" thick	40 S.F.	2.2	36.96
Shingles, asphalt, multi-layered, 285 lb. per square	3 Sq.	6.0	176.40
Siding, cedar, beveled, rough-sawn, stained	130 S.F.	4.3	469.56
Soffit and fascia, aluminum, vented	42 L.F.	6.1	108.86
Windows, double-hung, insulating glass, 3'-0" x 4'-6"	4 Ea.	3.6	873.60
Wall insulation, fiberglass, 3-1/2" thick, R-11, paper-backed	130 S.F.	0.7	35.88
Roof insulation, fiberglass, 6" thick, R-19	300 S.F.	2.1	122.40
Totals		99.2	$2,609.44

Project Size	20' x 9'	Contractor's Fee Including Materials	$8,557

Key to Abbreviations
C.Y.– cubic yard   Ea.– each   L.F.– linear foot   Pr.– pair   Sq.– square (100 square feet of area)
S.F.– square foot   S.Y.– square yard   V.L.F.– vertical linear foot   M.S.F.– thousand square feet

SUMMARY

If the ridge height and roof area are adequate, a relatively large dormer like this one will reclaim wasted floor space. Although the project is fairly costly and time-consuming, it remains an economical way of adding space that will return years of service while increasing the value of your home.

For other options or further details regarding options shown, see

Gable dormers – 4' and 7' wide

Shed dormers – 12' and 30' wide

Standard window installation

30' WIDE SHED DORMER

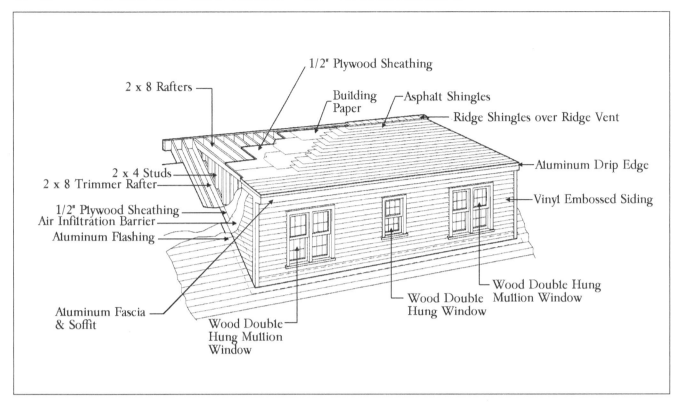

1/2" Plywood Sheathing

2 x 8 Rafters

Building Paper

Asphalt Shingles

Ridge Shingles over Ridge Vent

2 x 4 Studs

2 x 8 Trimmer Rafter

1/2" Plywood Sheathing
Air Infiltration Barrier

Aluminum Flashing

Aluminum Drip Edge

Vinyl Embossed Siding

Wood Double Hung Mullion Window

Wood Double Hung Window

Aluminum Fascia & Soffit

Wood Double Hung Mullion Window

Like other attached exterior structures, dormers should be carefully planned to meet the needs of the homeowner without harming the exterior appearance of the dwelling. For a large structure like this 29'-4" x 8' full shed dormer, the planning and design become even more critical, to ensure the efficient use of space and the appropriate architectural impact. Although shed dormers are usually placed on the rear roofs, they can often be seen from the front and side of the house. For this reason, the pitch and location of the new shed dormer should complement the roof lines and style of the house. This project is a major home improvement, requiring competence in carpentry and a significant amount of construction know-how.

MATERIALS

Construction grade 2 x 4, 2 x 6, and 2 x 8 lumber is needed to frame the roof opening and dormer. Standard sheathing covers the frame and serves as the subsurface for the roof and finished siding. Prefabricated window units and other finish and trim products are employed to complete the exterior of the structure.

The most important task involved in the project, after the layout and location of the dormer have been determined, is the cutting of the roof opening and the placement of the proper support for the structure. Time is an important factor because of the large opening that temporarily exposes the house's interior to the elements. Be sure to have plenty of help on hand, as there is a considerable amount of work to be done in a short period of time. Before opening the roof, double check to see that all supplies, tools, and equipment are readily available and in good working order. In short, all precautions should be taken to ensure expedient opening of the roof and closing in of the new dormer. Make sure that an emergency cover-up plan is in place.

Once the trimmer rafters and other support members have been placed, the new dormer is framed and sheathed. Because of its length, the face should be assembled and raised into position in sections with properly placed rough openings framed for the new windows. Double check all measurements for the locations and dimensions of the rough openings; correcting mistakes will be difficult and time consuming once the

partition is in place. After the face wall has been installed and temporarily braced, the 2 x 6 rafters and ceiling joists are placed. Because of the large area of this dormer roof, consistent angle cuts, notching and squaring of the rafters are paramount to ensure rapid placement of the roof sheathing and finish material. Accuracy of measurement and cutting is also important for the framing of the side walls, as the studs require two different angle cuts and have to be placed individually. The 1/2" plywood sheathing that covers the frame should be applied with the seams staggered to add strength to the roof and walls. Apply an air-infiltration barrier over the sheathing, and insulate the roof with fiberglass batts.

The finish materials and coverings that are installed over the sheathing will have the greatest impact on changes in the estimated cost for this project. The roofing and siding materials and the window units, especially, must be selected to match or complement those already in place on the house. The asphalt-strip shingle material suggested for the roofing, for example, is a commonly used product, but it may not be an appropriate choice for your roof. Remember, too, that roll roofing should be used if the new shed roof has little slope. The double-hung, vinyl-clad windows suggested in this plan may be replaced with thermopane casement units if house design features will allow. The embossed vinyl siding is low in maintenance, as are the aluminum soffit and fascia, but they may not be in keeping with the style of a house that features wood siding and trim. A contractor may have to be called in for the application of the roofing and siding, as these materials require particular skills and, often, specialized tools and equipment. The restoration of a weathertight condition to the roof and new dormer is vital to the prevention of leaks caused by rain and icing. Particular care should be given to the flashing of the old roof where it meets the dormer walls. A gutter with downspouts may have

to be included at extra cost if the location of the dormer roof poses potential drainage problems. The insulating of the dormer ceiling and walls should be accomplished as part of the exterior work.

LEVEL OF DIFFICULTY

This project is a difficult undertaking for all do-it-yourselfers because of the large size of the structure, the inconvenience of its location, and the expediency with which it must be completed. Also, many different carpentry skills must be employed during the project, including advanced framing tasks and precise fitting and angle cutting of framing, sheathing, and finish materials. The general inconvenience of the roof location often

adds time and extra effort to even basic jobs, and special equipment for roof work usually has to be employed to ensure safe and efficient working conditions. Generally, beginners should not attempt this or any other dormer installation, primarily because the project includes too many tasks requiring advanced carpentry skills within a very limited time span. Ambitious intermediates and experts should seek professional advice before beginning the project and arrange to have plenty of help on hand during the roof opening and framing operations. A dormer project of this magnitude involves a considerable effort just to open the roof, aside from the new building structure. Intermediates should add 75% to the professional time for all tasks, and more for specialized jobs like the installation of the flashing, vinyl siding, and aluminum soffit and fascia. Experts should add 25% to all of the labor-hour estimates throughout the project.

WHAT TO WATCH OUT FOR

The venting of any attic is an important part of its construction. This plan recommends the installation of a vented aluminum soffit to handle the attic's ventilation needs. If a wood soffit and fascia are placed, be sure to add enough vent area to ensure air movement through the attic space. Many different types of vents are available, from circular plug-in models of different sizes to larger rectangular fixtures and ventilation strips. The materials installation time required for additional vents has not been included in the plan.

SUMMARY

This 29'-4" x 8' shed dormer can add enough living area to the upper floor of your house to accommodate two moderate-size bedrooms and a full bath. With the correct planning and layout, the new dormer can also enhance your home's appearance and increase its value.

30' Wide Shed Dormer

Description	Quantity/ Unit	Labor- Hours	Material
Removal of roofing shingles, asphalt strip	400 S.F.	4.6	
Roof cutout and demolition, sheathing to 1" thick, per 5 S.F.	48 Ea.	64.0	
Framing, rafters, 2 x 8 x 12', 16" O.C.	288 L.F.	6.1	300.67
Trimmer rafters, 2 x 8 x 18'	36 L.F.	1.1	37.58
Studs and plates, 2 x 4 x 8', 16" O.C.	340 L.F.	6.0	146.88
Sub-fascia, 2 x 6	32 L.F.	0.9	21.89
Headers, double, 2 x 8 x 4' long	24 L.F.	1.1	25.06
Headers, double, 2 x 12 x 20' long	60 L.F.	3.2	122.40
Ceiling joists, 2 x 6 x 12'	288 L.F.	3.7	196.99
Sheathing, roof, 1/2" thick, 4' x 8' sheets	400 S.F.	4.5	232.85
Sheathing, walls, 1/2" thick, 4' x 8' sheets	180 S.F.	2.6	105.84
Housewrap, spun bonded polypropylene	500 S.F.	1.1	96.00
Building paper, 15 lb. felt	180 S.F.	0.4	4.32
Drip edge, aluminum, 0.016" thick, 5" girth	54 L.F.	1.1	12.96
Flashing, aluminum, 0.019" thick	50 S.F.	2.8	46.20
Shingles, asphalt strip, organic, class C, 235 lb. per square	5 Sq.	6.9	195.00
Ridge vent strip, polyethylene	30 L.F.	1.5	91.80
Siding, vinyl-embossed, 8" wide	180 S.F.	5.8	142.56
Soffit and fascia, aluminum, vented	52 L.F.	7.6	134.78
Windows, double-hung, insulating glass, vinyl-clad, 2'-6" x 3'	1 Ea.	0.8	264.00
Window, double-hung, insulating glass, vinyl-clad, 3' x 4'-6"	4 Ea.	3.6	1,464.00
Wall insulation, fiberglass, 3-1/2" thick, R-11, paper-backed	180 S.F.	0.9	49.68
Roof insulation, fiberglass, 6" thick, R-19	386 S.F.	2.7	157.49
Totals		133.0	$3,848.95

Project Size	29'-4" x 8'	Contractor's Fee Including Materials	**$11,953**

Key to Abbreviations
C.Y.– cubic yard   Ea.– each   L.F.– linear foot   Pr.– pair   Sq.– square (100 square feet of area)
S.F.– square foot   S.Y.– square yard   V.L.F.– vertical linear foot   M.S.F.– thousand square feet

For other options or further details regarding options shown, see

> *Gable dormers – 4' and 7' wide*
>
> *Shed dormers – 12' and 20' wide*
>
> *Standard window installation*

Section Three
EXTERIOR DOORS AND WINDOWS

Any projects that involve opening up interior space to the outdoors deserve some special considerations. It is wise to have the complete collection of needed tools and materials assembled in advance in order to prevent delays in installing the new door or window once the old one has been removed. Also, pre-hung doors can be heavy and awkward and windows are fragile. Having someone available to help you with these projects is an excellent idea.

- Measure the wall thickness before you purchase windows or doors. This is done by measuring the jamb width of another window or exterior door in your house. The manufacturer can make frame jambs to match your wall thickness.

- All wood doors in this book are priced pre-mortised for hinges and predrilled for cylindrical locksets. If you are planning to re-use hardware from an existing door, be sure to specify when ordering that you want a blank door, as opposed to a pre-drilled door ready to receive hardware.

- Many pre-hung doors come with average-quality hardware. Specialty knobs and hinges, such as solid brass or porcelain, can add considerably to the appearance, as well as the price, of a door. Locksets are purchased separately.

- Depending on the type and style of the exterior wood door you choose, a steel substitute could cost about half the price. Both steel and fiberglass doors provide a better thermal value; steel requires less maintenance.

- New wood doors should be sealed (primed, or primed and painted) before they are installed as per manufacturer's recommendation. Don't forget to cover all six sides, including the top and bottom edges.

- When deciding on a style of window for an older home, look at books that show homes from the same period for ideas on how to stay within the appropriate style. Major window manufacturers offer special lines of architecturally styled windows, and can provide well-illustrated catalogs to review.

- When replacing windows, beware of cracked or broken glass. You may want to apply masking tape to the glass before removing the sash.

- Most good windows must be ordered several weeks in advance. Don't remove existing windows before delivery of the new units, since shipment could be delayed.

- Skylights should let in as much light as possible. The goal is to distribute the light evenly inside the room without creating a glare. Skylights should ideally open for ventilation and to minimize heat gain in summer, but offer thermal protection to inhibit heat loss in winter.

- If a ventilating skylight is out of reach, motorized controls should be considered. When pricing skylights, also inquire about the cost of hardware extensions and handles to open and shut not only the windows, but also blinds and shades.

- Talk to your door or window dealer about glass types and options, such as tempered glass for strength, R-values for energy savings (4.0 is optimum), and "low-E" glass, which prevents sunlight from damaging interior furnishings.

Bow/Bay Window

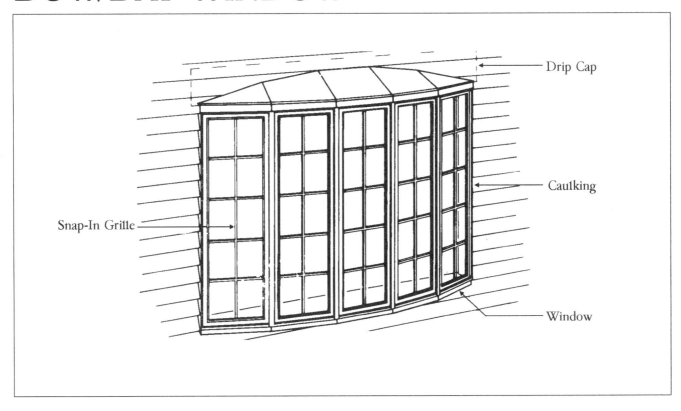

Drip Cap

Caulking

Snap-In Grille

Window

Of all the types of windows available today, none has as much dramatic visual impact as a bow or bay. While other windows might be higher and wider, or come with Palladian-style arches, the bay adds the third dimension – depth. On the inside, the effect is to make the room seem to expand outward, and the angles of the side lights take in the view like a wide-angle camera lens. The bay also adds diversity to the exterior walls. Do keep in mind that not every room is suitable for this type of window treatment, nor is every home. If your house and budget are right for a bay window addition, the end results can be impressive.

MATERIALS

One should begin this project by assessing the visual impact such a window will make on both the interior and exterior of the house. Factors to consider include: size, light (its orientation to the sun), view, structural constraints (e.g., load-bearing walls), interior appearance, insulation (heat loss is greater than with regular windows), exterior appearance, technical and functional features (these vary among brands), and the cost (of both

the unit and the installation). Before making a final purchasing decision, talk to someone with design and/or building experience, and educate yourself on the range of products available and how the various brands rate with contractors and remodeling professionals.

Bay windows are available in four basic styles, based on the angles they form with the plane of the house wall: 90°, 45°, 30°, and bow (made up of a number of sashes in a curved configuration). Because a bay projects outward, forming, in effect, a separate, tiny "room," it requires some sort of roof, as well as a support system to prevent sagging from its own weight.

The most common bay roof, used when the main roof is high above the window, is a hip-style matching the slope of the main roof, and shingled, either with red cedar or to match the house. Copper roofs are also popular. Most manufacturers offer roof kits for their windows, or you can customize your own. If your home has a narrow soffit, and if the bay is located close under it, adding an extension to the main roof is a way to cover the bay so that it blends into the existing exterior features and doesn't look like a poorly designed afterthought.

An adequate support system is essential to ensure that the window's weight doesn't cause it to pull away from the house, or to sag and rack the frame out of square, thereby causing the moving side sashes to bind. The most obvious way to provide this support is to install angled members under the window that will transfer the load to the main frame of the house. These can be simply two or more short pieces of 2 x 4 with 45° cuts on each end. They can be hidden by building an angled skirt, trimmed out to blend into the main wall. By extending this skirt straight down to the top of the foundation, you can give the unit the appearance of a walk-out bay (which, of course, it could actually be if you choose to expand the scope of the project by adding cantilevered extensions to the floor joists). Decorative brackets can also be used for support or designed and made on the job, but they should, in either case, match the style of the house. The number of brackets needed depends on the size and weight of the window.

Perhaps the best support system, and certainly the most ingenious, calls for suspending the bay from overhead using steel cables. The cables, which have threaded bolts on their lower ends, are fed up through holes in the seat- and head-boards, and fastened to cleats anchored to the framing of the house. The bay can be leveled by adjusting the nuts on the cable bolts. The cables themselves are hidden by the window's inside corner trim. With this system, any exterior angled skirt, or brackets, are merely decorative, and do not offer any necessary structural support. Be sure that any lower trim work is built so as to allow access to the cable bolts if future adjustments become necessary.

Be sure to check with your local building inspector to learn the code requirements for headers. There may be limitations on the height of the rough opening, and thus limits regarding the size of the window.

LEVEL OF DIFFICULTY

Installing a bay window is a major undertaking. Planning the job and selecting a unit that is appropriate to your home's size and style requires some experience and an ability to visualize the finished product. Experience is also needed to cut and frame the rough opening. If care is taken during this phase to make the rough sill dead level, it will greatly facilitate the installation. At least two people are needed to put the window in place, and for a very large and heavy unit, more help will be required.

Building a roof, whether from a kit or from scratch, is always tricky. The bay roof should be built with as much care and regard for proper flashing and sealing as would be given to any roofing job. Of the two support systems, the suspension cable type is generally the easier to install, and it has the added advantage of being adjustable. Just be sure the overhead cleat is anchored solidly to a framing member of the house wall or roof. Insulation and vapor barriers must be adequate, and correctly installed. Finally, good-quality interior and exterior finish work is essential to make the bay look like a feature that has always been where it is, not something merely "pasted" on later.

This is not a project for a beginner to attempt unaided, and the intermediate should also have experienced advice and help. The former should add at least 200% to the time estimates, and the latter should add at least 75%. The expert should figure on 25% more time than the hours listed, and should have a helper for putting the window in place.

WHAT TO WATCH OUT FOR

Careful planning makes every job easier. Remember that you are cutting a hole in your house; have everything – tools,

materials, and help – at hand so that this hole can be closed in and made weathertight as quickly as possible.

If your window is located in a load-bearing wall, temporary supports are needed to bear the load while the studs are being cut out and the header installed. These supports are put up in the manner of a framed wall with vertical studs between horizontal plates. This temporary "wall" should, of course, go at right angles to the floor joists, above and below. If the ceiling sheetrock is attached to furring strips, be sure the top plate of the support wall is located on a strip (furring strips are usually 1 x 3s spaced 16" on center); otherwise all the overhead weight will be "supported" by about 1/2" of sheetrock, which will collapse when the studs are cut and the load transferred to the support wall.

Also have temporary supports outside, 2 x 4s cut to length, and ground pads of wider 2x lumber for them to rest on. These will serve to bear the weight of the window once it is set in the rough opening. Setting the unit in place for a trial fit will allow you to identify and correct any problems, and once the window is centered and plumbed, you can mark exactly where to cut the siding to get a clean, tight joint where it meets the trim.

Bay windows lose a lot of heat through the glass, and can be cold spots in the winter. You can minimize heat loss by purchasing the option of double-glazed sashes, and by taking care to adequately insulate above, below, and around the window. A combination of rigid board insulation and standard fiberglass batts can be used to achieve the highest possible R value (R = resistance to heat flow). This is especially important in cold climates, and for a north-facing window. Be sure to install a vapor barrier between the insulation and the living space.

SUMMARY

A bay window, of whatever style, can add a note of substance and elegance to a home, but only if it is carefully designed to fit the proposed location and to blend with the existing features of the house, both interior and exterior.

For other options or further details regarding options shown, see

Standard window installation

Vinyl replacement windows

Bow/Bay Window

Description	Quantity/ Unit	Labor- Hours	Material
Demolition, cut opening for window, per 5 S.F.	8 Ea.	2.7	
Blocking, misc. to wood construction, 2 x 4	60 L.F.	1.9	25.92
Headers over opening, 2 x 12	24 L.F.	1.3	48.96
Casement bow/bay window, wood, vinyl-clad, 8' x 5' insul. glass	1 Ea.	1.6	1,410.00
Trim, interior casing, stock pine, 11/16" x 2-1/2"	33 L.F.	0.1	0.94
Paint, int. trim, incl. putty, primer, oil base, brushwork	33 L.F.	0.1	0.02
Int. trim, 2 coats, oil base, brushwork	33 L.F.	0.1	0.04
Drip cap, vinyl	8 L.F.	0.1	0.32
Caulking	32 L.F.	0.1	0.18
Snap-in grilles	5 Ea.	1.3	189.00
Totals		9.3	$1,675.38

Project Size	8' x 5'	Contractor's Fee Including Materials	$2,826

Key to Abbreviations
C.Y.– cubic yard   Ea.– each   L.F.– linear foot   Pr.– pair   Sq.– square (100 square feet of area)
S.F.– square foot   S.Y.– square yard   V.L.F.– vertical linear foot   M.S.F.– thousand square feet

BULKHEAD DOOR

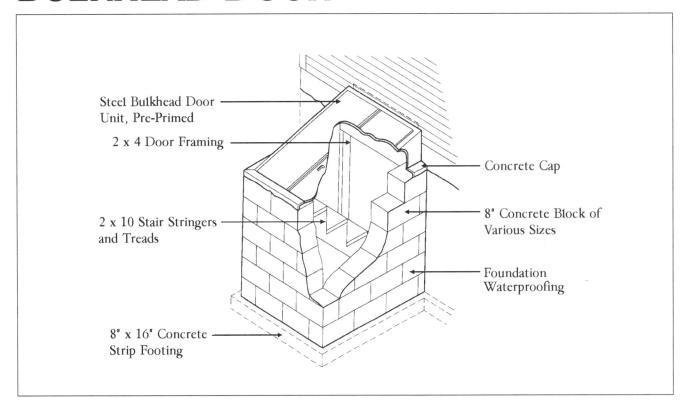

Steel Bulkhead Door
Unit, Pre-Primed

2 x 4 Door Framing

2 x 10 Stair Stringers
and Treads

8" x 16" Concrete
Strip Footing

Concrete Cap

8" Concrete Block of
Various Sizes

Foundation
Waterproofing

A house with a full basement should have exterior access via a walk-out door or a stairway under a sloped bulkhead. This might seem obvious, but in fact a great number of houses do lack such access. Builders and house movers have been known to avoid the extra foundation work expense involved in providing for an outside basement door. As a result, the homeowner must put up with the inconvenience and mess caused by moving every item bound to or from the basement, through the living area of the house. This includes everything from household tools and maintenance supplies to seasonal furniture. When moving a large object such as a hot-water heater, this inconvenience is compounded by the problem of having to negotiate, at the very least, two doorways and one set of narrow basement stairs. Adding a bulkhead door is not inexpensive, but in a house whose occupants make regular use of the basement for work, play, and storage, direct exterior access comes close to being a necessity.

MATERIALS

When locating your new bulkhead entrance, you must deal with both interior and exterior constraints and considerations. On the inside, the opening into the basement must be free of any obstructions and, ideally, related to the inside cellar stairs and other features in such a way as to make for a convenient traffic pattern and to allow plenty of room for maneuvering very large or long items that might need to be moved in or out. On the outside, the best locations are in the back or on the side of the house. While most homeowners would probably choose not to have the bulkhead show, practicality must take precedence over aesthetics. Installing a bulkhead requires excavation, and that means a backhoe has to be able to get to the site. Also, the usefulness of this door is limited if it is going to be difficult to get at from the outside. Look around at houses that have bulkhead doors; you will see that their low profiles make them unobtrusive, especially if they are partially screened by shrubbery. Taking everything into account, the number of possible locations for this new entrance will, no doubt, be reduced to one or two.

When you have decided on the best location, you must determine the dimensions of the areaway opening, and thus the size of the bulkhead door unit

and the length of the stair stringers. Door manufacturers provide brochures with tables for helping you make these calculations, based on the height of grade (ground level) above the basement floor. The area to be excavated should be staked off, allowing at least 18" all around for the footing and for working space. The hole should be dug to a depth equal to about 4" below the basement floor, with a perimeter trench to the depth of the existing foundation footing. The areaway opening is then cut through the foundation wall. This opening will later be framed with 2 x 8 lumber to provide the rough opening dimensions for the vertical door. Once a form is constructed, the footing can be poured for the areaway floor. Stucco the outside of the walls and, after letting the mortar dry, apply a heavy coat or two of foundation sealant. Once this is done, the excavated area surrounding the hole can be backfilled and tamped.

The metal bulkhead door frame is assembled and mounted by bolting it to the house and anchoring it to the top of the block walls. Once the doors are attached to the frame, the bulkhead is weathertight, needing only two coats of metal enamel paint, matched to the color of the house, to be finished. The next step is to set the stair stringers in place and secure them to the side walls with masonry nails. Treads are cut from standard 2 x 10 lumber and slid into the stringer slots. This type of stair system is not only economical in terms of materials and labor, but is also very functional, in that the treads can be removed at any time to allow the full areaway to be opened up for lowering very large or heavy items into the basement.

The vertical door leading into the basement should be of exterior grade, either solid wood or insulated, and weatherstripped to seal it against heat loss. Use 2 x 8 lumber to make a door frame, and secure it to the foundation opening with lead masonry anchors and countersunk lag screws. Once the door is in place, it can be cased off with appropriate lumber or moulding. In an unfinished basement where looks don't matter much, simple, square-cut 1 x 6 #2 pine will do. Be sure to stuff the cracks and crevices around the door frame with insulation before nailing on the casing. Once the door is primed and painted, the lockset can be installed to complete the job.

LEVEL OF DIFFICULTY

Considering the relatively small space it occupies, a bulkhead door installation is a fairly involved remodeling project with a number of distinct phases. Choosing a suitable location and determining the job specifications are easy using the information provided in manufacturers' brochures. You merely have to take measurements, then look them up in the tables, to know the dimensions of the areaway, the size of the bulkhead unit, the length of the stringers, and the number of stair treads. You will need to hire an excavating contractor for the excavation work.

Breaking through the foundation requires the use of a concrete saw. One can be rented, but it is a pretty formidable tool, heavy and loud, that is used for a pretty unpleasant job. Most homeowners would be better off hiring a contractor for this, and those who choose to do it themselves will probably wish they didn't.

Pouring the footing and building the block walls requires masonry tools and skills. Putting up plumb, square, and level walls with masonry blocks is not an easy job and, therefore, not one to be undertaken by anyone less than an expert.

The rest of the installation tasks – bulkhead, stairs, and vertical door – are not very difficult, though they can be time consuming.

Beginners and intermediates should hire professionals for the excavation, breakthrough, and masonry work, and should add at least 150% and 75%, respectively, to the time estimates for all other tasks. The expert should hire an excavator and will regret not hiring a concrete sawing contractor; for all other tasks he or she should add 50% to the times given.

WHAT TO WATCH OUT FOR

Scheduling is important during any remodeling job that requires the services of professional contractors. Once the hole is excavated, try to get the remaining work done as quickly as possible. Keep the hole covered with a large plastic tarp to prevent rain from creating muddy and unstable soil conditions, and from running into the basement. Also, set up barriers around the hole to keep children and others from getting too close to the edge.

This is not a good cold-weather project. Not only do concrete and mortar take longer to cure, but it is difficult to completely seal the foundation opening against heat loss. Moreover, most backhoe operators will not subject the teeth of their buckets to the abuse of trying to break through frozen topsoil.

Before the cutting begins on the foundation wall, do everything you can to seal off the work area from the rest of the basement and the living space of the house. Close all the windows near the hole when the saw is in operation.

SUMMARY

A basement's usefulness is reduced if it lacks direct access to the outside. This fact, combined with the mess and inconvenience the basement's traffic causes in the main living area of the house, makes the expense and labor involved in a bulkhead door installation easy to justify. You might be forced to choose a less than ideal location for the bulkhead; but any outside basement access is better than none at all. You probably cannot avoid the expense of hiring professionals for tasks other than the actual door installation, but you will have the assurance of knowing that the work was done correctly.

Bulkhead Door

Description	Quantity/ Unit	Labor- Hours	Material
Rent backhoe-loader	1 Day		225.60
Demolition, cut opening for door, per 5 S.F.	5 Ea.	1.7	
Footing, 9" thick x 18" wide, 3000 psi concrete	5 C.Y.	14.6	582.04
Slab on grade, 4" thick, incl. textured finish, no reinforcing	23 S.F.	0.5	24.84
Vapor barrier, 6 mil polyethylene	420 S.F.	90.8	1,476.72
Block wall, 8" thick, conc. block 8" x 16" x 8", reinforced	96 S.F.	10.7	278.78
Dampproofing, bituminous coating, 1 coat	96 S.F.	1.2	6.91
Anchor bolts, 1/2" diameter, 8" long, 4' O.C.	8 Ea.	0.3	5.95
Bulkhead cellar door with sides	1 Ea.	1.9	366.00
Door framing, 2 x 8	18 L.F.	0.4	10.80
Residential stl. door, pre-hung, insul. ext., flush face 3' x 6'-8"	1 Ea.	1.0	240.00
Stair stringers and treads, 2 x 10	42 L.F.	5.2	62.50
Lockset, standard duty, cylindrical, keyed, single cylinder	1 Ea.	0.8	78.60
Door trim, 1 x 6 pine, #2	17 L.F.	0.5	22.64
Paint, door, incl. frame & trim, exterior, primer & 1 coat, latex	1 Ea.	2.3	19.20
Totals		131.9	$3,400.58

Project Size 6' x 4'-7"

Contractor's Fee Including Materials **$11,408**

Key to Abbreviations
C.Y.– cubic yard Ea.– each L.F.– linear foot Pr.– pair Sq.– square (100 square feet of area)
S.F.– square foot S.Y.– square yard V.L.F.– vertical linear foot M.S.F.– thousand square feet

MAIN ENTRY DOOR

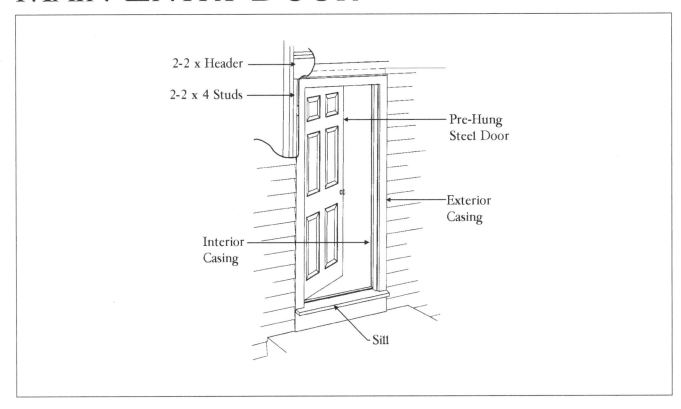

2-2 x Header

2-2 x 4 Studs

Pre-Hung
Steel Door

Exterior
Casing

Interior
Casing

Sill

Few working components of a house get more use than a main entry door, and with the additional deterioration caused by climate and neglect, such a door can clearly show its age after only a few years, as the sill rots and sags; wood panels split; hinges loosen; the lockset rattles; stiles and rails warp; and the door fails to hug the jamb and seal against the weather. Some of these problems can be solved with repair work or by adding a storm door, but this approach often results in a vain attempt to save the unsalvageable. Replacing the door is sometimes the best alternative.

MATERIALS

Which should it be – wood or steel? At one time the answer would have been obvious: steel doors were for institutional buildings or big city high-rises, and surely had no place overlooking the neat lawns of suburbia. Modern technology and design have made such an opinion no longer valid, even among traditional builders. This is what master carpenter Jim Locke has to say in his book The Well Built House: "Doors must be easy to open and close, to get through, and to pay for. They must close tightly against the cold,

and open with a light touch. They must appeal to our sense of being welcomed into the house, and resist attempts on them by the unwelcome. They should be easy to install and easy to maintain. All this points us directly to a hinged, insulated steel door." Such is the considered judgment of a very tradition-minded New England builder. Restoration of antique, period style houses is one obvious case where wood should be the material of choice, and there are certainly others (such as metal, fiberglass or vinyl clad, which can resemble wood and offer a significantly higher R-value), but for most entry door replacements in most houses, Mr. Locke's logic and conclusion would seem to apply.

Before making your final choice of a door, be prepared to be bewildered; the variety and options available today are truly amazing. Visit building suppliers and home centers; talk to knowledgeable salespersons; collect and study descriptive brochures, catalogs, and other data; seek the advice of designers and remodelers. In short, educate yourself so you are prepared to purchase the best door for your home. The entry door system shown here is quite basic, and it can serve as a point of departure when you

come to consider other designs and decorative features. An aluminum storm door may be added as well. Aluminum storm doors come pre-hung, as do most entry doors. The storm door comes with a screen and storm window panel, which can be easily removed in warm weather. These doors are available in both standard and custom sizes. Most are set up to swing a particular way but most are reversible so they can open on either side.

The main entrance door of a residence is usually 6'-8" high, and either 2'-8" or 3'-0" wide. Transoms and sidelights, fixed or operating, are optional and will add to the size of the rough opening. Another factor affecting dimensions is whether your house has wood or masonry siding.

A door comes hinged in one of four ways: left or right in-swing, or left or right out-swing. Most exterior doors swing into the house. This is somewhat a traditional convention to allow for an outside storm or screen door. An insulated steel door does not need a storm but it is a nice feature to have for ventilation on a nice day. A swing door, when open, takes up space in the entry or hallway, but most houses are designed in such a way that this is not a problem. If your existing door's swing presents no difficulties, it's probably best to keep it the same for the new unit.

Modern steel door construction is an impressive example of applied technology. A polyurethane insulating core (R-factor of about 12) is surrounded by 1-1/8″ kiln-dried wood stiles and rails, and all covered by rigid, 24-gauge steel facings, which are in turn given coatings of electrogalvanizing, phosphate, acrylic primer, and stainable, wood-grained, plastisol vinyl. Embossed panels or moldings are stamped into the metal, giving the door the look of wood with the strength of steel that will not warp, split, or rot. The jambs are weatherstripped with a magnetic, refrigerator-type system that locks out cold air and water. The oak threshold is adjustable, enabling it to form a weatherproof seal with the door's bottom sweep.

If your new door is exactly the same size as the old one, it should fit the existing rough opening. Center the unit in the opening with the casings removed and the door closed. Level the sill and shim between the side jambs at the bottom, tacking through both jambs and shims into the trimmer studs. Plumb the side jambs, shim at the top, and tack. Shim at several intermediate points along the side jambs and tack through the shims, making sure that the proper joint between the door and jamb is maintained. Check the operation of the door to see that it opens and closes without sticking or binding. Make any final adjustments, and complete the job by driving and setting all nails.

LEVEL OF DIFFICULTY

Removal of the existing door is very easy and not too messy. Carefully remove the casings inside and out, take the door off the hinges, cut through the side jambs with a reciprocating saw, and pry the frame out of the opening. There is more work involved if you are attempting to salvage the unit. You'll then have to cut the nails behind the jambs, push the frame out of the opening at the top, and carefully pry up the threshold. Installing a pre-hung door is a fairly straightforward operation, especially if the rough opening does not need to be enlarged or reduced. Entry door replacement systems such as this one come with detailed instructions that need only be carefully followed. The same is true for mounting the hardware.

A storm door is a fairly simple add-on to an entry door, and can be installed by a beginner with some knowledge about installing doors. A beginner should have experienced assistance during all phases of the main entry door project, and should add 100% to the time estimates. Both intermediates and experts should find little difficulty handling this job; they should add 20% and 10%, respectively.

WHAT TO WATCH OUT FOR

An entry door is an expensive, precision-made item of cabinetwork. Its finish appearance and proper functioning can be compromised by rough handling during transit and installation. A steel door is tough, but carelessness around it, for example with hammers swinging from tool belts, can cause scratches and dents that can be all but impossible to repair or hide. Store the door in a safe place, and leave on the protective plastic sleeve while installing it.

Take care when shimming the jambs not to push the frame out of alignment. The door's smooth operation depends on the frame being level and plumb. Also be careful, when driving the nails into the jambs, not to leave any hammer "smiles," those tell-tale marks of the unskilled hacker. Don't drive the nails all the way home with the hammer; leave the heads "proud," a tiny bit above the surface of the wood, and use a nailset to countersink them. Fill the nail holes with vinyl spackle after the wood has been primed.

Some manufacturers warn that dark colored paint and/or storm doors should not be used with any steel door that has plastic trim. Sun exposure can cause excessive heat build-up, which may distort the trim. Read all the printed material that comes with your door for information such as this.

SUMMARY

Great care should be taken in choosing a main entry door that matches the house's overall look and style, both inside and out. When properly installed, an insulated steel door will last many years, retaining its good looks, ease of operation, and weather-sealing efficiency.

For other options or further details regarding options shown, see

> Entryway steps
> Security measures
> Patio & sliding glass doors

Main Entry Door

Description	Quantity/ Unit	Labor- Hours	Material
Remove existing door, 3′-0″ x 6′-8″	1 Ea.	0.5	
Remove existing frame, including trim	1 Ea.	0.5	
Pre-hung door, insul, metal face, raised molding, 3′-0″ x 6′-8″	1 Ea.	1.0	186.00
Door trim set, 1 head and 2 sides, 4-1/2″ wide	1 Opng	1.5	28.80
Entrance lock, cylinder, grip handle	1 Ea.	0.9	126.00
Dead bolt	1 Ea	1.0	152.40
Paint, door and frame, primer	1 Ea.	1.3	2.11
Paint, door and frame, 2 coats	1 Ea.	2.7	6.84
Paint, trim, primer, including puttying	20 L.F.	0.3	0.48
Paint, trim, 2 coats, including puttying	20 L.F.	0.4	1.20
Totals		10.1	$503.83

Project Size	3′ x 6′-8″	Contractor's Fee Including Materials	$1,193

Key to Abbreviations
C.Y.– cubic yard   Ea.– each   L.F.– linear foot   Pr.– pair   Sq.– square (100 square feet of area)
S.F.– square foot   S.Y.– square yard   V.L.F.– vertical linear foot   M.S.F.– thousand square feet

PATIO & SLIDING GLASS DOORS

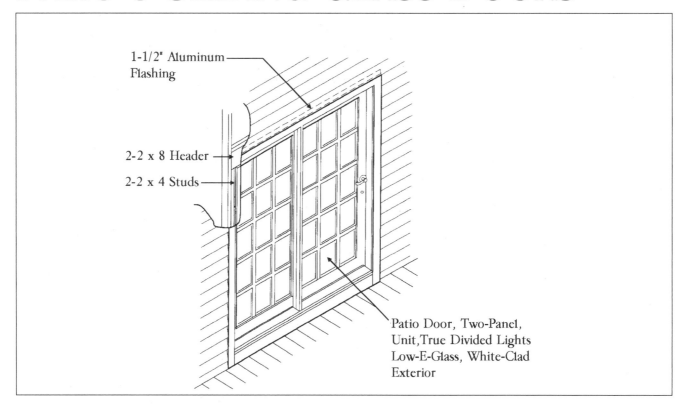

1-1/2" Aluminum Flashing

2-2 x 8 Header

2-2 x 4 Studs

Patio Door, Two-Panel, Unit, True Divided Lights Low-E-Glass, White-Clad Exterior

Recently builders and architects have been inclined to use fully glazed, swinging patio door units where, formerly, sliders would have been the automatic choice. Sliders were heavy and awkward to open and close; latches were often flimsy and insecure; bottom tracks were prone to clogging with dirt and debris; and the constant use tended to wear out the weatherstripping and cause it to leak cold air. In spite of their drawbacks, glass sliders continue to serve a useful purpose. Their aluminum construction makes them almost maintenance-free, they admit plenty of light, they are generally less expensive than patio units, and they take up no swing space.

Patio doors combine a light, airy, yet substantial look with highly efficient insulation and weatherproofing. The best quality doors are engineered to operate smoothly and easily on four or more hinges, and to latch securely with mortise locksets. Exterior protective cladding, of vinyl or polyester, makes modern units virtually maintenance-free. Manufacturers offer such a variety of panel sizes and shapes, in combination with transom and side-lite options, that a patio door can function as either a main or secondary entry, or as an interior passage between

any two rooms, upstairs or down. In addition, patio doors offer light weight and ease of operation that sliding doors notoriously lack.

MATERIALS

As is the case with all building products and materials, it hardly ever makes sense to try to economize by purchasing cheap goods. Shopping around for the best price on a quality door unit is smart; choosing an inferior door based solely on price is not.

A patio door is made up of two glazed door units, or panels, in one wide frame. The operating door of a patio unit is usually hinged to the mullion between the panels, with the lockset located at the outside jamb. Manufacturers offer a wide range of decorative options in terms of size, shape, and finish, as well as more technical options that deal mainly with glazing, insulation, and protective coatings.

The patio door unit in this project is constructed of clear pine, with true divided lights. The doors are glazed with low emissivity (low E) glass, which reduces the radiative emission of heat from the

glass surfaces, resulting in greater energy-saving efficiency than triple-glazing without the latter's weight and reduced light transmission. More insulation is provided by foam-filled weatherstripping that seals tightly and almost completely eliminates drafts. The exterior wood is protected from the elements by a .055" silicone-polyester-coated extruded aluminum cladding with a baked-on white enamel finish that makes the door practically maintenance-free. The sill is made of a high-tech composition, trade named Lexan, which combines good insulating properties with virtual indestructibility.

The patio door hardware is solid brass – more expensive, but much more durable than lightweight, plated brass. Four heavy-duty hinges provide smooth movement and ensure that the door will never sag.

Any kind of door needs a well-framed rough opening, but the unique characteristics of a glass slider call for extra care in the framing phase of this project. Remember that a glass sliding unit is quite heavy. You want to ensure the door's ability to slide as smoothly as possible, and this means a perfectly level sill and bottom track. Try to frame the opening to close tolerances (the rough opening for this unit is 96" x 80-3/4") to keep shimming of the aluminum frame to a minimum.

Cutting out and framing the opening for either unit is essentially the same as for any door. For a load-bearing wall installation, follow the procedure for erecting a temporary supporting wall, as outlined in the *Bay Window* project in this book.

Detailed instructions are generally provided with sliding door units to guide you step by step through the installation process. Once the frame is secured in the rough opening level, square, and true to plane, the fixed panel is put in place. The slider unit features a stainless steel sill track cover that is designed to prevent wear on the sill and rollers, and to ensure smooth operation. Once it is slipped on the track, the sliding panel can be installed and the rollers adjusted. If the rough sill is level and square to the sides, the amount of roller adjustment needed should be minimal.

LEVEL OF DIFFICULTY

If this is a same-size replacement project, it should be fairly easy and straightforward. When replacing a slider – either with a patio door or another sliding door – the old unit is lifted out and, with casings removed, the aluminum frame is unscrewed and taken from the opening in reverse order of installation. Cutting and framing a rough opening is a somewhat more serious operation, demanding experience in both planning and execution. A beginner could undertake a simple remove-and-replace project, but would need experienced help and guidance with a new installation. Intermediates and experts should encounter few difficulties here. Transporting the heavy glass panels and lifting them into place is easier and safer with at least two people, and very large units definitely require two strong backs.

Time estimates for both types of door should be increased by 100% for beginners. For the sliding glass door installation, add 25% for intermediates,

and 10% for experts. For the patio door, an intermediate should add 50% to the time estimates, and an expert, 20%.

WHAT TO WATCH OUT FOR

A common design flaw of decks is to make them virtually flush with the inside floor. This means that rain can bounce off the decking and drench the lower portions of the door, filling the track with water that can work its way under the panels. If you are installing a sliding door that will open onto a new deck or landing, make sure that there is a full step down from the door sill to the deck.

Local building codes should be consulted to determine header specifications and whether safety glass is required. Safety glass is tempered to break into small, gravel-like bits, rather than sharp shards and spears. The problem is, safety code takes precedence, so you may be forced to opt for triple-glazed or other types of panes in order to be in conformity. Check out all safety requirements before you begin shopping for a door.

SUMMARY

Sliding glass doors can provide a good entryway option where occasional access is needed to a terrace, deck, or patio, or where space restrictions preclude the use of a swinging door. A well-designed, good-quality patio door unit, properly installed, will provide years of trouble-free use and add significantly to the appearance and value of your home. The finest door can, however, be compromised by poor installation. Take care to frame a rough opening that is square and level, and to read and follow the manufacturer's instructions.

For other options or further details regarding options shown, see

Bow/Bay window

Security measures

Patio Doors

Description	Quantity/ Unit	Labor- Hours	Material
Wood French door, 6'-0" x 6'-8", w/ 1/2" insul. glass and grille	1 Ea.	2.3	1,320.00
Interior casing, raised molding	22 L.F.	0.7	52.80
Drip cap, aluminum	7 L.F.	0.3	2.52
Entrance lock, cylinder, grip handle	1 Ea.	0.9	126.00
Paint, trim, incl. puttying, primer, oil base, brushwork	22 L.F.	0.3	0.53
Trim, 2 coats, oil base, brushwork	22 L.F.	0.4	1.32
Door, 2 sides, incl. frame & trim, primer & 2 coats, brushwork	1 Ea.	2.7	4.81
Totals		7.6	$1,507.98

Project Size	6'-0" x 6'-8"		Contractor's Fee Including Materials	**$2,484**

Key to Abbreviations
C.Y.– cubic yard   Ea.– each   L.F.– linear foot   Pr.– pair   Sq.– square (100 square feet of area)
S.F.– square foot   S.Y.– square yard   V.L.F.– vertical linear foot   M.S.F.– thousand square feet

FIXED BUBBLE SKYLIGHT

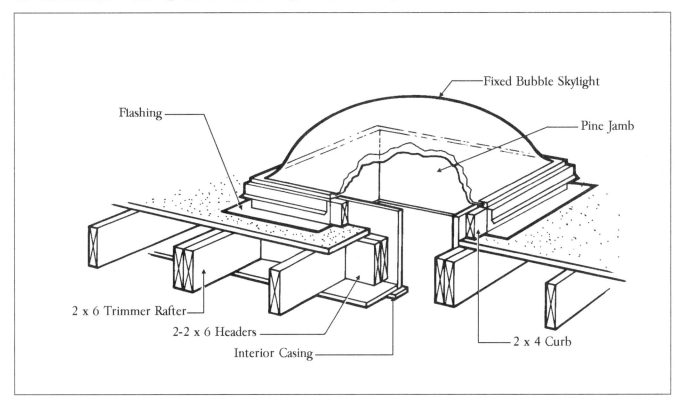

Fixed Bubble Skylight

Flashing

Pine Jamb

2 x 6 Trimmer Rafter

2-2 x 6 Headers

Interior Casing

2 x 4 Curb

As in other exterior home improvements, the overall planning layout and design of a skylight installation must be determined with care. This plan features a small fixed bubble skylight, which can provide an economical solution to lighting problems in an interior room or porch. Although this project involves general inconveniences of roof work and the complications of roof hole-cutting, many do-it-yourselfers can complete the installation in a surprisingly short time.

MATERIALS

Other than the skylight itself, the materials required to complete this project consist of standard construction products. These basic materials include lumber for adding strength to the rafters, flashing, incidental items for making the roof weathertight, and various interior trim products to finish the shaft and to refurbish the ceiling area affected by the installation. Because each project will present a different set of conditions, the type and amount of these materials may vary from the estimate. Slight increases or decreases in cost, therefore, are to be expected in installations of this type.

Before the prefabricated skylight unit can be placed, an extensive amount of preliminary work must be completed. The roof and ceiling areas to be opened must be carefully laid out, the holes cut, and support restored to the rafters and, if necessary, the ceiling joists. Determining the location of the skylight is a critical procedure for both practical and aesthetic reasons. Generally, a central location in the ceiling between existing rafters is ideal, as the hole-cutting is easier and trimmer rafters do not have to be placed. However, these conditions apply only when the unit you are installing is small enough to fit within your rafter spacing (usually 16"). In most cases, at least one rafter must be cut and trimmers and headers installed to restore roof support. This operation can often be difficult and time-consuming, as the trimmers may have to span a major portion of the length from the ridge to the top of the outside wall. Cutting the holes in the roof and ceiling, placing the trimmer and header rafters and joists, and installing extra support can be difficult jobs that require know-how and moderate to advanced carpentry skills. Inexperienced do-it-yourselfers are advised to seek help if they are unfamiliar with these operations.

Additional costs may be incurred during the preliminary phase of the project if extensive roof reconditioning is required. All do-it-yourselfers should have a professional contractor evaluate and, if necessary, correct any deficiencies in the structure of the roof before proceeding with the project.

The design and the size of the skylight unit should also be carefully selected so that it meets practical and aesthetic needs. This plan calls for a small fixed bubble, but a larger fixed unit or a ventilating model may be more appropriate for a given situation. In addition to the amount of light and ventilating capacity of the skylight, factors like the roof pitch, exposure, and exterior design of the fixture should also be considered. Generally, fixed dome skylights like this one are a good choice for well-ventilated rooms, high ceilings above stairwells and hallways, and screened-in porches. As with other prefabricated materials that will be exposed to the weather for many years, the quality of the skylight unit should be taken into consideration. Double-layered plastic bubbles like this one must be hermetically sealed to prevent condensation between the layers of plastic and to ensure efficient energy control. Most models include a base flange or provision for a curb and self-flashing. Be sure to follow precisely the manufacturer's guidelines when installing the skylight unit.

The time and material required to finish the interior of the skylight shaft will vary, as each installation will present a different set of conditions. This plan assumes that the skylight shaft extends only the thickness of the roof frame, as in attic rooms or basic porch locations where the ceiling follows the line of the roof. A longer shaft will be required, however, if the roof and ceiling are separated by attic space. In these cases, more materials and installation time will be necessary to build, insulate, and finish the shaft. The installation can also be more difficult when flared or angled shaft designs are needed or desired. Where long shafts are required, the extra cost and time will be considerable.

LEVEL OF DIFFICULTY

With good conditions such as 16″ rafter spacing and no extra shaft requirements, most do-it-yourselfers can handle this skylight installation. Small, fixed plastic bubbles of this type tend to be relatively inexpensive, so there isn't the worry of working with very costly or complicated components. The units are often installed in porches and attics, on flat or gently sloping roofs, and while the hole-cutting is never easy, it is less difficult in these locations. As long as conditions are favorable, beginners can complete this installation with the addition of 150% to the professional time; but they should proceed cautiously and seek instruction beforehand from a knowledgeable person. For skylight placement on a steep roof or where long shafts and extensive preliminary work are necessary, beginners should consider hiring a professional installer. Intermediates and experts should add 75% and 25%, respectively, to the labor-hour estimates for all tasks. All do-it-yourselfers should build in extra time for the general inconvenience of roof work.

WHAT TO WATCH OUT FOR

Although they do not allow for increased ventilation, fixed skylights like this one provide many other advantages that make them the right choice for certain installations. One is that they are ideal for flat roofs and for roofs that tend to drain slowly, ice up, or collect snow. Once the sealed unit has been correctly placed and made weathertight, the chances of its leaking are minimal. A second advantage is the bubbled shape of the exterior, which exposes more domed surface area to natural light. This feature makes them a good choice for roofs with northern exposures. A third advantage is that they are economically priced. For locations like porches and breezeways, they function as well as more expensive operable windows.

SUMMARY

This bubble skylight will dramatically increase the amount of light and sense of spaciousness in a bedroom, attic area, porch, or breezeway. If the conditions are right, it can be installed by most do-it-yourselfers for just the cost of the unit and a few basic construction materials.

For other options or further details regarding options shown, see

Fixed sky window

Operable sky window

Ventilating bubble skylight

Fixed Bubble Skylight

Description	Quantity/ Unit	Labor- Hours	Material
Roof cutout and demolition incl. layout, sheathing per 5 S.F.	2 Ea.	2.7	
Demolition, roofing shingles, asphalt strip	4 S.F.	0.1	
Demolition, framing rafters, ordinary 2 x 6	4 S.F.	0.1	
Trimmer rafters, 2 x 6 x 14′	28 L.F.	0.7	37.97
Headers, double, 2 x 6 stock	10 L.F.	0.4	6.84
Curb, 2 x 4 stock	10 L.F.	0.3	4.32
Skylight, fixed bubble, 24″ x 24″	1 Ea.	1.0	268.80
Flashing, aluminum, 0.013″ thick	9 S.F.	0.5	3.78
Trim, interior, casing, 9/16″ x 3-1/2″	10 L.F.	0.3	13.68
Jamb, 1 x 8 clear pine	10 L.F.	0.4	16.32
Paint, trim, including puttying, primer	20 L.F.	0.3	0.48
Trim, including puttying, 1 coat	20 L.F.	0.3	0.72
Totals		7.1	$352.91

Project Size	32″ x 46″	Contractor's Fee Including Materials	$841

Key to Abbreviations
C.Y.– cubic yard   Ea.– each   L.F.– linear foot   Pr.– pair   Sq.– square (100 square feet of area)
S.F.– square foot   S.Y.– square yard   V.L.F.– vertical linear foot   M.S.F.– thousand square feet

VENTILATING BUBBLE SKYLIGHT

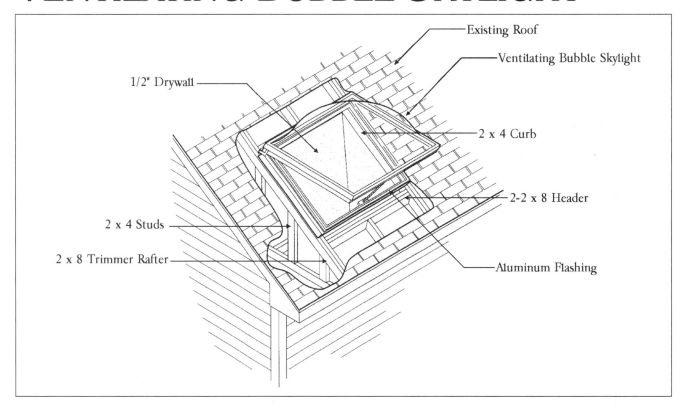

1/2" Drywall

Existing Roof

Ventilating Bubble Skylight

2 x 4 Curb

2-2 x 8 Header

2 x 4 Studs

2 x 8 Trimmer Rafter

Aluminum Flashing

When they are installed under the right conditions, ventilating skylights can greatly improve the appearance and air quality of an interior space. The large 52″ x 52″ ventilating bubble featured in this plan can considerably increase air circulation while providing nearly 18 square feet of window area. When the skylight is closed, the benefits of natural light continue, while the unit's double-layered insulating dome helps to keep the interior warm in winter and cool in summer. Because of the large size of the skylight unit and the considerable preparation required to install it, this project is challenging. Beginners and intermediates with limited remodeling experience should consider hiring a professional, at least for the more difficult exterior phase of the project.

MATERIALS

The primary component in this remodeling project is the skylight unit, but several basic construction materials are also included in the plan. These are needed to restore the roof structure, to make the installation weathertight, and to finish the interior around the new skylight. There are many styles and brand

names of skylights available to the consumer in a wide range of quality and price. Therefore, it pays to do some research before you decide on the unit that meets your needs and budget requirements.

The most challenging task faced in this skylight project is that of locating the fixture on the roof and ceiling, and the subsequent hole-cutting. Although only a few new materials are included in this preliminary work, the procedure is time-consuming and demands considerable know-how. To install a skylight of this size under normal conditions, about 20 square feet of roofing must be stripped, an equal area of sheathing removed, and two or possibly three rafters cut. If a finished ceiling is involved, a comparable area of interior covering will have to be removed and an equal number of joists cut. Temporary bracing may be required for the severed rafters and joists until the trimmers and doubled headers are placed. This plan suggests 2 x 8 trimmer rafters and headers, but 2 x 6s and even 2 x 10s may be needed if the existing rafters match these dimensions. Extra time and cost may be required for the preliminary work if extensive

reconditioning and strengthening of roof framing members is needed.

Once the hole-cutting and rafter trimming have been completed, the skylight unit can be placed in the opening and fastened into position. A unit of this size is heavy and awkward enough to cause problems in getting it to the site. You should have several workers available to help in getting it onto the roof and to position it from inside and outside of the opening. Remember that these expensive units are preframed, often prefinished, and fragile. Dropping or banging one can cause cosmetic damage at the very least and, worse, misalignment or breakage. After the skylight has been placed, the surrounding roof and perimeter seams are made weathertight with flashing, caulking, and reroofing. Reusing some of the old roofing can be a bit of a challenge, but it can also help to blend the new installation into the roof. Follow the manufacturer's guidelines and instructions throughout the placement operation, including the recommendations for proper flashing and weatherstripping.

The amount of interior finish material and installation time will vary according to the location and arrangement of the ceiling. If the finished covering is applied directly to the rafters, as in some attic and cathedral ceiling designs, the amount of finish material and the labor-hours to install it will be fairly limited. In such cases, the hole is closed in with finished wood and then faced with trim. If, as in this plan, the finished ceiling lies below several feet of attic space and joists, the job becomes more difficult, costly, and time-consuming. A light shaft must be built to span the attic space and joists to the level of the finished ceiling. A frame of 2 x 4s is normally used for the walls of the shaft. Finish material like sheetrock, pine boards, or paneling can then be applied to the inside of the shaft and appropriate casing used to trim the opening. Insulation of the recommended R-value for your area should be placed on the outside of the shaft walls. Further expense will be incurred for the shaft construction if special effects like flashing and angling are desired. Remember that trimming and heading off the ceiling joists are necessary preliminary tasks for the installation of a light shaft.

LEVEL OF DIFFICULTY

This large skylight installation is a major project for do-it-yourselfers of all ability levels. If the project is complicated by the need to build a light shaft, the undertaking becomes even more challenging. Remember, too, that because the skylight unit is an expensive product, the risks of damaging it have to be weighed against the savings in labor costs when you install it on your own. Experts and skilled intermediates can handle all procedures in the project, but they should add 25% and 75%, respectively, to the professional time estimate. They should allow even more time for placing the skylight itself, as careful handling and workmanship are important considerations. Any do-it-yourselfers who are inexperienced in roofing should get some advice on the task of replacing the disturbed roofing and applying the flashing. Beginners and less skilled intermediates should leave all of the exterior work to a professional installer; but they may be able to handle the interior finish if the required ceiling work is minimal. They should double the labor-hour estimates for all basic tasks, including painting.

WHAT TO WATCH OUT FOR

Determining the size, shape, and type of the skylight or sky window deserves ample consideration before the master plan for the project is finalized. For example, a large unit like this one can overwhelm a small room and, conversely, skylights that are too small can look out of place on an expansive ceiling. Often, several medium-sized units will look better and provide more flexibility for your ventilation, lighting, and aesthetic needs. Of course, more expense is incurred for more units; but if they are small enough to be placed between the roof framing members, installation costs for each one will be reduced. Remember that the exterior appearance of the skylight, especially on front roofs, is also important. The choice between fixed and ventilating models should be given careful consideration as well, since each type has qualities that make it more or less suitable to a particular location and set of conditions.

Ventilating Bubble Skylight

Description	Quantity/ Unit	Labor-Hours	Material
Roof cutout and demolition, roof sheathing, per 5 S.F.	20 S.F.	5.3	
Demolition, roofing shingles, asphalt strip	18.50 S.F.	0.2	
Demolition, framing rafters, ordinary 2 x 6	18.50 S.F.	0.1	
Trimmer rafters, 2 x 8 x 14'	28 L.F.	0.7	37.97
Headers, double, 2 x 8	18 L.F.	0.8	12.31
Curb, 2 x 4 stock	18 L.F.	0.6	7.78
Skylight, ventilating bubble, 52" x 52"	1 Ea.	2.7	642.00
Flashing, aluminum, 0.019" thick	21 S.F.	1.2	19.40
Shaft construction, 2 x 6 x 12' trimmer joists	24 L.F.	0.6	32.54
Headers for joists, 2 x 6 stock	20 L.F.	0.9	13.68
Framing, 2 x 4 stock	64 L.F.	2.1	27.65
Drywall, 1/2" x 4' x 8' sheets, taped and finished	64 S.F.	1.1	19.20
Paint, shaft walls, primer, oil base, brushwork	64 S.F.	0.5	3.84
Shaft walls, 1 coat, oil base, brushwork	64 S.F.	0.4	3.84
Insulation, foil-faced batts, 3-1/2" thick, R-11	64 S.F.	0.3	26.11
Trim, interior casing, 9/16" x 4-1/2"	18 L.F.	0.7	28.51
Paint, trim, primer, oil base, brushwork	18 L.F.	0.2	0.43
Trim, 1 coat, oil base, brushwork	18 L.F.	0.2	0.65
Totals		18.6	$875.91

Project Size	52" x 52"	Contractor's Fee Including Materials	$2,142

Key to Abbreviations
C.Y.– cubic yard   Ea.– each   L.F.– linear foot   Pr.– pair   Sq.– square (100 square feet of area)
S.F.– square foot   S.Y.– square yard   V.L.F.– vertical linear foot   M.S.F.– thousand square feet

SUMMARY

This large bubble skylight can improve the livability of your home by bringing in new light and ventilation. If the right design and roof location are chosen, this improvement can also enhance the exterior appearance of your home.

For other options or further details regarding options shown, see

Fixed bubble skylight

Fixed sky window

Operable sky window

FIXED SKY WINDOW

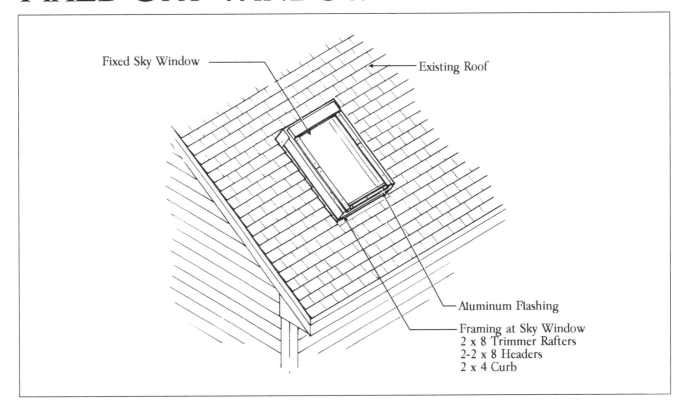

Fixed Sky Window — — Existing Roof

Aluminum Flashing

Framing at Sky Window
2 x 8 Trimmer Rafters
2-2 x 8 Headers
2 x 4 Curb

Sky windows, like bubble skylights, are available in square and rectangular shapes, and various sizes, levels of quality, and price. The homeowner should research the possibilities and then purchase the unit that best serves his or her needs. This model features a medium-sized fixed rectangular sky window with tinted insulated glass. Like other operating skylight installations, this project can be a challenge for do-it-yourselfers. Beginners and less-skilled intermediates are encouraged to hire a professional for the exterior work and, in some cases, for the interior part of the installation as well.

MATERIALS

Except for the skylight unit itself, the materials needed to complete this project under normal conditions are relatively inexpensive. Trimmer rafters, constructed from 2 x 6s or framing lumber that matches the existing roof rafters, and a 2 x 4 curb are needed for the opening. The other basic construction materials for this project are flashing and milled trim for the interior casing. All of these items can be purchased at most lumber yards and building supply outlets. The sky

window is a factory-prepared unit that includes a flashing kit, basic fasteners, and complete installation instructions, in addition to the preassembled window. These units are available under many brand names from supply outlets, door and equipment retailers, and specialty solar equipment stores.

Aside from the actual placement of the skylight, the most important operation in the project is the preparation of the roof and ceiling to receive the unit. In some situations, this preliminary procedure is basic, but in others it can be quite difficult and costly. This plan allows for the materials and installation time under basic conditions, as in an attic, loft, or winterized porch, where the finished ceiling follows the roof line and is attached to the rafters. In this case, the roof area affected by the skylight is first laid out, carefully squared, and stripped of its finish covering. The sheathing is then cut out with a reciprocating or circular saw and a section of the intermediate rafter is removed. The opening is trimmed with 2 x 6s and 2 x 8s, as required.

Although they will add slightly to the cost, joist hangers and angle brackets may facilitate the header installation and

add strength to the reconditioned support system. If the rafters are spaced at 24", as in some older homes, rafter cutting may not be required. However, additional cost may be incurred in this case to bolster the support of the old roof. If extensive roof reconditioning is required, the increase in cost may be substantial, and a professional contractor may have to be consulted.

If the conditions are not this accommodating, the interior operation can be considerably more difficult and expensive. This is because the ceiling must be opened and the joists trimmed in a procedure similar to that done on the roof. Also, a light shaft between the openings must be framed, sheathed, insulated, and finished. Because the dimensions and design features of the shaft will be different from project to project, the cost and time to install it will vary greatly. Remember that a longer shaft is required for a skylight located near the ridge than for one placed near the eaves.

LEVEL OF DIFFICULTY

As advised earlier, beginners and less-skilled intermediates will find much of this project beyond their experience. The exterior installation can be particularly challenging, especially on steep roofs and those that require extra reconditioning. Joist work and the building of a shaft, if needed, might also make the interior operations too difficult for inexperienced do-it-yourselfers. Beginners, therefore, should consider hiring a qualified carpenter to complete the exterior part of the installation and, possibly, the interior. For any basic tasks that they feel they can handle, beginners should double the estimated labor-hours. Experts and experienced intermediates should add at least 25% and 75%, respectively, to all tasks, with slightly more time allowed for the careful handling and placing of the expensive sky window unit. All do-it-yourselfers should take into consideration the general inconvenience and hazards of roof work before they tackle the project. Extra expense may be incurred if staging or other specialized equipment is required for high ceilings and steep roofs.

WHAT TO WATCH OUT FOR

Sky windows provide their unique lighting function during the day, but at night they may offer little more than large dark areas on the ceiling. Also, during the warm months, they may contribute unwanted extra heat to the house if they are located on sunny roofs. Accessories are available to alleviate both of these shortcomings. To cut down on an excess of summer sunlight, you might invest in one of the many shade and sun-screen options available from the skylight manufacturer. A shade may also be used at night to insulate this area of the ceiling. This accessory will add to the cost of the project, but the benefit derived will make the added expense worthwhile.

SUMMARY

Even a medium-sized sky window like the one in this plan can make a considerable impact on interior lighting. It is a challenging project, but one that can tremendously improve the appearance of a room.

For other options or further details regarding options shown, see

Fixed bubble skylight

Operable sky window

Ventilating bubble skylight

Fixed Sky Window

Description	Quantity/ Unit	Labor- Hours	Material
Roof cutout and demolition incl. layout, sheathing per 5 S.F.	8 S.F.	2.1	
Demolition, roofing shingles, asphalt strip	8 S.F.	0.1	
Demolition, framing rafters, ordinary 2 x 6	8 S.F.	0.2	
Trimmer rafters, 2 x 8 x 14'	28 L.F.	0.7	37.97
Headers, double, 2 x 8 stock	16 L.F.	0.8	16.70
Curb, 2 x 4 stock	12 L.F.	0.4	5.18
Sky window, fixed thermopane glass, metal-clad wood, 24" x 48"	1 Ea.	3.2	618.00
Flashing, aluminum, 0.013" thick	13.50 S.F.	0.8	5.88
Trim, interior casing, 9/16" x 4-1/2"	12 L.F.	0.5	19.01
Jamb, 1 x 8 clear pine	12 L.F.	0.4	19.58
Paint, trim, primer, oil base, brushwork	24 L.F.	0.3	0.58
Trim, 1 coat, oil base, brushwork	24 L.F.	0.3	0.86
Totals		9.8	$723.76

Project Size	24" x 48"	Contractor's Fee Including Materials	$1,491

Key to Abbreviations
C.Y.– cubic yard   Ea.– each   L.F.– linear foot   Pr.– pair   Sq.– square (100 square feet of area)
S.F.– square foot   S.Y.– square yard   V.L.F.– vertical linear foot   M.S.F.– thousand square feet

OPERABLE SKY WINDOW

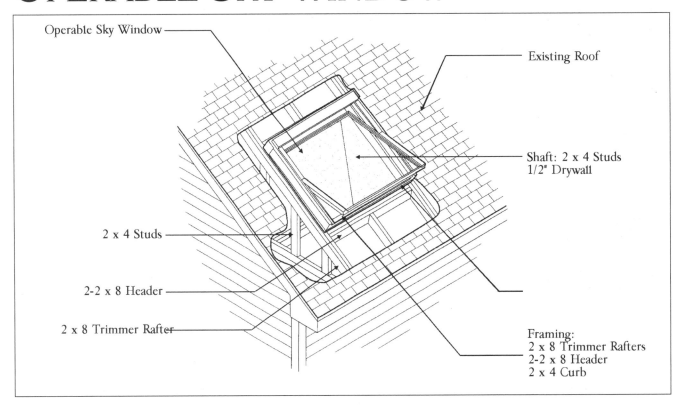

Operable Sky Window

Existing Roof

Shaft: 2 x 4 Studs
1/2" Drywall

2 x 4 Studs

2-2 x 8 Header

2 x 8 Trimmer Rafter

Framing:
2 x 8 Trimmer Rafters
2-2 x 8 Header
2 x 4 Curb

This project features a 44" x 57" top-of-the-line sky window with an operable sash. Units like this are equipped with insulated tinted glass and factory-prepared curbing. A wide selection of accessories is also available to suit the particular needs of each installation. This window addition can dramatically improve the lighting and ventilation in an interior space.

MATERIALS

Only a small number of materials are required for this project, as the prefabricated skylight unit is designed for convenient installation. A considerable amount of preliminary roof work is necessary, however, and interior and exterior finish work will be required after the unit has been installed. The materials needed to complete these operations include inexpensive home building products, such as framing lumber, flashing, and interior wood trim. These items represent only a small percentage of the total cost of the project, but they are necessary components. The trimmer joist/rafters, especially, are vital structural elements that will be used to restore ceiling and roof support after the opening has been cut.

Before you begin the task of opening the ceiling and the roof, take time to lay out the location of the window unit. Most of the popular brands of skylights are manufactured for easy placement between standard-sized rafter spacings. The frame of this 57" skylight unit can be located in the space opened when two 16" rafters are cut and partially removed. Whenever possible, locate the skylight between existing rafters, as you will save time and disturb the roof structure only slightly. Be sure to square the layout and double-check all measurements before the hole is cut. After trimmers and headers are installed, a curb may have to be fastened to the perimeter of the opening, depending on the manufacturer's recommendations. The instructions will also provide specific information for proper fastening and flashing of the window frame. Use accepted roofing methods for replacing the disturbed area with finished materials. Reusing salvaged pieces of the old roofing will neatly blend the new work with the existing roof.

In many instances, sky windows are placed in attic rooms, family rooms with vaulted ceilings, and other rooms with slanted ceilings. The installation process is easier in these cases because the roof rafters and ceiling joists are often one and the same. After the skylight unit has been secured in its opening and made weathertight on the outside, the interior finishing is a relatively easy undertaking. The inside of the opening is trimmed with appropriate material, usually wood, and matching trim is applied for the casing.

In some instances, the interior finishing may be more difficult and costly. If, as in this plan, the roof and ceiling are separated by unused attic space and separate holes are required for the exterior and interior openings, you will have to build a light shaft between the two. The cost of this procedure and its degree of difficulty will vary with each situation, depending on the shaft's dimensions, finish, and design features. Usually, a 2 x 4 frame, sheathing, and insulation are required for the basic structure. Trim work and an appropriate covering are then applied to the shaft's interior to complete the procedure. If the bottom of the shaft is joisted on a horizontal ceiling the opening could be flared to get the maximum benefits of light and ventilation. Remember to figure in some extra time and cost for cutting and trimming the ceiling opening and joists as part of the shaft installation.

LEVEL OF DIFFICULTY

If the basic conditions are right, the installation of this sky window can be a manageable project for do-it-yourselfers experienced in the use of tools and routine roof work. Beginners and less-skilled intermediates, however, should not rush into the undertaking without considering the risks involved in roof hole-cutting and related procedures.

Steep roofs and inaccessible attic locations can provide dangerous and troublesome conditions for inexperienced workers. Remember that you must also use power tools in awkward positions such as from a roof ladder. Because nearly 20 square feet of roof will be opened, rapid and effective work is required to cut the hole, trim it, place and flash the unit, and restore the finish roofing to weathertight condition within a few hours' time. Experts and skilled intermediates should be able to complete these tasks with 25% and 75%, respectively, added to the professional time. Inexperienced do-it-yourselfers should not attempt the exterior work and should double the estimated labor-hours for finishing the interior. All do-it-yourselfers are reminded to allow extra time for handling this expensive and fragile skylight unit and for the general inconvenience of roof work.

WHAT TO WATCH OUT FOR

If a long light shaft is required for the installation of your sky window, you might want to put some extra effort into planning its design features. Although the length and width of the shaft's top opening are fixed, the sides, bottom, and length of its run can be creatively arranged to provide maximum practical operation and aesthetic appeal. One of the best ways to increase ventilation and light penetration is to flare the shaft by increasing the length and width of the opening at the ceiling end. This design requires more time and expertise to install because the rafters, joists, 2 x 4s for the frame, and sheathing must be cut at precise angles; but any additional cost for materials is small. Although this modification to the conventional shaft design will add some time and expense to the project, it can significantly improve its appearance and function.

SUMMARY

Installing a large sky window like the one featured in this plan can transform a gloomy interior area into attractive and pleasant living space. With the addition of a clever shaft design and the appropriate accessories, the installation can produce spectacular results.

For other options or further details regarding options shown, see

Fixed bubble skylight

Fixed sky window

Ventilating bubble skylight

Operable Sky Window

Description	Quantity/ Unit	Labor- Hours	Material
Roof cutout and demolition incl. layout, sheathing per 5 S.F.	18 S.F.	4.8	
Demolition, roofing shingles, asphalt strip	20 S.F.	0.2	
Demolition, framing rafters, ordinary 2 x 6	20 S.F.	0.4	
Trimmer rafters, 2 x 8 x 14'	28 L.F.	0.7	37.97
Headers, double 2 x 8	16 L.F.	0.8	16.70
Curb, 2 x 4 stock	14 L.F.	0.5	6.05
Sky window, operating, thermopane glass, 44" x 57"	1 Ea.	1.3	678.00
Flashing, aluminum, 0.019" thick	14 S.F.	0.8	12.94
Shaft construction, 2 x 6 x 12" trimmer joists	24 L.F.	0.6	32.54
Headers for joist, 2 x 6 stock	20 L.F.	0.9	13.68
Framing, 2 x 4 stock	64 L.F.	2.1	27.65
Drywall, 1/2" x 4' x 8' sheets, taped and finished	64 S.F.	1.1	19.20
Paint, shaft walls, primer, oil base, brushwork	64 S.F.	0.5	3.84
Shaft walls, 1 coat, oil base, brushwork	64 S.F.	0.4	3.84
Insulation, foil-faced batts, 3-1/2" thick, R-11	64 S.F.	0.3	26.11
Trim, interior casing, 9/16" x 4-1/2"	18 L.F.	0.7	28.51
Paint, trim, primer, oil base, brushwork	18 L.F.	0.2	0.43
Trim, 1 coat, oil base, brushwork	18 L.F.	0.2	0.65
Totals		16.5	$908.11

Project Size	44" x 57"	Contractor's Fee Including Materials	$2,099

Key to Abbreviations
C.Y.– cubic yard   Ea.– each   L.F.– linear foot   Pr.– pair   Sq.– square (100 square feet of area)
S.F.– square foot   S.Y.– square yard   V.L.F.– vertical linear foot   M.S.F.– thousand square feet

STANDARD WINDOW INSTALLATION

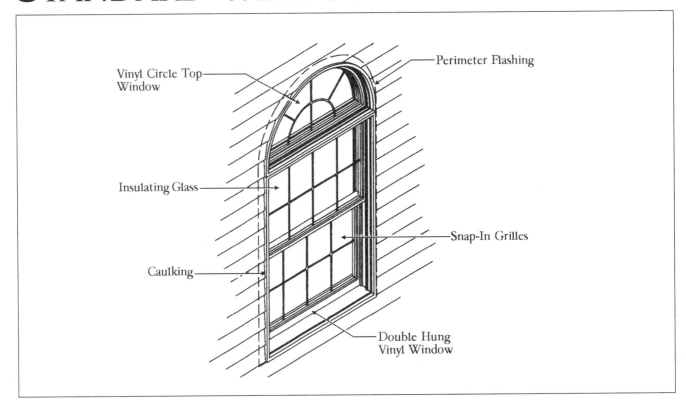

Vinyl Circle Top Window

Perimeter Flashing

Insulating Glass

Snap-In Grilles

Caulking

Double Hung Vinyl Window

Adding a window requires a good deal of careful planning to ensure that the end result is indeed a home *improvement* and not an unsightly and costly liability. It is important that the new window be selected to match or blend with the existing windows as to location, type, size, and shape. *Location* is usually predetermined by the available wall space, but consideration must be given to the look of the window both inside and out. For example, the preferred interior location that accommodates furniture placement might cause it to be out of balance with other architectural elements on the exterior.

Choose a *type* of window that will complement the age and style of your home. A vinyl-clad, casement unit would be an inappropriate choice for an antique colonial with nine-over-nine double-hung windows, but perfectly acceptable for a ranch-style contemporary.

Assuming that the new window is not simply to be an exact match of existing ones, its *size* and *shape* should be of pleasing proportions and, again, in keeping with the overall interior and exterior look of the house. To help envisage what it will look like, cut a piece of roofing felt (tar paper) to represent

the new window and tape or staple it in place. Then view it at various distances, from different angles, squinting your eyes to eliminate distracting details and help focus your attention on how the shapes of windows, door, roof lines, and walls balance and blend. Take your time and make the right choice. Windows are the "eyes" of a house and, perhaps more than any other single element, they greatly determine its external character and beauty.

MATERIALS

There are many types of windows available: double-hung, casement, fixed pane, sliding pane, awning, half-round, elliptical, and many other shapes. Many manufacturers offer these with a variety of options such as primed wood or vinyl-clad casing; true divided lights with removable screen and storm panels; thermopane or double insulated glass with clip-in grills; and crank or lever-type opening mechanisms. A clerk at a reputable supply house can provide you with information and descriptive brochures for the various brands and can answer your questions. Expect to pay for quality. Inexpensive windows look

insubstantial, often perform poorly, and offer you little in the way of guarantees. Remember, it costs as much to install a cheap window as a high-grade one. Trying to save a few dollars on a feature of your house as important and as highly visible as a window is false economy.

You may discover that none of the manufacturer's catalogs shows exactly the window you need. A supplier can usually refer you to a mill where you can have a unit custom-made to your specifications. Again, don't skimp on quality.

Regardless of the type of window required, preparation and installation is similar for all. Typical framing would be 2 x 4 studs with a 2 x 6 header. In older houses, the dimensions of the framing lumber, inside finished wall, and outside sheathing can vary widely from what is standard today, and in order to match them some creativity will be called for.

LEVEL OF DIFFICULTY

Because of the various skills involved, the cost of the window unit, and the potential for leaks, this type of project is not recommended for beginners. If you are a beginner and can find a carpenter who will let you serve as a helper, you can learn a lot from actively participating in the job. Don't be surprised, however, if your participation costs, rather than saves, money. You will, no doubt, make a few mistakes and slow the carpenter down.

A window installation is within the capabilities of an intermediate, but a helper will be needed when leveling the unit and nailing it in place – one person on the outside and the other inside.

Any installation above the ground floor should be undertaken only by an expert with an experienced helper. The obvious difficulties and hazards involved are challenges for even a seasoned professional. The risks to a merely competent intermediate far outweigh the cost of hiring a skilled carpenter. An intermediate should add 75% and an expert 30% to the estimated time. For an upper-story installation, the expert should add another 20%.

WHAT TO WATCH OUT FOR

Before deciding where you want to place the window and, most especially, before cutting into the wall, check for the presence of electrical wiring, water pipes, and heating ducts. There is, literally, no more shocking surprise than encountering a live wire with the blade of a reciprocating saw. Cut the opening in the inside wall first, making exploratory cuts with a drywall saw. Only when you have determined that your line of cut is free of hazardous obstructions should you go to a power saw. If wires are encountered, they can easily be rerouted around the new window. You may be able to do this yourself. Pipes, and particularly heating ducts, are harder to deal with and will require a plumber. Most problems can be solved with the help of skilled and creative tradespeople and your willingness to pay the added expense. The framing members used to create an opening in a wall provide support for the unit being installed.

Be sure to check the building codes for header size requirements. A large picture window in a load-bearing wall calls for more header support than a small ventilation window in an attic gable.

Under no circumstances should a window frame be used to support a vertical load. Care should be taken to follow manufacturers' recommendations for correct clearance and fastening sequence.

Windows are always susceptible to leaks. Be sure that all water-shedding components – building paper, drip cap, flashing, siding, caulking – are properly installed. When in doubt, seek expert advice.

Before closing up the inside wall and trimming out, be sure any insulation you may have removed during installation is replaced and all crevices around the new window are stuffed with insulation to prevent heat loss.

If the window is to be painted, take care you don't paint it shut. Also, install an adequate locking device, especially if the window is easily accessible from the outside.

Care must be taken to frame the rough opening level, square, and plumb. Failure to do so will cause a number of problems during installation and could, in fact, result in the window not functioning properly.

Installing a window means cutting a hole in a wall. The best time to schedule this project would be during the months of mild weather, but it can be undertaken at any time of the year as long as you are prepared to make the opening weathertight in the event of a storm or an unforeseen delay. Even if all goes smoothly, the simplest window project can take a few days to complete. An upper-story installation, for example, will involve working from a ladder and possibly even building a scaffold. Handling tools at heights and in awkward positions will add time to the job.

Stripping and replacing the exterior siding may involve a contractor if your house has aluminum or vinyl, and most certainly if it is of brick or stone. Wood siding such as clapboard or shingle makes it easier to do yourself. You will need a shingle bar to strip the siding back from the rough opening in a staggered pattern so that the vertical joints don't line up and result in leaks. Be sure the sheathing is covered with building paper properly overlapped.

SUMMARY

A new window provides increased visibility, natural lighting, and possibly ventilation to the interior while enhancing the exterior of the house by adding balance and proportion to the existing facade.

For other options or further details regarding options shown, see

> *Bay window*
> *Skylights*
> *Vinyl replacement window*

Standard Window Installation, 3' x 4'

Description	Quantity/ Unit	Labor- Hours	Material
Demolition, cut opening for window, per 5 S.F.	2.50 Ea.	0.8	
Blocking, misc. to wood construction, 2 x 4	40 L.F.	1.3	17.28
Headers over opening, 2 x 8	8 L.F.	0.4	11.90
Window, wood, vinyl-clad, premium, 3' x 4', insulated glass	1 Ea.	0.9	296.40
Trim, interior casing	15 L.F.	0.5	36.00
Paint, interior, trim, primer and 2 coats, enamel, brushwork	2 Side	2.7	4.90
Caulking	14 L.F.	0.5	2.52
Snap-in grilles	1 Set	0.2	130.80
Totals		7.3	$499.80

Project Size	3' x 4'	Contractor's Fee Including Materials	$1,061

Key to Abbreviations
C.Y.– cubic yard   Ea.– each   L.F.– linear foot   Pr.– pair   Sq.– square (100 square feet of area)
S.F.– square foot   S.Y.– square yard   V.L.F.– vertical linear foot   M.S.F.– thousand square feet

VINYL REPLACEMENT WINDOW

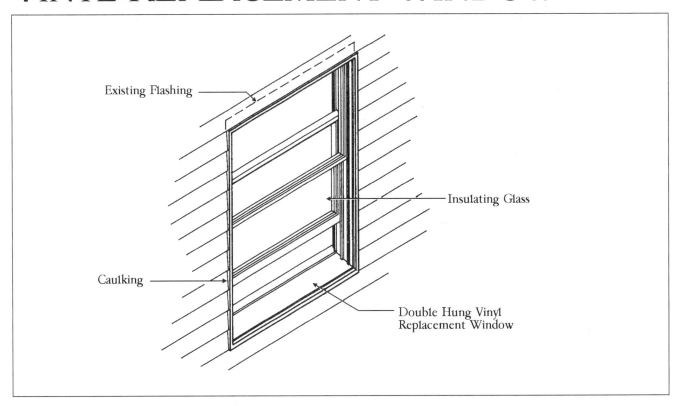

Existing Flashing

Insulating Glass

Caulking

Double Hung Vinyl
Replacement Window

Replacing an older window with a new energy-efficient unit is a simple way to improve your home's appearance while decreasing your heating and cooling costs. The most popular choice for replacement windows has traditionally been vinyl, which is estimated in this project. However, major manufacturers have recently introduced solid paint-grade wood (interior and exterior), and aluminum-clad or vinyl-clad wood exteriors, in a range of styles and sizes.

MATERIALS

The three most common materials used in replacement windows are vinyl, wood, and aluminum. The new windows come with a frame that includes new balances and hardware. They are almost always custom made to fit a particular window opening. These windows should last a long time, so do not skimp on quality just to save a few dollars.

Be sure to measure twice before ordering your replacement window. Once a window is made, it's yours. You may want to request that a sales representative measure the opening for you. If you hire

a contractor, he or she will measure and order the window for you.

Replacement windows generally consist of what is called the sash. The sash is the operable portion of the whole window assembly. Depending on the age of your home or the vintage of the window itself, this project can be relatively easy and involve minor preparation of the frame to receive the new window, or it can require careful planning, scheduling, and tedious trim removal and replacement. The test of a good installation is that an observer can't tell that you have done anything major other than clean up existing windows.

LEVEL OF DIFFICULTY

Basic carpentry skills are involved in installing a replacement window. Although not recommended for beginners, this project is within the capabilities of an intermediate do-it-yourselfer, but a helper may be needed when positioning the replacement unit.

Depending on the type of window unit and the amount of preparation required for the opening, experts and experienced intermediates should plan on 50% more

time than that estimated here; beginners should plan on 100% more time.

WHAT TO WATCH OUT FOR

If your home is in a historical district or other area where there are restrictions on the types of home improvements you may undertake, be sure to check with your local building department before replacing any windows. Also be sure to choose a style of window that will complement the style of your house, no matter what its age.

Installation above the ground floor should be taken on only with careful preparation. Installing upper-story windows involves working from a ladder or even scaffolding. Working at heights and in awkward positions can add time and difficulty to the job.

Because windows are susceptible to leaks, ensure that your existing window frames are clean and in good condition and seal them well with caulking when the new window is installed. Before replacing the window stops on the interior, be sure to fill any open areas with insulation to avoid heat loss.

This type of project is best done in times of mild weather to avoid problems related to poor weather conditions. Be prepared to make the opening weathertight in the event of a storm or other unforeseen delay.

The final step in the installation of any replacement window requires some form of caulking. If you are doing the installation yourself, be sure to do a thorough job. All your efforts could be diminished by poor caulking.

SUMMARY

Replacement windows will enhance the appearance of your home and provide better insulating properties than older windows. Today's vinyl replacement windows come with many options forcolor, style, and architectural appeal. Other features include energy-efficient low-E/argon glazing, tilt-in systems for cleaning, and grille systems in a variety of configurations. Care should be taken to choose a unit that best suits your budget and the look of your home. Shop carefully and make sure you understand the measuring procedure and the tools and equipment you will need.

For other options or further details regarding options shown, see

Bay window

Standard window installation

Vinyl Replacement Window

Description	Quantity/ Unit	Labor- Hours	Material
Demolition, remove existing window sashes and interior trim	1 Ea.	0.4	
Window, vinyl replacement unit, 3' x 5', insulated glass	1 Ea.	1.0	224.40
Caulk exterior perimeter	16 L.F.	0.5	2.88
Insulate voids as necessary	16 S.F.	0.1	4.03
Install window trim	1 Opng.	0.6	17.82
Paint, interior, trim, primer and 2 coats, brushwork	16 L.F.	0.4	1.34
Totals		3.0	$250.47

Project Size	3' x 5'	Contractor's Fee Including Materials	$502

Key to Abbreviations
C.Y.– cubic yard   Ea.– each   L.F.– linear foot   Pr.– pair   Sq.– square (100 square feet of area)
S.F.– square foot   S.Y.– square yard   V.L.F.– vertical linear foot   M.S.F.– thousand square feet

Section Four
GARAGES AND CARPORTS

Garages and carports should be carefully planned around such factors as their intended use (storage, workshop space, and/or vehicle shelter), the size and style of your house, the available and buildable (within requirements of the local building department) land and the characteristics of the site. Following are some additional considerations.

- Be sure to design your garage for full size cars even if your car(s) are subcompacts. Not only will you appreciate having space to move around (on foot and in your car), but it won't present future potential problems to buyers of your home, should you ever decide to move.

- Prior to finalizing plans, mark off the proposed garage on the ground with stakes and string. Get a feel for the size of the project.

- Garage floors can not be flush with an existing floor. A minimum difference is required for protection from gasoline spills. Check with your local fire and building departments.

- Unless you have a reason to call attention to your garage, the doors should be unobtrusive and harmonize with the rest of your house.

- If home security is a concern, avoid garage doors with windows.

- When deciding between a carport and a garage, consider that carports are less expensive and easier to build than conventional garages. Do-it-yourselfers can do much of the work to construct a carport on their own.

- Roll-up garage doors have more moving parts than any other door in your house. Therefore, they require more maintenance to keep them working properly. Oil the roller bearings, pulleys, lock mechanism, and cables twice a year, and check track alignment once a year.

- If you decide to shop for an automatic garage door opener, ask how loud the opener will be and ask for a demonstration if possible. Generally a 1/4 horsepower motor can handle a single door, and a 1/2 horsepower motor can operate a double door. Inquire about safety features as well, particularly for a home with children.

- Be sure to check the local building code if you are adding an attached garage to your house. There are important requirements to ensure adequate fire separation that cannot be ignored.

- Garages require the same standards for seismic bracing as your building code requires for dwellings.

- If a project requires excavation (e.g., to install a foundation), be alert to the dangers of encountering and damaging underground electrical, gas, water, or septic lines. You or your contractor should contact your utility companies to verify the location of these services underground.

ATTACHED SINGLE CARPORT

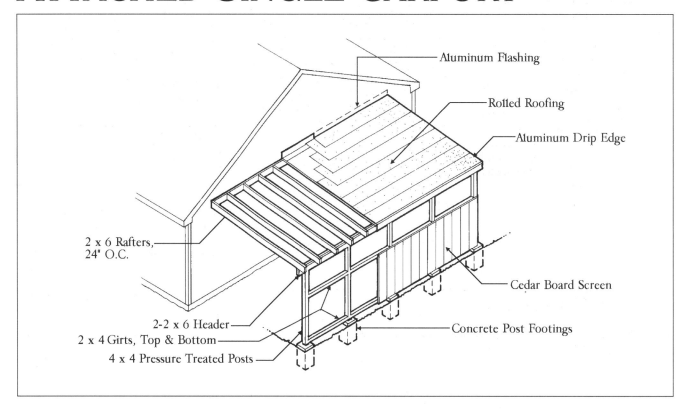

Aluminum Flashing

Rolled Roofing

Aluminum Drip Edge

2 x 6 Rafters, 24" O.C.

Cedar Board Screen

2-2 x 6 Header
2 x 4 Girts, Top & Bottom
4 x 4 Pressure Treated Posts

Concrete Post Footings

Carports offer a simple and economical alternative to the more costly framing and finish work required for conventional garages. In addition to requiring less material, carports are also easier to build than garages. Their uncomplicated design and structure offer the opportunity for do-it-yourselfers to do all or much of the work on their own. Freestanding carports require considerably more skill and know-how to erect than those that are attached to the house, but both projects can be completed by intermediates who have basic construction skills and knowledge. This plan demonstrates how a one-car facility can be built for modest expense. Many do-it-yourselfers can complete the job for the cost of the materials.

MATERIALS

The components of this carport project are standard construction materials that can be purchased at building supply outlets. Try to buy all of the materials at the same place, and be sure to make arrangements for curb delivery as part of the purchase price. Because the carport is attached to the house on one side and open at either end, materials are needed

only for the side opposite the house and the roof. Both of these sections are basic in their design and thus are fairly easy to build, if you lay them out correctly, follow good carpentry practices, and work slowly on tasks that are new to you.

The side of the structure serves two purposes: providing the support for one end of the roof and supporting the girts to which the siding is fastened. Because the roof structure is large and heavy, the five 4 x 4 posts that support it must be set in concrete footings at least 3' deep. Precise placement of the posts is important, so take time to space them at even intervals, align them with a stringline, and plumb them with a level. If you place the 4 x 4s in concrete, be sure to use pressure-treated lumber as recommended in the plan. Standard grade 4 x 4s can also be used at reduced cost, but they should not be set in concrete and require fastening to anchors preset in the footings.

After the posts have been placed, the roof can be framed with 2 x 6s set 24" on center, and doubled headers. This roof framing task is the trickiest operation in the project, so get some help if you have not done it before. Be sure to cut the joists at the correct angle on the house

end and at the reverse angle on the other end. Also, prepare the house wall to receive the roof assembly by cutting back the siding to the sheathing and arranging for the roof flashing. The method of installing the flashing will depend on the type of siding, so costs may vary for this operation. At the other end of the roof frame, temporary support should be provided for the first few joists before the headers are positioned and fastened. Be sure to square the frame before laying the 1/2" plywood roof sheathing and to stagger the termination seams of the plywood as you are placing it. Rolled roofing material should be used in place of asphalt shingles, because of the flat pitch of the shed roof. The installation of this type of roofing can go quickly with the right methods and equipment. With a little advice and instruction beforehand, and the right tools, most intermediates can undertake the job, even if they have no prior experience with rolled roofing. Be sure to place the drip edge on all open sides of the roof before you apply the roofing material and install flashing on the house side.

Once the roof is completed, the girts can be placed between the posts, and a

5'-high screen attached to them to make up the side of the carport. This plan suggests rough-sawn 1 x 12 cedar siding boards, but many options are available to suit the style and decor of the house, including various wood sheet goods, and fiberglass and metal panels. The cost of these products varies widely, as does their installation time.

No consideration has been given in the plan for the cost of the floor surface within the carport because it is assumed that a driveway is already in place. If a floor surface is needed, the cost of asphalt pavement, concrete, or gravel, and the time to install it, will have to be added to the project estimate.

LEVEL OF DIFFICULTY

This carport plan is a manageable exterior project for most homeowners, including beginners who have the time to work slowly and are willing to learn as they go. Some of the tasks involved in the building process require knowledge that beginners may not have, but the required skills can be learned by trial and error and proper guidance from an experienced builder. Two of the jobs within this project are critical: the precise setting and leveling of the support posts and accurate layout and placement of the roof frame. If these two installations are correct, the rest of the job should go smoothly. If you are a beginner, be sure to get some assistance, particularly on these and other unfamiliar tasks. Intermediates and experts should be able to handle all of the tasks required in the project, but may need to learn the details of rolled-roofing installation before they begin that part of the job. Generally, beginners should double the professional time for all tasks; intermediates should add 40%; and experts, about 10%.

WHAT TO WATCH OUT FOR

With the investment of a little more time and money, carports can be made to double as screened-in sitting and eating enclosures. Once the carport structure, as described in this plan, has been completed, screen panels can be made up to fit the open areas in the rear end and side of the enclosure. The front end of the carport can be fitted with roll-down screening that can be raised when the facility is used to shelter the car. These modifications will increase the cost of the project, but they can be done at any time after the carport itself is finished and will add considerably to the practical use of the enclosure.

SUMMARY

A carport can serve as a practical alternative to a full garage at a fraction of the cost for both materials and labor. Because of the basic design of this project, most intermediates can complete this home improvement for only the cost of the materials.

For other options or further details regarding options shown, see

Attached single garage

Driveway

Freestanding double carport

Freestanding double garage

Attached Single Carport

Description	Quantity/ Unit	Labor- Hours	Material
Excavate post holes, incl. layout	5 Ea.	5.0	
Concrete, field mix, 1 C.F. per bag, for posts	5 Bags		37.20
Post base, 4 x 4	5 Ea.	0.3	28.50
Post cap, 4 x 4	5 Ea.	0.3	13.98
Posts, pressure-treated, 4 x 4 x 10'	50 L.F.	1.6	67.20
Headers, 2 x 6, doubled	96 L.F.	3.1	129.02
Joists/rafters, 2 x 6 x 12', 24" O.C.	132 L.F.	4.2	177.41
Sheathing, 5/8" plywood, 4' x 8' sheets	288 S.F.	3.6	196.99
Roofing, rolled, 30 lb. asphalt-coated felt	3 Sq.	2.4	97.20
Drip edge, aluminum, 5"	48 L.F.	1.0	11.52
Girts, 2 x 4, for screening panel	40 L.F.	1.2	24.00
Siding boards, cedar, rough-sawn, 1 x 12, board on board	110 S.F.	3.4	266.64
Paint, all exposed wood, primer	625 S.F.	7.7	67.50
All exposed wood, 2 coats	625 S.F.	12.3	112.50
Totals		46.1	$1,229.66

Project Size	11' x 21'	Contractor's Fee Including Materials	**$3,898**

Key to Abbreviations
C.Y.– cubic yard   Ea.– each   L.F.– linear foot   Pr.– pair   Sq.– square (100 square feet of area)
S.F.– square foot   S.Y.– square yard   V.L.F.– vertical linear foot   M.S.F.– thousand square feet

ATTACHED SINGLE GARAGE

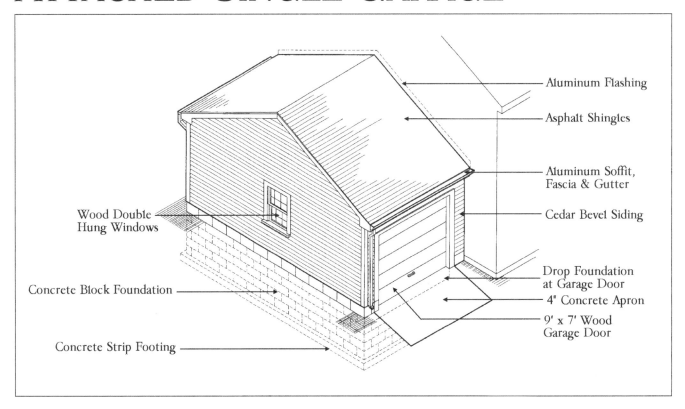

Aluminum Flashing

Asphalt Shingles

Aluminum Soffit, Fascia & Gutter

Cedar Bevel Siding

Wood Double Hung Windows

Drop Foundation at Garage Door

4" Concrete Apron

9' x 7' Wood Garage Door

Concrete Block Foundation

Concrete Strip Footing

A garage can boost the value of a house, as it provides protection for vehicles and sheltered work and storage space. Like carports, garages can be attached or freestanding, with single or double bays. This model is a basic single-car garage, attached to the house and built on level ground. Many do-it-yourselfers can accomplish this project with limited assistance and subcontracting.

MATERIALS

The components of this garage plan include materials for the foundation and floor, three walls, and the roof, as well as some specialized products like the gutters and garage door unit. Because all of these items are usually in good supply, prices are competitive and it pays to shop around. You probably will get a better "package" price if you buy all of the materials from the same retailer.

Before the foundation and floor are placed, the site for the new garage must be cleared and a perimeter trench excavated to a depth of about 4'. The site conditions of the garage location will have an impact on cost. For example, a severe slope in the grade may require a

higher foundation or backfill to raise the grade. If needed, such operations will add to the cost of the project. Generally, if the terrain is level and free of trees and other obstructions, no extra cost will be incurred in the site preparation. The excavation of the 12' x 22' foundation trench should be dug to a level below the frost line and should be deep and wide enough to allow for an 8" x 16" footing and a 3' to 4' foundation. Where no frost line exists, it might be possible to decrease the foundation depth. The estimate for this garage provides for hand excavation and the installation of 8" concrete blocks for the foundation. It is probably best to hire specialty contractors for the foundation work. At the very least, they can complete the process more rapidly and with less aggravation. The concrete slab can be placed at any time after the foundation has been completed. When excavating for the slab, be sure to allow for several inches of sub-base, 4" of concrete, and at least one exposed course of block above the floor surface, except in the section that will include the door. If you are installing the floor on your own, be sure to allow a gradual slope toward the door opening for drainage.

Once the foundation has been placed and backfilled, the three walls can be framed and sheathed. Standard grade 2 x 4s placed 16" on center, with a doubled plate on the top and a single shoe plate on the bottom, make up the walls. Be sure to fasten the shoe plate firmly to the 2 x 6 sill, which is anchored to the top of the foundation wall. The rear wall of this garage is solid, but the side and front partitions require openings for the window and door. The cost for 2 x 8 headers and 2 x 4 jack studs and their installation time have been included in the plan. Remember to allow additional time for removing some or all of the siding from the house wall before the framing begins. The other walls can then be framed, 2 x 6 joists can be placed, and the roof framing can be completed.

The framing of the roof is more difficult than that of the partitions; but this roof is still small enough to provide a good chance for do-it-yourselfers to pick up experience in placing a ridge board and rafter system. With some advice before you start and guidance as you go, your skills will improve as you progress. Consistency of rafter length and accurate angling and notching are the key to a square and level roof frame. Once the rafters have been laid out and cut from a template, the placement and fastening operations usually move quickly. After the frame is complete, the sheathing, drip edge, and roofing are installed. The shingling can be done by most do-it-yourselfers because the roof is straight and basic in its design. Get some assistance on placing the flashing, however, if the process is unfamiliar to you. A poor or incorrect flashing job can cause water damage to the interior of the garage and to the adjacent house wall and ceiling.

There are several options for the exterior finish products to be used on this garage. An economical, standard double-hung model is a good choice for the window. This plan suggests low-maintenance aluminum fascias, soffits, and gutters, but wood products could also be used. Similarly, garage door units are available in expensive wood-paneled models, as well as in economical fiberglass and hardboard designs.

LEVEL OF DIFFICULTY

Foundation, framing, and roofing operations are challenging tasks, regardless of the type of structure being built. Inexperienced do-it-yourselfers should assess their level of skills and the amount of time they can give to the project before taking it on. Because this structure is relatively small, with one side already in place, intermediates can improve their building skills while working on the project; but they should seek assistance when they need it. The concrete and masonry work may be out of the reach of many do-it-yourselfers. Digging trenches, constructing forms, and laying concrete blocks are physically demanding, time-consuming operations. Ambitious beginners can complete most of the project once the foundation is in, but they should get instruction at the start and guidance along the way. They should double the labor-hours for the tasks they attempt. Intermediates and experts should add 40% and 10%, respectively, for the carpentry work, and more for specialty operations like the foundation, floor, and garage door installations.

WHAT TO WATCH OUT FOR

An automatic garage door opener is a convenient accessory that can be installed by most homeowners. These devices can be adapted to all types of door units and garage interiors and installed in just a few hours' time. The cost of an automatic opener has not been included in this plan, but some of the money saved by doing the project on your own might be applied to this convenience item.

Attached Single Garage

Description	Quantity/ Unit	Labor- Hours	Material
Site clearing, layout, excavate for footing	1 Lot	7.0	
Footing, 8" thick x 16" wide x 44' long	2 C.Y.	5.5	218.40
Foundation, 6" concrete block, 3'-4" high	264 S.F.	23.2	497.38
Edge form at doorway	10 L.F.	0.5	4.92
Floor slab, 4" thick, 3000 psi, with sub-base, reinforcing, finished	265 S.F.	5.6	292.56
Wall framing, 2 x 4 x 10', 16" O.C.	38 L.F.	6.1	170.54
Headers over openings, 2 x 8	14 L.F.	0.7	14.62
Sheathing for walls, 1/2" thick, 4' x 8' sheets	384 S.F.	4.4	225.79
Ridge board, 1 x 8 x 12'	12 L.F.	0.4	17.28
Roof framing, rafters, 2 x 6 x 14', 16" O.C.	280 L.F.	4.5	191.52
Joists, 2 x 6 x 22', 16" O.C.	220 L.F.	2.8	150.48
Gable framing, 2 x 4	45 L.F.	1.3	19.44
Sheathing for roof, 1/2" thick, 4' x 8' sheets	352 S.F.	4.0	206.98
Sub-fascia, 2 x 8 x 12'	24 L.F.	1.7	25.06
Felt paper, 15 lb.	340 S.F.	0.7	8.16
Shingles, asphalt, standard strip	4 Sq.	6.4	180.00
Drip edge, aluminum, 5"	52 L.F.	1.0	12.48
Flashing, aluminum	21 S.F.	1.2	19.40
Fascia, aluminum	21 S.F.	1.2	53.93
Soffit, aluminum	38 S.F.	1.5	45.14
Gutters and downspouts, aluminum	44 L.F.	2.9	63.89
Wood, window, standard, 34" x 22"	1 Ea.	0.8	246.00
Garage door, wood, incl. hardware, 9' x 7'	1 Ea.	2.0	450.00
Siding, cedar, rough-sawn, stained	400 S.F.	13.3	1,444.80
Sheetrock, 5/8" fire resistant	484 S.F.	8.0	151.01
Painting, door and window trim, primer	85 S.F.	1.7	5.10
Painting, door and window trim, 2 coats	85 S.F.	1.7	5.10
Totals		110.1	$4,719.98

Project Size	12' x 22'	Contractor's Fee Including Materials	$12,190

SUMMARY

Adding a garage improves the value of your home while providing shelter for your vehicle and increasing storage and protected work space. If you do some or all of the work on your own, this garage can be built at a surprisingly low cost.

For other options or further details regarding options shown, see

Driveway

Freestanding double carport

Freestanding double garage

Single attached carport

Skylights

Standard window installation

Key to Abbreviations
C.Y.– cubic yard   Ea.– each   L.F.– linear foot   Pr.– pair   Sq.– square (100 square feet of area)
S.F.– square foot   S.Y.– square yard   V.L.F.– vertical linear foot   M.S.F.– thousand square feet

FREESTANDING DOUBLE CARPORT

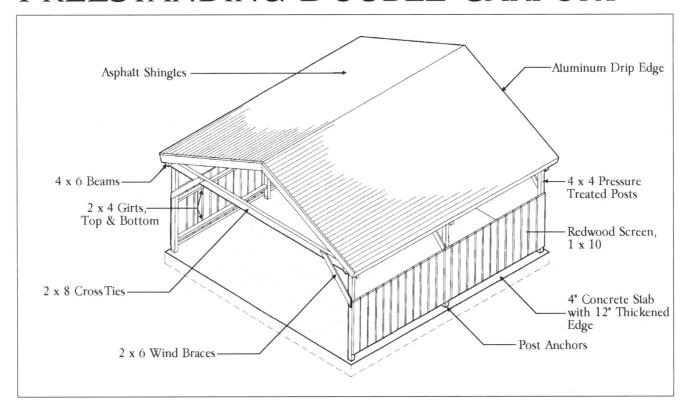

Asphalt Shingles

Aluminum Drip Edge

4 x 6 Beams

2 x 4 Girts,
Top & Bottom

4 x 4 Pressure
Treated Posts

Redwood Screen,
1 x 10

2 x 8 Cross Ties

4" Concrete Slab
with 12" Thickened
Edge

Post Anchors

2 x 6 Wind Braces

Carports can be relatively small, simple enclosures, like the one-car, attached model on previous pages, or they can be constructed on a larger scale in a freestanding design. This plan demonstrates how a two-car facility can be economically built by using post-and-beam construction methods. The simplicity of its design makes it a reasonable undertaking for intermediates who are experienced in the use of tools and have a fundamental knowledge of carpentry. Because of the size of this structure, beginners should leave much of the post-and-beam support installation and roof framing to the professional, but they are encouraged to tackle some of the other tasks in the project. Generally, skilled intermediates and experts can complete the entire project, with the exception of the concrete floor.

MATERIALS

The materials used in this project start with ready-mixed concrete for the floor and the conventional components of post-and-beam construction, including 4 x 6s and 4 x 8s (or doubled 2 x 8s). Standard grade 2 x 6 rafters, plywood sheathing, and asphalt shingles are

needed for the roofing, and siding material is required for screening. All of these products, except the ready-mixed concrete, are readily available at lumber yards and building supply outlets, so do some comparison shopping before you buy. Also, before you finalize the purchase agreement, make sure arrangements have been made for curb delivery. Although this structure is simple in design, the materials used in its construction are heavy and require trucking.

The preparation for the wood structure begins with the excavation for and placement of the concrete floor. There are several important aspects in this procedure that, when closely monitored, can help the rest of the project go smoothly. The first is to double-check the square and level of the form before the concrete is placed. The second is to be sure that the post anchors for the 4 x 6s are set in precise position and alignment as they are embedded in the concrete. Remember that the wood structure will be supported by the slab and aligned according to the position of the anchors, so measure and square the layout carefully. If you are unfamiliar with this type of concrete placement,

get some assistance from an experienced person or subcontract this part of the project to a professional.

The post-and-beam construction of the roof support requires basic carpentry know-how and some engineering skills as well. The three post-and-crosstie assemblies can be laid out and fabricated on the slab before being raised and positioned on the anchors and temporarily braced. Be sure that they are plumb, level, and square before the 4 x 6 plate beams are placed and fastened. Some skilled intermediates and experts can tackle this operation, but beginners and intermediates with limited building skills should seek the help of someone who is more experienced or hire a professional.

The roof frame is constructed of 2 x 6 rafters, set 24" on center, and a 2 x 8 ridge board. Like the post-and-beam support system, the roof framing requires a fundamental knowledge of carpentry and advanced building skills. If you are unfamiliar with the process of cutting and setting rafters, get some instruction beforehand, and be sure to have some helpers available. Remember, too, that you will be working 8' to 12' off the ground when placing the rafters, so make arrangements to rent or borrow some staging, trestle ladders and planks, or other means of working safely and efficiently at the 12′ ridge board height. An alternative to installing the rafters is to purchase manufactured 2 x 4 trusses and place them 24″ on center. The cost for materials will be higher, but the roof will go up faster.

The 1/2″ plywood sheathing is placed on the rafters after the roof frame has been squared and fastened. The aluminum drip edge, which is installed before the roofing is laid, is an inexpensive but important item, as it provides run-off clearance for rain water and creates a straight, neat, finished edge for the roofing. Experts and most intermediates should be able to handle this roofing job, as it consists of straight runs with no valleys, flashing, or other challenging obstacles.

The finished sides of the carport should conform to the style of your house. You can select from many different materials, including plywood products, siding boards, and metal and fiberglass panels. This plan suggests that the 1 x 10 redwood tongue-and-groove siding be placed on the 2 x 4 girts as screening and then stained or painted. The post-and-beam supports, girts, rafters, underside of the roof deck, and the fascia board are covered with two coats of stain or paint.

LEVEL OF DIFFICULTY

This carport plan calls for a moderate to advanced level of building expertise and carpentry skills. In addition, the size of the structure and the post-and-beam layout requirements involve complex planning that may put parts of the project out of reach for beginners and many intermediates.

After the supports and roof frame are in place, most intermediates can finish the job, including the installation of the roof sheathing and shingles. Be sure to get some instruction in the roofing process if you have not laid shingles before. Straight roofs are not that hard to install, and a good start can make it a fairly simple operation.

Beginners should hire a professional for the concrete slab installation. Experts and intermediates should allow extra time for a careful layout of the forms and should arrange to have several helpers on hand when the concrete is placed. Accomplished intermediates and experts should add 100% and 50%, respectively, to the professional time for the slab installation, including the form work. They should add 40% and 10%, respectively, for all other tasks. Beginners should not attempt the slab, post-and-beam support, or roof-framing jobs. They should double the professional time for all other tasks, provided they are given ample instruction before they start.

WHAT TO WATCH OUT FOR

The concrete surface that serves as the floor of the carport is an important feature and must be correctly placed. If you are installing it on your own, set the forms accurately so that the edges of the slab are level and square. Metal-rod or wire reinforcement, a sub-base of granular material, and a plastic vapor barrier are also necessary for correct installation. Remember, too, that provisions for surface-water drainage should be made by crowning the slab. Wind-driven rain or snow and water from wet or snow-covered vehicles can collect and puddle in low spots on the surface, but these problems can be avoided with proper finishing of the floor. Seek help from a cement mason or concrete contractor if you are inexperienced and need assistance in placing the slab.

SUMMARY

Although this double carport is fairly basic in design, it is a challenging undertaking for most do-it-yourselfers. It is not out of reach, however, with help and guidance. This project is a good way to gain experience while creating a practical addition to your home.

For other options or further details regarding options shown, see

Driveway

Single attached carport

Freestanding Double Carport

Description	Quantity/ Unit	Labor- Hours	Material
Grading, by hand, for slab area	53 S.Y.	1.8	
Edge forms for floor slab	86 L.F.	4.6	42.31
Vapor barrier, polyethylene, 6 mil	4.20 Sq.	0.9	14.77
Concrete floor slab, 4″ thick, 3000 psi complete	420 S.F.	8.8	463.68
Post anchors, embedded in slab	6 Ea.	0.2	4.46
Posts, pressure-treated, 4 x 6 x 8′	48 L.F.	2.2	121.54
Crossties, 2 x 8 x 22′	132 L.F.	4.7	229.68
Plate beams, 4 x 6 x 12′	48 L.F.	2.2	121.54
Ridge board, 2 x 8 x 12′	24 L.F.	0.9	41.76
Fascia board, 2 x 8 x 12′	48 L.F.	1.7	83.52
Rafters, 2 x 6 x 12′, 24″ O.C.	288 L.F.	9.2	387.07
Sheathing, 1/2″ plywood, 4′ x 8′ sheets	544 S.F.	6.2	319.87
Drip edge, aluminum, 5″	96 L.F.	1.9	23.04
Shingles, asphalt, 235 lb. per square	6 Sq.	8.7	205.20
Braces, 2 x 6 x 4′	40 L.F.	1.3	53.76
Girts, 2 x 4, for screening panels	84 L.F.	2.4	50.40
Screening, redwood channel siding, 1 x 10	210 S.F.	5.9	561.96
Paint, primer	1,100 S.F.	13.5	118.80
Paint, 2 coats	1,100 S.F.	21.7	198.00
Stain, 2 coats, brushwork on screen	420 S.F.	7.1	45.36
Totals		105.9	$3,086.72

Project Size	20′ x 21′	Contractor's Fee Including Materials	$9,425

Key to Abbreviations

C.Y.– cubic yard   Ea.– each   L.F.– linear foot   Pr.– pair   Sq.– square (100 square feet of area)
S.F.– square foot   S.Y.– square yard   V.L.F.– vertical linear foot   M.S.F.– thousand square feet

FREESTANDING DOUBLE GARAGE

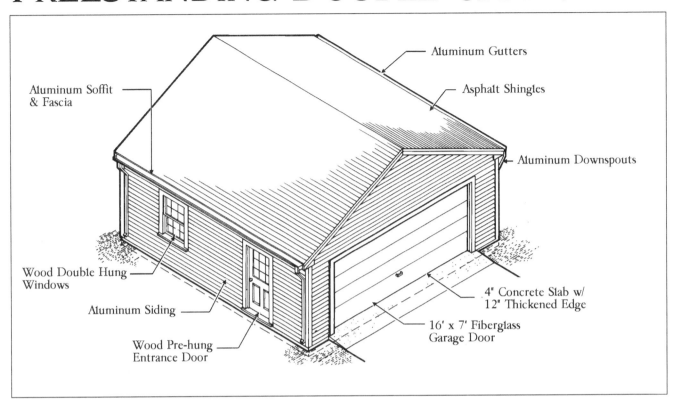

Aluminum Gutters

Asphalt Shingles

Aluminum Soffit & Fascia

Aluminum Downspouts

Wood Double Hung Windows

4" Concrete Slab w/ 12" Thickened Edge

Aluminum Siding

16' x 7' Fiberglass Garage Door

Wood Pre-hung Entrance Door

If your house lot is large enough to accommodate the addition of a freestanding garage, you might consider using this model for a two-car facility. This plan is basic in both design and materials; but it is still a major project, and one that will challenge do-it-yourselfers of all ability levels. With patient and careful work, however, intermediates can save on labor costs by completing some or all of the project on their own.

MATERIALS

The basic components of this two-car garage are commonly available building materials used in wood-frame structures. A 4" reinforced concrete slab with a thickened perimeter comprises the foundation and finished floor of the structure. Standard 2 x 4 framing materials, plywood, and asphalt shingles are employed for the roof. Several specialty products are also included to finish the structure, including aluminum gutters, soffits, and fascias, and a double-width fiberglass garage door. Bear in mind that all of these materials are competitively priced at building supply outlets. With this size order, you have some leverage in getting the best possible

price. Arrange for local delivery in the final purchase agreement, as most of these materials require trucking. Cost, quality, and convenience are factors to consider before purchasing your materials.

If the site for the new garage is level and dry, the suggested 4" reinforced concrete slab with a thickened perimeter is generally all that is required for a foundation. Poor soil conditions, frost, and the possibility of future attachment to an existing structure may require the placement of a footing and foundation wall at extra cost. The site clearing and excavation will take some time because of the large surface area, so plan accordingly, particularly if you intend to clear and dig by hand. If you are unfamiliar with concrete installations, seek advice from a knowledgeable person or subcontract the job to a professional. Incorrect slab placement can be the cause of wasted time and materials later in the construction.

Installing the wall materials for this four-sided structure requires a command of basic carpentry skills. If you have not framed a four-sided structure before, you will need some assistance in the layout, erecting, squaring, and securing of the

partitions. After the wall framing is in place, 1/2" exterior plywood sheathing should be installed for support and as a base for the finish siding.

The roof framing is the most difficult task in the entire project, as the 2 x 6 rafters and 1 x 8 ridge board have to be positioned with accuracy. Poor workmanship can cause wasted time and material, and result in variations in the finished roofing. Like the straight roof in the freestanding carport project, this shingling job can be handled by most do-it-yourselfers. Roofing felt is an extra weatherproofing material that is applied under the asphalt shingles in this project. It is not an essential item, however, if you are looking for a way to cut costs.

Once the roof has been completed, the structure is ready to receive the windows, entrance door, siding, garage door, and other finish products. A wide variety of standard pre-hung window and door units can be used, depending on the style of your home, the garage's visibility, and your budget. The extra expense for insulating glass or deluxe airtight units is unnecessary unless you plan to insulate and heat the facility. This plan features low-maintenance aluminum siding, soffits, and gutters, but standard wood products can be used in their place at varying costs. If you choose the suggested aluminum material, a specialty contractor should be consulted or called in, as these products require particular tools and expertise for fast and efficient installation. Many different types of garage doors are available in a wide range of prices, in both wood and manmade materials. The large fiberglass door suggested in this plan is recommended for its light weight, durability, and maintenance-free properties. A specialty door contractor should also be consulted or hired before the installation.

LEVEL OF DIFFICULTY

If you are familiar with framing operations and reasonably skilled in exterior home improvements, you should be able to complete this garage project. As noted earlier, it is a major undertaking and will take considerable time to complete, but if you are willing to work slowly and carefully, the results will be worthwhile. Two critical procedures, the slab installation and roof framing, can be costly if they are not done correctly. Seek professional assistance for these tasks if you feel that they are beyond your abilities. Beginners should not attempt the concrete, framing, and specialty installations, and they should add 100% to the professional time for other jobs like the sheathing and roofing. Intermediates and experts should add 40% and 10%, respectively, to the labor-hour estimates for the jobs they attempt. They should add more for the slab placement and roof framing if they are inexperienced in these operations. Intermediates and experts should also consider hiring professionals for the specialty work, like the aluminum siding and garage door installation.

WHAT TO WATCH OUT FOR

This two-car garage project provides an excellent opportunity for the homeowner to create storage and workshop space in addition to shelter for vehicles. If the peak of the structure is high enough, over 300 square feet of usable storage space exists in the attic above the ceiling joists. Although it will add to the cost and time of the project, decking can be laid over the joists, and an access door can be framed into the gable above the garage door opening. Remember that the ceiling joists suggested in this plan are only 2 x 6s; 2 x 8s should be used for the joists if heavy materials are to be stored in this area. Electrical service can also be added to the plan for safety or convenience lighting, or for workshop use.

SUMMARY

If your lot and dwelling can accommodate a large freestanding structure, this design might be the right choice. It is a major project, but intermediates who have the time can complete much of the work on their own, for a substantial savings in the cost of construction.

For other options or further details regarding options shown, see

Attached single garage

Driveway

Freestanding double carport

Main entry door

Single attached carport

Skylights

Standard window installation

Freestanding Double Garage

Description	Quantity/ Unit	Labor- Hours	Material
Grading, by hand, for slab area	576 S.F.	19.8	
Edge forms for floor slab	88 L.F.	4.7	43.30
Concrete floor slab, 4" thick, 3000 psi, complete	484 S.F.	9.4	522.72
Anchor bolts, embedded in slab	20 Ea.	0.8	14.88
Wall framing, 2 x 4 x 8', 24" O.C., studs and plates	80 L.F.	10.2	239.04
Headers over openings, 2 x 12	28 L.F.	1.5	57.12
Sheathing for walls, 1/2" thick, plywood, 4' x 8' sheets	736 S.F.	10.7	768.38
Ridge board, 1 x 8 x 12'	24 L.F.	0.7	34.56
Roof framing, rafters, 2 x 6 x 14', 24" O.C.	336 L.F.	5.4	229.82
Joists, 2 x 6 x 22', 24" O.C.	264 L.F.	3.4	180.58
Gable framing, 2 x 4	75 L.F.	2.2	32.40
Sheathing for roof, 1/2" thick, plywood, 4' x 8' sheets	756 S.F.	8.6	444.53
Sub-fascia, 2 x 8 x 12'	48 L.F.	3.4	50.11
Felt paper, 15 lb.	7 Sq.	1.5	17.89
Shingles, asphalt, standard strip	7 Sq.	11.2	315.00
Drip edge, aluminum, 5"	100 L.F.	2.0	24.00
Fascia, aluminum	50 S.F.	2.8	128.40
Soffit, aluminum	70 S.F.	2.7	83.16
Gutters and downspouts, aluminum	64 L.F.	4.3	92.93
Windows, wood, standard, 2' x 3'	2 Ea.	1.6	492.00
Garage door, incl. hardware, hardboard, 16' x 7', standard	1 Ea.	2.7	714.00
Entrance door, wood, prehung, 2'-8" x 6'-8"	1 Ea.	1.0	237.60
Siding, aluminum, double, 4" pattern, 8" wide	680 S.F.	21.1	971.04
Painting, entrance door and window trim, primer	50 S.F.	0.6	1.20
Painting, entrance door and window trim, 2 coats	50 S.F.	1.0	3.00
Totals		133.3	$5,697.66

Project Size	22' x 22'	Contractor's Fee Including Materials	$14,579

Key to Abbreviations
C.Y.– cubic yard   Ea.– each   L.F.– linear foot   Pr.– pair   Sq.– square (100 square feet of area)
S.F.– square foot   S.Y.– square yard   V.L.F.– vertical linear foot   M.S.F.– thousand square feet

Section Five
LANDSCAPING

Much of the visual appeal of older, established neighborhoods lies in the abundance of their greenery. Such landscape features as "mature plantings," meaning fully developed trees and shrubs, increase your property's value. Following are some tips for planning your landscape project.

- In planning landscape improvements, read through books, magazines, journals, and catalogs. Local nurseries or county agricultural extension offices can also provide up-to-date advice on which plant varieties thrive in your area, and which ones pose special problems.

- Large-scale landscape projects are often best left to contractors with properly equipped and experienced crews.

- When planning your project, try to create a three-to-five year plan to allow for growth and development of select shrubs and plantings. In particular, be sure to place shrubs a proper distance from the house to allow for mature growth.

- When planting shrubs or other plants close to the house, be sure to leave plenty of room for access to oil tanks, meters, fills, and so on. Also, make sure foliage is *at least* 6–8" away from the house, in order to avoid mildew and other moisture damage, and to allow access to the exterior walls and windows for maintenance.

- Follow planting and transplanting directions provided by lawn and garden experts. Do not overdo fertilizers. Also, do not ignore regular maintenance requirements.

BOARD FENCE

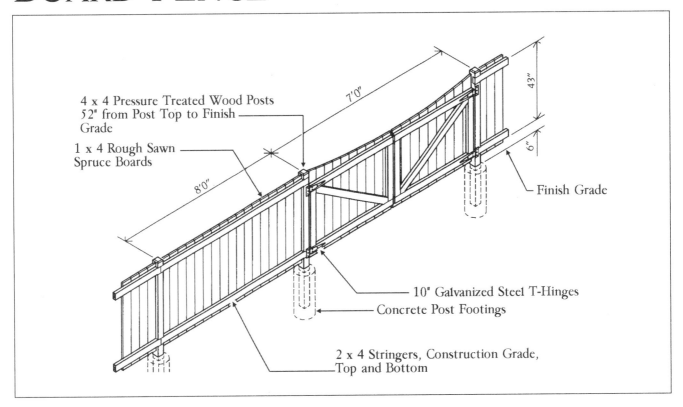

4 x 4 Pressure Treated Wood Posts 52" from Post Top to Finish Grade

1 x 4 Rough Sawn Spruce Boards

7'0"

43"

6"

8'0"

Finish Grade

10" Galvanized Steel T-Hinges

Concrete Post Footings

2 x 4 Stringers, Construction Grade, Top and Bottom

Fence-building with prefabricated panels saves you the labor involved in constructing them yourself and can reduce the time needed to install the fence, but these savings can be offset by the cost of the panels themselves. Also, your choice of design and size is limited to the panels your local fencing supplier provides. Site-building your own fence gives you the freedom and flexibility to design the style of fence that best conforms to your needs and to the look of your home's architecture and landscaping. Building on-site also allows you to choose the most appropriate materials for the job and to purchase them most economically. Keeping in mind the dictum "form follows function," custom building can be the best and most cost-efficient way to get a fence that combines good looks with practicality.

MATERIALS

The most difficult task in a custom fence project like this may well be designing it. The freedom to build a fence of any style, using any combination of materials, can cause a certain paralysis of the will, for the very reason that the choices are so unlimited. You must determine exactly

why you need or want the fence, what its purpose is to be, what style of fence would harmonize with your house and landscape, and where it is to be located on your property. Answering these questions makes the design process a matter of narrowing choices by eliminating fencing styles and materials that are not appropriate to your stated needs. You may find that preassembled panels matching your choice can simplify your project. And don't overlook that method of research which involves getting in your car and driving around to observe the fences that others have built as well as those that are available at local fence suppliers. Picture how those fences that most appeal to you would look in your own yard. Be sure you know your boundary lines, and are familiar with any codes, covenants, or restrictions (such as historical preservation or conservation commission guidelines) that may affect your plan.

An ornamental fence — that is, one intended to enhance the looks of your house — should be in keeping with its architecture and landscaping. For example, a split-rail fence that would look fine with an antique farmhouse would look out of place in front of a

gingerbread Victorian. Most homeowners would recognize and avoid such obvious mismatches, but many homes are not so easily categorized, and they may require the experienced eye of a professional designer to select a fence that will indeed be an ornament, not a detriment to the home.

A fence that is intended to serve a practical need — for example, confining children or pets — is generally easier to design because the functional requirements help to eliminate the inappropriate. Thus a ranch-style fence can confine cattle, but not the children or dogs, and a low picket fence can beautify the perimeter of a yard, but provides no privacy.

This solid board fence is designed as an economical way to enclose a backyard, keeping your small children and pets in, and those of others out. Being just 4' high, it is visually unobtrusive, yet provides enough vertical surface to serve as an attractive backdrop for shrubs and flowers. The material components have been selected for their workability, practicality, and reasonable cost.

The posts are pressure-treated 4 x 4s, rated for ground contact, which means they can be set directly in the ground and are guaranteed against rot for up to 30 years. The stringers are construction grade 2 x 4s. These are much easier to cut, screw, and nail than pressure-treated stock and cost about half the price. The same reasons govern the choice of 1 x 4 rough-sawn spruce for the fence boards, rather than pressure-treated lumber. A good quality preservative stain does not take long to apply, especially if a sprayer is used, and the wide range of available colors allows you to choose one to complement your home or to blend the fence into the surrounding landscape. Every few years an application of the same stain, or a clear preservative, will keep the fence looking good and help protect it from the elements. The double gate is about 7' wide, enough room to allow access for vehicles and lawn care equipment.

Lay out the line of the fence with stakes and mason's cord. If the fence attaches to your house, you can work out from the house, using it as a fixed point of reference for determining measurements and angles. If the fence is freestanding, it is best to establish the position of the corner posts and lay out 90° angles to connect them. Site-building allows you to locate the posts where convenient, and enables you to avoid obstacles to digging, such as large roots or immovable rocks. Eight-foot spacing between the posts is the general rule, but if necessary, shorter or longer spans can be used, the difference being hardly noticeable once the fence is complete. This is an especially useful expedient where the fence must follow the contours of a sloping or hilly yard.

The stringers can be joined to the posts in a number of different ways: with facenails or screws; toenails or screws; dadoes on the post sides; dadoes on the post fronts or backs (as illustrated); or with galvanized steel post brackets. The post tops can be left square-cut and plain, or beveled in various ways to dress them up a bit and to help them shed water. The gates swing on heavy T-hinges, and shut with an external mount latch, which can be padlocked for additional security.

LEVEL OF DIFFICULTY

A custom-designed, site-built fence is a challenging project. By its very nature a fence is a "public" construction that greatly affects the appearance of your house and yard for better or for worse. Beginners should not undertake the job without qualified help in design, layout, and construction. An intermediate should have no trouble completing the work independently. It calls for a few basic construction skills: measuring, cutting rough dadoes and angles, plumbing posts, and nailing or screwing the parts together. It also involves the drudgery of digging a number of post holes between two and three feet deep. Local conditions can make this chore either routine and fairly easy, as in loose, sandy soil, or a real pain in the back, as in hard-packed clay full of roots and rocks.

Beginners should have guidance and assistance throughout all phases of this project, and for any tasks undertaken, should add 150% to the times estimated. Intermediates and experts could build this fence unaided, and should increase the estimates by 75% and 40%, respectively.

WHAT TO WATCH OUT FOR

For digging the post holes, a long-handled shovel is the one indispensable tool. You might also find a clamshell-type post hole digger useful. A long pry bar and an axe or a hatchet may be needed to help remove rocks and tree roots. A rented power auger can make quick work of the hole-digging chore if the soil is fairly free of such underground obstacles. A two-man machine works best, but be sure you understand how to operate it safely before pulling the starter cord.

Once the posts are firmly in place and the stringers secured to them, the boards can be nailed on. They should be butted tight together because some shrinkage is bound to occur as the boards dry out. The resulting gaps – generally less than 1/4" – will be narrow enough not to be noticeable. Remember the effect of your fence's overall appearance. Take time to first tack the boards in place and stand back to see how they look before nailing them permanently.

To accommodate a slope, the fence sections can either follow the slope or be stepped. In the first instance, the stringers are run parallel to the slope; in the other, the stringers remain level. In both cases the posts and boards must be perfectly plumb.

A very easy and attractive way to set off the gate is to cut a curve across the top. Lay the gate sections side by side and scribe a large radius arc, using a string-and-pencil compass, a plywood template, or a flexible wood batten. Make the cut with a saber saw. Prior to installing a fence, be sure you know your boundary lines and are familiar with any codes, covenants, or restrictions (historic boards) that may impact your decision.

SUMMARY

Prefabricated fencing can limit your choice to the designs, sizes, and materials offered by local dealers. The price of the panels includes, of course, the cost of shop labor, which is considerably more than the cost of the materials alone, but will cut down on your assembly time. A custom-designed, site-built fence gives you the freedom to get exactly the size and type of fence you want, within your budget.

Board Fence

Description	Quantity/ Unit	Labor-Hours	Material
Board fence, 1 x 4 boards, 2 x 4 rails, 4 x 4 posts, treated	320 L.F.	56.9	2,304.00
Gate, general wood, 3'-6" wide, 4' high	2 Ea.	5.3	122.40
Excavation, hand pits, sandy soil for post anchors	1 C.Y.	1.0	
Pre-mixed concrete, for post anchors, 70-lb. bags	13 Bags		96.72
Placing concrete, footings, under 1 C.Y.	1 C.Y.	0.9	
Totals		64.1	$2,523.12

Project Size	4' high x 320' long	Contractor's Fee Including Materials	**$6,048**

Key to Abbreviations
C.Y.– cubic yard   Ea.– each   L.F.– linear foot   Pr.– pair   Sq.– square (100 square feet of area)
S.F.– square foot   S.Y.– square yard   V.L.F.– vertical linear foot   M.S.F.– thousand square feet

BRICK WALKWAY

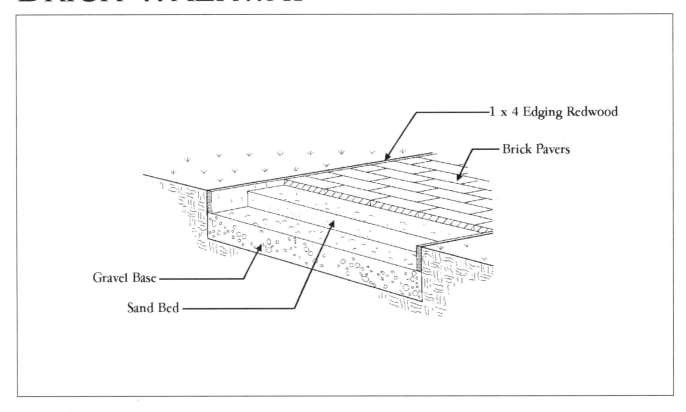

1 x 4 Edging Redwood

Brick Pavers

Gravel Base

Sand Bed

Walkways and other landscape features are often installed long after a house is built and occupied. The exception occurs when "spec" builders landscape as soon as the siding is on, to dress up the lot and make the house attractive and marketable. The backfill and grading around the house, however, will continue to settle over the course of many months, and this can cause annoying and unsightly problems, from dead shrubs to eroded and sunken walks, which eventually need repair or, in the worst cases, replacement. Avoiding such "instant landscaping" is generally recommended, but postponing the work often means it gets done piecemeal, sometimes remaining incomplete for years.

A walkway is both attractive and practical: it provides a visual and physical link between elements in the landscape; it directs visitors to your home's entrance, and makes them welcome; it offers clear and safe footing; and it protects grass from the trampling of traffic. If your walkway has deteriorated, or if one was never built, this is a fairly easy and economical plan for replacing or installing one.

MATERIALS

The simplest and most common way to build a walkway is to lay the paving material in a bed of sand or stone dust, without mortar. This is called "dry" construction. It can be used with a variety of paving materials: bricks, concrete pavers, or flagstones; it drains well; and it allows you to correct mistakes, replace broken pieces, or remove either a section or the whole of the walk at any time.

Flagstones are cut from any type of a stone that splits into flat pieces, such as sandstone or slate, and are available in irregular or rectangular shapes, ranging in thickness from 1/2" to 2". Bluestone, granite, and marble are beautiful, but expensive, and look best in a formal setting. Stones that are multiple-cut, cut to size, or are of special shape (e.g., round), are the easiest to arrange. Irregular pieces of stone require time and care in laying, because they look best when fitted fairly close together, like a jigsaw puzzle. The dry construction method lets you work out and alter the arrangement of the pattern as you go.

Concrete pavers come in hundreds of sizes, colors, shapes, and textures. They can be used either alone or in combination with other paving and border materials, such as brick, wood, or loose aggregates (e.g., wood chips or crushed stone). Some concrete pavers interlock, which helps make them resistant to frost heaves. They are often used to create elegant-looking driveways.

Bricks are the most versatile and easily laid paving material. A visit to a quarry or masonry supplier will reveal the surprising variety of colors, textures, and shapes available in this most ancient and durable building material. Individually, bricks are small and relatively lightweight, which makes them easy to handle, to arrange in interesting and attractive patterns, and to lay in conformity with gradual changes of direction and ground level.

Bricks are graded according to specifications established by the American Society for Testing and Materials (ASTM). Common bricks are graded as follows: SW (severe weathering) – used, regardless of climate, where the bricks will be in contact with the ground; MW (moderate weathering) – able to withstand freezing weather, but not if in contact with the ground; and NW (nonweathering) – for interior work only. Pavers are a special category of brick, being harder and more resistant to wear than common bricks. Bricks can be laid in a number of different arrangements known as "bonds." The illustration shows the popular, economical, and easy to lay "running bond." Other bonds are "jack-on-jack," "herringbone," and "basket weave." The latter two bonds require the use of "modular" pavers; that is, bricks that are exactly twice as long as they are wide. The walkway can be edged with bricks in one of several configurations, termed "stretcher," "header," "soldier," or "sailor," or with other materials such as stone or treated wood.

Mark off the edges of the walkway with string and stakes, and excavate to a depth of about 6". Dampen and tamp the earth in the walkway area, and put down a layer of synthetic landscaping fabric. This will help block the growth of weeds without inhibiting drainage. Cover the walkway with a layer of sand or stone dust to approximately 2" from grade level. Lay the bricks in the pattern you have chosen, tamping down each brick with a rubber mallet. Add or remove sand to keep all the bricks in line and on the same plane. When all the bricks are laid, spread sand over them and sweep it into the joints, hosing down the walk to compact the sand. Repeat this procedure as many times as necessary to completely fill the joints. Laying the bricks tightly together keeps the joint spacing to a minimum; this helps the sand stay in place and makes the walkway almost maintenance-free.

Like highways, walkways should be crowned; that is, slightly pitched from a high point at the center down to the edges. This eliminates the possibility of ponding surface water by creating a natural drainage system. To crown the walkway, simply sculpture the base material to the desired shape. The bricks will then follow this contour when placed.

LEVEL OF DIFFICULTY

The physical demands of a landscape project can be accurately gauged by the materials used and the tools needed: bricks and sand, pick and shovel, hammer and cold chisel (brick chisel or set) all conjure images of sweaty work and aching muscles. If one is willing and able to pay this physical price, even a beginner can successfully complete a dry construction walkway of flagstone or brick. You can largely control the amount of labor involved in laying bricks by the bond you choose. A running bond is easy to lay and requires few cuts. More complex bonds require more effort to arrange, and more cutting, which, of course, takes more time and creates more waste.

A beginner should get experienced advice in choosing the most appropriate type of brick, and assistance in laying out the walkway, adding 150% to the estimate listed. An intermediate can complete this job alone, and should add about 50% to the time. An expert should add about 10%.

WHAT TO WATCH OUT FOR

Both sand and stone dust are used as a leveling base and filler in dry construction walkways. Stone dust is grayish and gritty, producing a lot of airborne dust when dry, and forming a sort of clingy paste when wet. This makes it somewhat messy to work with, but it compacts very firmly to an almost cement-like hardness over time. Sand is a cleaner, "softer," and more pleasant material. In fact, if there are children around when it is delivered, they won't be able to resist leaping into the sand pile to play. They should, however, be directed elsewhere until you have finished the job, at which point they can play with whatever is left.

The only specialized tools you need are a rubber mallet for tamping the bricks and a 3" masonry chisel for cutting them. Cutting a brick involves holding the chisel to the brick and hitting it sharply with a hammer. A heavy hammer, such as a 25-ounce framing hammer or a small sledge deliver a solid, dead blow and will not bounce back off the chisel as a lighter hammer will. Your first few attempts may produce more rubble than usable brick, but you will improve with practice. A short piece of wide 2" lumber makes a good cutting board. Protect your eyes with safety glasses.

Brick walks generally cause few drainage problems, but to ensure good surface run-off, the walkway should be very slightly crowned, curving gradually to either edge from a 3/4" to 1" high point in the center. The 3' width of this walk allows two people to walk comfortably abreast. After the job is completed, observe the walkway on a rainy day to see if there are any depressions that cause puddling. On a good day, remove the low bricks and add sand to raise them.

The secret to successful brick paving is properly constructing the border. It must be strong and lock the paving field in place. Any place that the brick is terminated, as at a step, the final course should be laid in mortar.

SUMMARY

Brick, one of the oldest and most durable building materials, is compatible with most styles of architecture. Adding a walk is an appealing project for a do-it-yourselfer.

For other options or further details regarding options shown, see

> *Brick or flagstone patio in mortar*
> *Brick or flagstone patio in sand*
> *Driveway*

Brick Walkway

Description	Quantity/ Unit	Labor- Hours	Material
Sand, base fill, 4" deep	90 S.F.	1.7	64.80
Compaction in 6" lifts, hand tamp	1 C.Y.	0.4	
Hand grade, fine grading of base	3 C.Y.	2.1	55.08
Brick pavers, laid flat	90 S.F.	14.4	232.20
Edging, redwood, 1 x 4	60 L.F.	2.9	192.24
Totals		21.5	$544.32

Project Size	3' wide x 30' long	Contractor's Fee Including Materials	$1,731

Key to Abbreviations
C.Y.– cubic yard   Ea.– each   L.F.– linear foot   Pr.– pair   Sq.– square (100 square feet of area)
S.F.– square foot   S.Y.– square yard   V.L.F.– vertical linear foot   M.S.F.– thousand square feet

DRIVEWAY

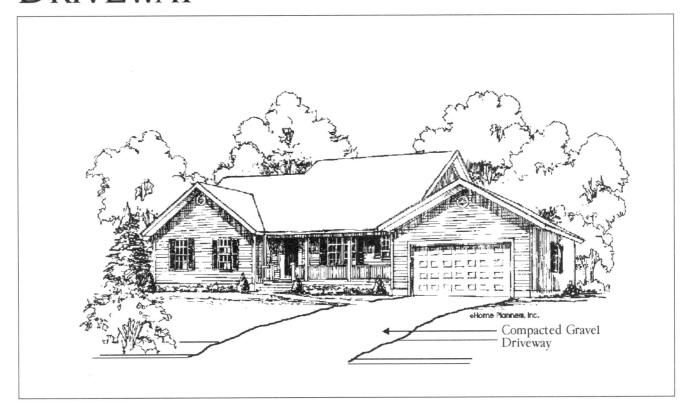

©Home Planners, Inc.

Compacted Gravel
Driveway

In today's mobile society, the driveway has become a basic requirement for most homeowners. The addition of a new car to the family can raise havoc with the drivers and their schedules if the existing driveway is not large enough. In addition to space for cars, the driveway also often serves as a setting for such activities as basketball, street hockey, and car washing.

MATERIALS

When locating your driveway you must consider local zoning/building codes, which will determine the required set-back distance from your property lines. Other considerations include the size driveway you need, whether there are utility lines or a septic system running under the proposed driveway location, whether you have enough room for a driveway on your property, and how and where the new driveway will drain.

Once you have decided on the size and location of the driveway, you will need to decide on the type of surface material. The choices include asphalt, brick, concrete, and gravel. Gravel is economical and is considered by many to be more attractive than other materials because of

its natural look and the wide range of available colors and stone sizes. A gravel driveway is something most homeowners can install with some careful planning and a few tips from an expert. The others should be installed by a professional, as they require special equipment and experienced installers.

Installing a driveway requires excavation, which means a skid steer loader will need to be able to maneuver easily in the area of the proposed driveway. The area to be excavated should be staked off so that the installer always has a layout line to follow. Topsoil should be excavated to a minimum depth of 6″, and a base of bank run gravel should be placed to a minimum depth of 3″. Hand grade the base and compact it using a gas-powered vibratory plate compactor. Once the base has been compacted, lay down a top layer of pea gravel and compact this in the same way as the base was done. Finally, backfill any open areas around the driveway using the top soil that was excavated.

The preparation for brick, concrete, gravel (stone), or concrete pavers is the same as that for asphalt. Each requires proper grading and base preparation.

LEVEL OF DIFFICULTY

A gravel driveway can be installed by most homeowners. The most difficult part of the driveway installation is the excavation, which can be done using a rental skid steer loader (such as a Bobcat or Case) or by hiring a professional excavator. Most of the work involves the physical labor of spreading the gravel with a rake. A compactor can also be rented and used by a beginner.

Beginners and intermediate do-it-yourselfers could install a brick, concrete, or interlocking paver driveway, but should consider complicating factors such as the weight and storage location of the materials, and requirements for cutting materials. Refer to the patio projects for more information.

Experienced do-it-yourselfers may want to rent all of the aforementioned equipment and try the work on their own. They should add 100% to the estimated time. Beginners should probably restrict themselves to the basic preparation and layout, and the final grading and landscaping.

A professional should be hired for any type of surface other than the gravel.

WHAT TO WATCH OUT FOR

Scheduling is important during any job that requires the services of a professional contractor. Schedule equipment rental in such a way as to avoid paying for equipment while it sits idle.

In climates where snow is common, gravel can be displaced by shoveling. After several years, a gravel driveway may need to be regraded or improved with another layer of stone.

When considering paving, keep in mind that prices for asphalt and concrete are generally higher during the cold seasons because of installation complications in cold temperatures.

A well-installed driveway should have no puddles or surface water remaining on the surface after a rainstorm or car washing session.

DRIVEWAY OPTIONS
Cost per Square Foot, Installed

Description	
Asphalt	$ 0.79
Concrete	$ 2.85
Crushed Stone	$ 0.48
Brick	$ 7.41
Paving Stones	$12.30

SUMMARY

A driveway is a basic feature taken for granted by everyone who has one. If you are planning to undertake a new driveway project, be sure you understand the difference between merely laying it on top of existing ground and excavating, preparing a base, and applying multiple layers of finish material. And remember, since water runs downhill, the pitch of the driveway is critical.

For other options or further details regarding options shown, see

> *Attached single carport*
> *Attached single garage*
> *Brick or flagstone patio in sand*
> *Brick or flagstone patio in mortar*
> *Concrete patio*
> *Freestanding double carport*
> *Freestanding double garage*

Driveway

Description	Quantity/Unit	Labor-Hours	Material
Rent wheeled skid steer loader	1 Day		168.00
Rent vibratory plate compactor, 13" plate, gas	1 Day		46.20
Bank run gravel	273 S.F.	0.1	130.56
Pea gravel	2 C.Y.	1.7	39.48
Backfill perimeter of driveway	47 L.F.	1.1	
Totals		2.9	$384.24

Project Size	13' x 21'	Contractor's Fee Including Materials	$688

Key to Abbreviations
C.Y.– cubic yard   Ea.– each   L.F.– linear foot   Pr.– pair   Sq.– square (100 square feet of area)
S.F.– square foot   S.Y.– square yard   V.L.F.– vertical linear foot   M.S.F.– thousand square feet

GARDEN POND

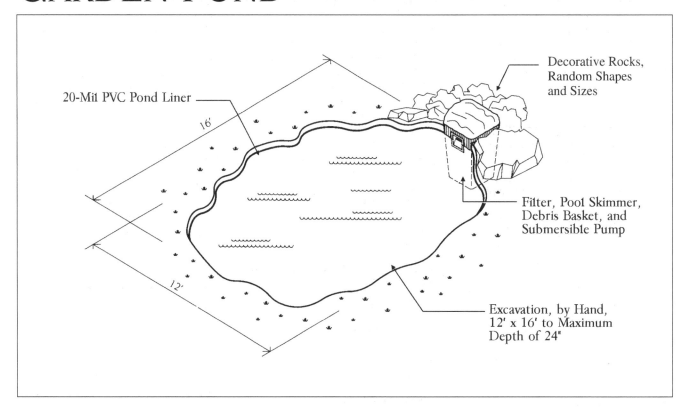

20-Mil PVC Pond Liner

16'

12'

Decorative Rocks,
Random Shapes
and Sizes

Filter, Pool Skimmer,
Debris Basket, and
Submersible Pump

Excavation, by Hand,
12' x 16' to Maximum
Depth of 24"

Humans are strongly attracted to water – its look, feel, and sound. Some biological researchers and anthropologists interpret this as an urge of beings whose chemical makeup is 90% saline water to stay connected to the source of their existence, the primal sea. This helps to explain the premium placed on waterfront real estate and the popularity of swimming pools – that, and being able to take a cool dip on a hot day! For those who can neither move to seashore or lakeside, nor afford the expense of a pool, building a small garden pond can be a way of satisfying this curious, but undeniable, aquatic urge.

MATERIALS

Americans are willing to go to great lengths and considerable expense to create and maintain beautiful yards; and while anyone can plant petunias and pull weeds, the average person lacks both the ability to visualize major landscaping alterations and the knowledge of materials and techniques that make such changes possible. This is the service a landscape designer provides. Before undertaking this project there is a great deal of careful planning and research to

do, which may well include hiring this kind of professional help.

Your backyard doesn't have to be very large to accommodate a garden pond, but its layout is most important. The pond should look like a natural part of the landscape, and in the wrong setting it could appear incongruous. If your yard is not suited for the type or size of pond treatment described in this project, you might consider a small, fixed-shape pool or fountain. Some suppliers offer as many as 60 different styles of fiberglass pools that can be sunk in the ground and surrounded by bricks, tiles, or stone, along with suitable shrubs and flowers, to make a very attractive addition to virtually any yard.

The ideal setting for this pond is a low area with a natural backdrop. In nature, water flows to and settles in kettle holes and depressions to form ponds, and your aim, remember, is to replicate what nature does. Hillsides, ledges, trees, and shrubbery all serve the purpose of a backdrop, which is to block unsightly or distracting views and focus attention on the pond.

The lay of the land and the size of the pond will determine the amount of

excavation necessary, and whether you want to do the work by hand, rent a small machine, or hire a backhoe. The pond does not need to be very large or deep; an irregular bed, sloping to a maximum depth of 18" to 20", is sufficient. The bottom is lined with 20-mil PVC, which is thin enough to conform to any contour but thick enough to resist punctures and natural degradation. A 1/2"-thick layer of damp newspapers under the liner provides a protective cushion for it and eventually decomposes to form an almost solid and virtually impermeable substratum. Rocks and crushed stone spread randomly over the liner hold it in place and give the look of a natural pond bottom.

To prevent the pond from becoming a stagnant hatchery for mosquitoes, a filtered circulating system is required to skim off debris and aerate the water. This system allows you to create a natural ecosystem by growing aquatic plants, introducing fish, and attracting birds and dragonflies that feed on noxious insects and their larvae. A submersible pump draws the pond water through a swimming pool skimmer unit and pushes it out through a flexible plastic pipe that returns it to the pond, ideally rippling over rocks in a small stream. In most areas of the country you are within driving distance of a water garden specialty center. Consult with them for advice and supplies suited to your climate and local water conditions. If the skimmer and pump, electrical hook-up, and plastic liner are all artfully concealed and surrounded with appropriate plantings, a year's seasoning will give the pond the appearance of always having been there.

LEVEL OF DIFFICULTY

For many people, attacking the ground with a long handled shovel is the essence of drudgery. The fact is, it is the main task involved in this project. If your vision of the finished pond is strong and compelling, it can be enough to motivate you to literally "move the earth" to accomplish it. If the very idea of a shovel is enough to raise blisters on your hands, there are plenty of local landscapers who will gladly take on the job, but be prepared to pay a premium price. This type of improvement can add immeasurably to the beauty of your yard and the pleasure you derive from it, but it is unquestionably a luxury, and purveyors of such luxuries are generally well reimbursed for their efforts.

While the physical work of digging dirt and lugging stones is undeniably hard, the most difficult phase of this project can be the designing of it. If your yard does not provide an obvious setting for a pond, you might do well to hire a landscape architect to create one. This is a major undertaking, and if the outcome is to be a natural-looking small pond, as opposed to an artificial large puddle, paying for the services of a designer can be money well spent.

A beginner should not attempt this project without advice regarding design, layout, and construction, and should add 100% to the time given for any physical tasks performed. An intermediate may also need assistance in formulating an attractive and workable design, and should add at least 50% to the time. An expert should add about 25% for all tasks.

WHAT TO WATCH OUT FOR

In researching this project, you might want to consult books that describe Japanese landscaping ideas and techniques. The Japanese are the acknowledged masters of the art of shrinking nature to a compact, manageable size while, paradoxically,

increasing the impact of its beauty on the beholder. By subtly balancing natural materials, colors, and textures, and maintaining precise proportions, they manage to create the illusion of vastness in a small space. Consult your local library for other ideas as well.

If you choose not to hire a plumber or an electrician, be sure you follow all local codes regarding the underground piping and wiring your pond design requires. You will need to wire a circuit to a GFI breaker or an outdoor receptacle to provide power for the pump. If you live in a cold climate, shut down and drain the pond's circulating system before the first freeze.

As shallow as this pond is, it still represents a potential drowning hazard to very young children. Take whatever precautions are necessary or are mandated by law.

SUMMARY

Nothing fosters a greater feeling of peace and tranquility than a shaded garden nook beside a small pond or fountain. If your yard provides a proper setting, a garden pond can be made to look like a natural feature of the landscape. A yard lacking such a setting might be equipped with a fixed-shape fiberglass pool and fountain for a man-made but completely appropriate look.

This project is labor-intensive, whether you do the work yourself or hire a contractor. It is a luxury landscaping improvement, and as such, the professionals who can provide it tend to sell their services at a premium.

If your budget allows, it may be a good investment to hire a landscape designer to plan and lay out the pond. With the plans in hand, you can then arrange the purchase of materials and the construction schedule as economically as possible, undertaking those tasks within your competence and subcontracting those you are not equipped to do. If you are undaunted by the labor involved, this small water garden can turn your ordinary backyard into a show piece.

Garden Pond, 12' x 16'

Description	Quantity/ Unit	Labor- Hours	Material
Excavation, by hand, 12' x 16' to a max. 24" depth	7 C.Y.	14.0	
Rock, medium round stone, to line bottom	3 C.Y.	0.1	50.94
Liner, 20-mil PVC	200 S.F.	5.6	52.80
Filter and pump	1 Ea.	8.9	1,260.00
1-1/4" pipe with fittings	40 L.F.	7.6	113.76
Wire, type MC	100 L.F.	1.3	20.04
Outdoor receptacle, GFI, 15 amp	1 Ea.	0.8	52.20
Totals		38.3	$1,549.74

Project Size	12' x 16'	Contractor's Fee Including Materials	$4,029

Key to Abbreviations
C.Y.– cubic yard   Ea.– each   L.F.– linear foot   Pr.– pair   Sq.– square (100 square feet of area)
S.F.– square foot   S.Y.– square yard   V.L.F.– vertical linear foot   M.S.F.– thousand square feet

BRICK OR FLAGSTONE PATIO IN SAND

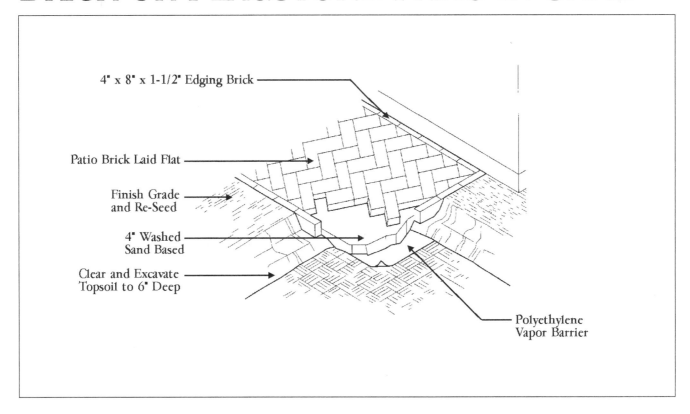

4" x 8" x 1-1/2" Edging Brick

Patio Brick Laid Flat

Finish Grade
and Re-Seed

4" Washed
Sand Based

Clear and Excavate
Topsoil to 6" Deep

Polyethylene
Vapor Barrier

There are many advantages to working with brick, concrete, and stone. They are durable, strong, and attractive, and can be used in a variety of installations, from interior fireplace facings to exterior walkway and patio surfaces. Once you have learned the basic skills, acquired the fundamental hand tools, and gained some experience, working with these materials can be very rewarding. If you are inexperienced with masonry tasks, it is best to start with a project that is small and not too demanding, such as these 6' x 12' plans. Because the bricks and stones are set in sand, mistakes can be easily corrected, with little or no waste of materials. This project can be completed by most do-it-yourselfers for the cost of the materials.

MATERIALS

The basic materials used for patio projects of this type are readily available from masonry suppliers, though the choice of stone may be limited. Flagstone, slate, and bluestone are the most common, but you may have to compromise on the size of the rectangular cut pieces if you plan to buy from the retailer's stock. The first step is arranging a form system to

set the limits and the perimeter design of the patio. Then, a subsurface of sand is put into place to establish a level base and support for the installation. Lastly, the surface material of brick or stone is applied and finished. Since these materials are readily available, it pays to shop for the best prices. In most cases, materials must be trucked, and delivery arranged as a part of the purchase agreement. The retailer may or may not charge for curb delivery, depending on the size of the order and the distance to the site.

In addition to the masonry supplies, some basic hand tools are needed for all brick or stone patio installations. These include a stonemason's hammer, flat stone chisel, 3' or 4' level, square, and long tape measure.

No special type of sand is required for the base, as long as it is not too fine to restrict drainage, nor too coarse to prohibit easy placement of the brick or stones. Sand that has been screened and washed, like mason's sand, works the best and is usually the most economical. Only a small amount is needed to cover this 6' x 12' area to a depth of 2" and to provide the grouting between the bricks

or pieces of stone. If you live in an area where clean sand (not sandy soil) is naturally available, it can be used for the base. Screen it first through 1/4" wire screen to remove pebbles and organic matter. Money saved by acquiring or transporting the sand yourself should be weighed against the time and trouble of preparing it.

Before the bricks or stone can be set, the area must be laid out, excavated, and prepared with a sand base. Minimal materials cost is involved in this stage of the installation process, but a significant amount of time is required to dig out the site to the required depth and to place the sand. If the area is level, the procedure is basic, but more time and expense may be involved if you have to contend with a slope. Be sure to completely dig out all vegetation, especially grass and other perennial plants that are hardy enough to reestablish themselves later on between the bricks. This plan suggests that a polyethylene barrier be placed below the layout of base sand to prohibit the growth of deep-rooting vegetation, but some hardy plants with shallow root systems may still establish themselves above the barrier if they are left unattended. The cost of the barrier is minimal and its benefits are many, so be sure to include it.

For the brick patio, there are many different types, sizes, and grades of brick available. In most cases, bricks that can be used for paving are referred to as all-purpose bricks, hard bricks, or pavers, and are rated as SW (severe weathering). Bricks with a rating less than SW may crack and crumble in time when exposed to water, icing, and constant freeze-thaw weather conditions. Verify the rating of the brick with your retailer.

Placement of flagstones is easier and faster if they are uniform in shape and thickness. Like bricks, stones of a generally consistent proportion can be placed on the leveled base with little fuss or readjustment. If there is much variation in thickness, the process takes longer, as each stone has to be leveled separately by adding or taking away sand. In most cases, rectangular bluestone flags maintain a consistent 2" thickness, but common flagstone and other paving materials are apt to be irregular. The trouble-free placement of regular stones might be worth their extra cost as you save on installation time. If you have not cut stones before, be sure to get some advice or have an experienced person show you the technique. You may have to invest in a stone chisel or brickset because a conventional cold chisel or other substitute will not do the job as well. Flags damaged by incorrect cutting

methods may cause considerable waste. Even with the right tools and technique, a 20% waste allowance should be expected for irregular flagstone.

The method and pattern for laying the bricks or stone on the sand base should be carefully considered. An advantage of setting them in sand is that trial and error will usually cost you only more time. As a rule, patterns that require little or no cutting install faster and create minimal waste. In brick work, simple jack-on-jack and basketweave patterns require little or no cutting. There is some cutting involved in the intricate running bond and herringbone patterns, like the one illustrated. A masonry saw can make the job easier.

After the brick or stone has been set and leveled in the desired pattern, the seams are grouted with screened sand or stone dust and packed solid. The same material used for the base sand can be used for the grouting, though there are alternatives. Preprocessed grouting, stone dust, and fine sand cost more, but can provide more effective packing for the stones. Grouting correctly takes time, but is a vital operation in the project. If the sand grout is not packed tightly, the bricks or stones will wobble and eventually become misaligned. Several treatments are usually required to solidify the surface. A light watering with a hose should follow each packing session. Be sure to kneel outside of the patio, whenever possible, when grouting. If you have to walk or kneel

on the newly-placed bricks or stones, use a piece of plywood to distribute your weight. Keep the new patio wetted down for a couple of weeks after it is finished, and regrout from time to time.

LEVEL OF DIFFICULTY

If you take time to lay out and prepare the site properly, and select a design that is commensurate with your ability, little can go wrong; and the result will be of professional quality. The advantage of a sand-set patio is that you can work at a comfortable pace and rectify mistakes easily. Those who are inexperienced should get instruction on brick or stone cutting before beginning the project and should allow an additional 75% to the professional time, and 100% for more intricate brick designs. Intermediates and experts should add 40% and 10%, respectively, to the labor-hours for all tasks included in the project.

WHAT TO WATCH OUT FOR

Patios that are set in sand are not as easy to maintain as those set in mortar. Stones may have to be reset periodically, and the sand replenished because of frost heaving or loosening from foot traffic. Though the plastic barrier prevents deep-rooted plants from becoming established, smaller surface plants will still grow between the stones. An inert grouting material like stone dust or professionally prepared exterior grout will reduce the problem. Careful application of a mild herbicide also helps. Extra material and installation time are required for special patterns or angled surface courses. This herringbone design uses about 20% more brick than a basketweave or jack-on-jack pattern.

SUMMARY

With their basic design, small size, and easy installation, these patios can be completed entirely on your own, at significant savings.

For other options or for further details regarding options shown, see

Brick or flagstone patio in mortar

Brick walkway

Brick Patio In Sand*

Description	Quantity/Unit	Labor-Hours	Material
Layout, clearing and excavation, by hand	1.33 C.Y.	2.7	
Polyethylene vapor barrier, .006" thick	72 S.F.	0.2	2.59
Washed sand for 4" base	1 C.Y.	0.1	14.34
Fine grade and level, by hand, sand base	8 S.Y.	0.3	
Compaction, vibratory, 8" lifts, sand base	1 C.Y.	0.1	
Edging brick, 4" x 8" x 1-1/2", laid on edge	36 L.F.	1.6	38.88
Patio brick, 4" x 8" x 1-1/2" thick, laid flat, herringbone pattern	72 S.F.	11.5	185.76
Re-seed grass, hand push spreader, 4.5 lbs. per M.S.F.	0.02 M.S.F.	0.1	0.31
Totals		16.6	$241.88

Project Size 6' x 12'

Contractor's Fee Including Materials **$1,054**

Key to Abbreviations
C.Y.– cubic yard   Ea.– each   L.F.– linear foot   Pr.– pair   Sq.– square (100 square feet of area)
S.F.– square foot   S.Y.– square yard   V.L.F.– vertical linear foot   M.S.F.– thousand square feet
For flagstone instead of brick, add $239 to the contractor's fee.

BRICK OR FLAGSTONE PATIO IN MORTAR

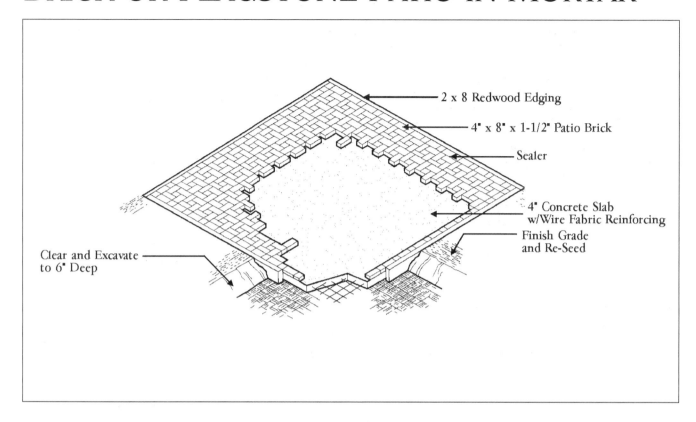

2 x 8 Redwood Edging

4" x 8" x 1-1/2" Patio Brick

Sealer

4" Concrete Slab
w/Wire Fabric Reinforcing

Finish Grade
and Re-Seed

Clear and Excavate
to 6" Deep

The process of placing bricks or stones in mortar on a concrete slab is more difficult than setting them in sand, but it is well within the reach of many do-it-yourselfers. If the slab is already in place and the laying of the brick or stone is all that has to be done, the job is fairly easy. However, if you begin from scratch and place the concrete slab first, the project becomes more difficult, costly, and time-consuming. Because of the specialized know-how required, inexperienced do-it-yourselfers should seek some assistance before they start and might also want to do some reading on the subject. Flagstones differ from brick in that they can be more expensive and often take longer to install because of their irregular sizes and shapes.

MATERIALS

This plan calls for the basic masonry materials used in most mortar-set patios. Either project is really two installations in one. The first involves excavating and preparing the site and then placing the slab; the second includes the placement of the flagstone or brick in the mortar bed and the finish work. Several basic masonry tools are needed. Some of these

tools may have to be purchased or rented, and this extra cost may have to be figured in.

Before the slab is placed, a substantial amount of excavating has to be done. The patio area should be carefully located, measured, squared, and then dug out by hand to a size slightly larger than that of the finished surface. Some additional cost may be incurred for a layer of gravel if the slab is being placed on poor subsoil. This precautionary measure will help to reduce cracking and heaving of the slab during freezes. Be aware that the edges may take some abuse during the slab placement. If you are fussy about its appearance, you might consider placing the border after the concrete has set up. Be sure to square the form to its precise dimensions and then brace the corners and stake the sides to hold it firmly in place while the slab is poured. A slight pitch off level in one direction away from adjacent structures will help to drain the surface.

Once the form has been placed and the reinforcing wire positioned, the slab is poured and finished. Because you have to work quickly and efficiently, this process can be challenging if you are inexperienced. Be sure to get some

advice beforehand and to have plenty of help available when the ready-mix truck arrives. Getting prepared for the truck's arrival requires much organization and a set plan to minimize the time of the delivery. If the truck is tied up for too long, you will be charged extra. After the delivery has been completed, the concrete is leveled and rough-finished in preparation for the placement of the bricks or stones.

The important aspect of this operation is to complete small, manageable sections at first, with increased production as the project progresses. Flagstone, however, does require more know-how, skill, and installation time than brick because its irregular shape necessitates cutting and fitting some pieces. Be sure to select flags of uniform thickness, as adjustments cannot be easily made for variations of thickness. Before laying a section of stones, place the pieces on the slab and position them by trial and error until you arrive at a combination that provides maximum coverage and joints that average about 3/4" in width.

The entire surface does not have to be completed in one session, so work very slowly at first with the anticipation of faster placement as you get better at it.

The process of laying the bricks on the finished slab also requires basic masonry knowledge. Never mix more mortar than you will use in an hour of brick laying. If any brick cutting is required, do it before mixing the mortar or starting the brick placement for that session.

As in all construction projects, keep an eye on level and square as you proceed. Investment in a stringline, long mason's level, and a good trowel will add to your efficiency and to the appearance of the finished patio. The time and money invested in an application of masonry sealer will be returned many times over in protection for the finished patio.

LEVEL OF DIFFICULTY

These projects are challenging for most do-it-yourselfers, but manageable when taken in small steps. With patient work, even novices can complete the patio with professional results. The most demanding part of either project is the placing of the slab. Novices may consider hiring a contractor to form and pour the slab and could then follow up with the brick work.

Experts and most intermediates should be able to complete either project completely on their own. They should add 10% and 40%, respectively, to the labor-hours estimate for the project. Beginners should add 75% to the professional time for all tasks.

WHAT TO WATCH OUT FOR

Although special border features and materials usually cost more and take longer to install, they can bring the design of the patio to life and add lasting protection to the edge of the surface. There are various options for rot-resistant wood edges, including the redwood shown here, cedar, locust, or pressure-treated pine. Patios that are recessed or terraced can be made even more attractive with appropriate wood edging. Bricks set in soldier, sailor, or sawtooth arrangements can also enhance the surface design while adding durability to the patio's edge.

Part of a neat job is minimizing the amount of mortar that gets on the brick or stones. Clean up any drops or smears with a damp cloth or sponge before they dry as you proceed with the grouting. A final cleaning can be done with muriatic acid after the grout has cured.

PATIO OPTIONS

Cost per Square Foot, Installed

Description	
Brick	$ 7.60
Redwood Plank	$ 7.80
Granite Pavers	$12.30
Bluestone	$10.85
Concrete	$ 2.52

SUMMARY

The patios presented in this plan will make an attractive and practical addition to the exterior of any home. Although the installation requires a moderate level of masonry know-how, the project can be completed in small steps and at a comfortable pace for do-it-yourselfers who are willing to learn as they go.

For other options or further details regarding options shown, see

> *Brick or flagstone patio in sand*
> *Brick walkway*

Brick Patio In Mortar*

Description	Quantity/ Unit	Labor- Hours	Material
Layout, clearing and excavation, by hand	2.66 C.Y.	5.3	
Compaction, vibratory, 8" lifts, common fill	2 C.Y.	0.1	
Edging, redwood, 2" x 8"	48 L.F.	2.3	153.79
Concrete, ready mix, 2000 psi, for slab	2 C.Y.		156.00
Reinforcing, welded-wire fabric, 6 x 6 - 10/10	160 S.F.	0.7	13.44
Place concrete, slab on grade, 4" thick	2 C.Y.	0.9	
Concrete finishing, broom finish	144 S.F.	1.8	
Patio brick, 4" x 8" x 1-1/2" thick, grouted, 3/8" joint	144 S.F.	25.6	426.82
Re-seed grass, hand push spreader, 4.5 lbs. per M.S.F.	0.10 M.S.F.	0.1	1.24
Sealer, silicone or stearate, sprayed on concrete,1 coat	144 S.F.	0.3	46.66
Totals		37.1	$797.95

Project Size	12' x 12'	Contractor's Fee Including Materials	$2,782

Key to Abbreviations
C.Y.– cubic yard   Ea.– each   L.F.– linear foot   Pr.– pair   Sq.– square (100 square feet of area)
S.F.– square foot   S.Y.– square yard   V.L.F.– vertical linear foot   M.S.F.– thousand square feet
For flagstone instead of brick, add $1101 to the contractor's fee

CONCRETE PATIO

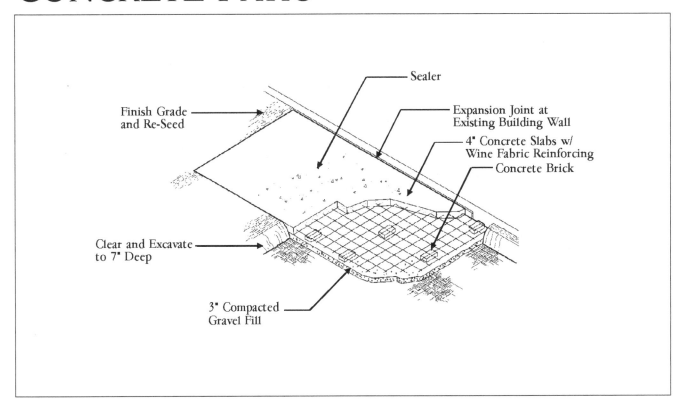

Sealer

Finish Grade and Re-Seed

Expansion Joint at Existing Building Wall

4" Concrete Slabs w/ Wine Fabric Reinforcing

Concrete Brick

Clear and Excavate to 7" Deep

3" Compacted Gravel Fill

MATERIALS

The primary advantages of concrete as a patio, walkway, or floor material are its strength and durability, but as an added benefit, it can also be shaped and molded into a variety of configurations and forms. It can take on many surface textures and can be colored before it is placed, or painted after it is installed. This project presents two concrete patios – one 6' x 12' and the other 12' x 16' – both of which are moderately challenging for do-it-yourselfers. If the work is done in stages, experts, intermediates, and many beginners can complete the project on their own.

MATERIALS

The materials and methods of installation for both patios are the same. The site has to be laid out and excavated, the forms built, the subsurface and reinforcement placed, and the concrete poured and finished. As long as the grade on which the patio is to be built is level or gently sloped, and the subsoil easy to dig into, the project can be completed as estimated with no extra costs.

All masonry work that involves mortar and concrete requires efficient organization and detailed preparation. Before you begin work, you will need some basic masonry hand tools: shovels, stiff rakes, a sturdy wheelbarrow, trowels, a float, and an edging or jointing tool. A stringline, square, 3' or 4' level, 16' tape measure, and hammer are required for the construction of the forms. Extra cost will be incurred if you have to buy the tools, but all of them are necessary to complete this project. The specialized concrete finishing tools, reinforcement materials, and premolded expansion joint can be purchased from a masonry supplier.

The excavation work required for both patios is substantial, as the area has to be dug out by hand to a depth of at least 7" to allow for 3" of gravel fill for a subbase under the 4" slab. Be sure to dig out the edges of the cavity thoroughly to permit easy placement of the form and to allow some leeway for squaring and leveling. The form should be pitched slightly away from any adjacent structures, squared, and braced firmly with stakes and corner ties to its final position. The subsurface material, reinforcement wire, and expansion joint should then be placed in preparation for the concrete delivery. The amount of subsurface

material will vary with the consistency of the soil, and the cost may increase or decrease proportionally. For example, if the soil is hard-packed and drains poorly, more gravel will be called for; if it is sandy and porous, less will be required. When in doubt, consult a professional.

The concrete for a relatively small job like a patio may cost more per cubic yard than it would for a complete foundation or cellar floor, because of the dealer's overhead for trucking costs. A minimum charge, therefore, is not unlikely. Be sure to work out all of the conditions of delivery with the retailer before the order is placed. The amount of time allowed for the delivery and the cost of additional truck tie-up time are important details that you should be aware of. Good organization and advance preparation will keep the costs down. If the site is accessible to the truck, and the concrete poured directly from the chute, the job will go quickly, efficiently, and most economically. If it has to be transported from the street or driveway, then more time, manpower, and, possibly, expense will be involved. Wheelbarrowing concrete is a physically demanding task that involves concentrated, rigorous work during the delivery period.

After the concrete has been put into place, several hours can elapse before the material hardens to an unworkable state. Always keep some fresh concrete handy during the screeding to add to low spots, and to offset settling as the surface is finished. On warm days, you may have to work a little faster in the finishing process, but there will still be enough time to create the desired texture. If the surface sets up too quickly, it can be enlivened with mist from a hose. Do not give it the full spray of a hose at this time, however, as this may expose the aggregate and damage the finish. Care should be taken when finishing the surface, as concrete can be troweled to a slick, even dangerous, smoothness. A broom finish provides a safe, non-skid textured surface. Other finishes are also possible, but they may require more time to complete. Imitation flagstone designs can be tooled into the surface, for example, or coarse ripples and swirls can be made by moistening the surface and reworking it with a sponge, float, or trowel. Generally, a broom finish is the most popular texture and the easiest to accomplish for do-it-yourselfers. Do not overwork the surface during finishing, as too much troweling can cause the aggregate to settle and separate from the rest of the slab. Be sure to tamp the edges only enough to fill voids and to create a solid side surface. Control joints and rounded edges should be added after the finishing process.

LEVEL OF DIFFICULTY

Either patio plan can be accomplished by expert do-it-yourselfers. The most difficult procedures are the actual placement of the concrete and its subsequent finishing, both of which can be troublesome for inexperienced workers. Remember that several hours of intense physically demanding work are involved in accomplishing this phase of the project, particularly in the larger patio. Beginners, therefore, should make arrangements to have at least one experienced person on the job during the pour and for some time after to oversee the operation. Novices should add 75% to the professional time for all of the preliminary tasks, like the setting of the forms and placement of the reinforcing. Intermediates with experience in concrete work and experts should add 40% and 10%, respectively, to the labor-hours estimated for either project.

WHAT TO WATCH OUT FOR

Even the smaller patio should have one or two joints placed in the surface to control the cracking that is part of all concrete installations. As a general rule, control joints should be placed at 4' to 6' intervals to a depth of about 3/4". When the surface has been finished and allowed to set up enough so that it will support your weight when distributed on a piece of plywood or lumber, use a jointing or edging tool to place the joint in a straight line across the slab. The three exterior edges of the slab should then be rounded or beveled with an edging tool before the finishing job is completed.

Many alternatives to the conventional square, single-level patio are possible if the surrounding terrain is suitable and if you are willing to exercise your imagination and ingenuity. Terracing the patio is one example that is not as complex as it may seem. The same basic techniques used for single-level slabs are employed in multi-level installations, only in modular arrangements. Where the land slopes away from the house, two- and three-tier patio designs are an option, with interconnecting pressure-treated pine or redwood steps. Railroad ties and other rot-resistant border materials can be used to shore cutouts in the embankment. Rounded corners and oval shapes are among the alternatives to conventional square and rectangular patio shapes. Masonite or 1/4" plywood can be used to form the curved edges. Custom-colored surface materials are also available on special order from concrete dealers and masonry suppliers if you want to mix them on your own. All of these variations will add expense to the project, but they can enhance the facility's appearance, and, in some cases, contribute to its practical functions.

SUMMARY

If you are unfamiliar with concrete construction, be sure to seek advice and arrange to have a knowledgeable person available to assist you during the slab placement and finishing. Your efforts will produce substantial savings and a home improvement that will give you and your family years of enjoyment.

For other options or further details regarding options shown, see

> Brick or flagstone patio in sand
>
> Brick or flagstone patio in mortar

6' x 12' Concrete Patio*

Description	Quantity/ Unit	Labor- Hours	Material
Layout, clearing and excavation, by hand	1.33 C.Y.	2.7	
Compaction, vibratory, 8" lifts, common fill	1 C.Y.	0.1	
Gravel fill, 3" deep	0.66 C.Y.	0.1	11.40
Compaction, vibratory, 8" lifts, bank run gravel	1 C.Y.	0.1	
Edge forms, 4" high	24 L.F.	1.3	11.81
Expansion joint at existing foundation, premolded bit. fiber	12 L.F.	0.3	5.47
Concrete, ready mix, 3000 psi, for patio	1 C.Y.		82.80
Reinforcing, welded-wire fabric, 6 x 6 - 10/10	72 S.F.	0.3	6.05
Place concrete, slab on grade, 4" thick	1 C.Y.	0.4	
Concrete finishing, broom finish	72 S.F.	0.9	
Re-seed grass, hand push spreader, 4.5 lbs. per M.S.F.	0.03 M.S.F.	0.1	0.44
Sealer, silicone or stearate, sprayed on concrete, 1 coat	72 S.F.	0.1	23.33
Totals		6.4	$141.30

Project Size	6' x 12'	Contractor's Fee Including Materials	$463

Key to Abbreviations
C.Y.– cubic yard Ea.– each L.F.– linear foot Pr.– pair Sq.– square (100 square feet of area)
S.F.– square foot S.Y.– square yard V.L.F.– vertical linear foot M.S.F.– thousand square feet
For 12' x 16', add $700 to contractor's fee.

PATIO ROOF (PERGOLA)

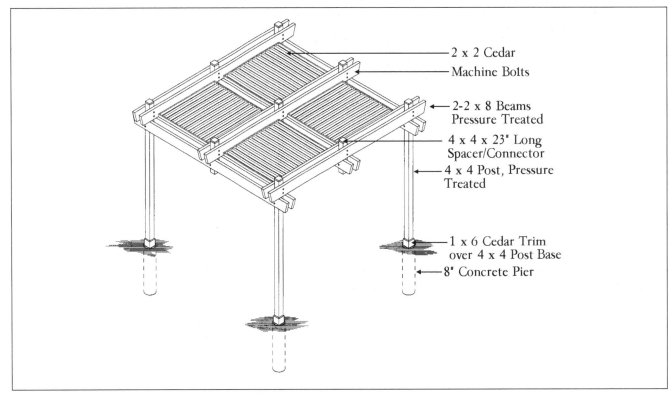

2 x 2 Cedar

Machine Bolts

2-2 x 8 Beams
Pressure Treated

4 x 4 x 23" Long
Spacer/Connector

4 x 4 Post, Pressure
Treated

1 x 6 Cedar Trim
over 4 x 4 Post Base

8" Concrete Pier

A patio roof, or pergola, can help turn an outdoor space into living space – whether it is a garden roof, a party pavilion or a shade structure, in a contemporary design or a traditional trellis. This pergola project can be freestanding or attached to the dwelling. The structure is square, with four equal roof segments. This design can be customized to fit an existing patio simply by adding or eliminating segments. Larger segments may require deeper horizontal members.

MATERIALS

A variety of woods can be used to build this structure, in total or in combination with each other. The choice depends on your personal taste, budget, and the style of your house. Pressure-treated wood has been selected for this project, as it is available at most retail lumber or home centers. No matter what species of wood you select, the same principles will apply to the layout, fabrication, assembly, and finishing. If you choose another type of wood, take the time to apply at least one coat of wood preservative prior to assembling the pieces, to ensure total coverage.

The location of the post bases is critical to the success of this project. The four segments are identical in size. This allows for the precutting of all the members from patterns that are laid out based on a plan. Using a pattern is recommended, to ensure that all the components are consistent in size. Patterns also permit you to select the best pieces before you cut, thus minimizing waste and ensuring the project's appearance.

It may be necessary to remove sections of the existing patio to place the footings for the posts. If the soil is stable, you may consider omitting the round, fiber tube forms included in this project, and simply place the concrete in the prepared holes. Before you place the anchor bolts in the concrete, check the dimensions and corners for squareness. The post bases will allow you to make minor adjustments horizontally and vertically when you set the posts; take care to make the tops of the footings level with each other. The final step in the construction process involves tightening the bolts at the footings after plumbing the posts one last time.

Once you have laid out the placement of the post bases, you will be able to precut the entire structure. The layout should be developed off of centerlines, to ensure the correct positioning of the members. Because this project is assembled with bolts at the main connection points, you will need to predrill the holes in the members. This is much easier to do on the ground than from a ladder. Once you have laid out and measured the wood for the first panel, you can use it as a pattern for the other three. When all the members have been cut, sand and prefinish them before assembly. Temporary bracing of the vertical posts is recommended when you are assembling the pergola components.

LEVEL OF DIFFICULTY

This is not a particularly difficult structure to build, because it is square and the four segments are identical. The work requires patience and a steady hand with the power saw. Because all of the joints are exposed, good square cuts will make a big difference in the finished product. Using the cut members as patterns should reduce errors and eliminate waste. Beginners should have little trouble with the basic structure. Intermediates might enjoy being more creative with the shaping of the ends of the structural members with a saber or scroll saw. The construction requires working off of ladders or scaffolding, which might slow down inexperienced do-it-yourselfers. The job also calls for two people working together, because the wood pieces are heavy. Intermediates and beginners should add 50% and 100%, respectively and slightly more for creating fancy end shapes. Experts should add 20% to the professional time for the total project.

WHAT TO WATCH OUT FOR

The key to building this structure is proper setting of the post bases. Be sure you have a tape long enough to check all dimensions. The easiest way to check the layout and the overhead structure for square is to tape the diagonals. If the opposite sides and the diagonals are equal, the structure is square.

A combination or planer blade is recommended for this project, because it makes a smoother cut. Be sure your drill bits are long enough to go through a minimum of 4" of wood. If you do not intend to apply any finish to the structure, make all your marks lightly with pencil. Because all of the wood is visible in the finished project, you may want to select your lumber for straightness and pleasing appearance.

SUMMARY

A pergola is a manageable project for homeowners who enjoy remodeling work. You should have someone available to help lift and install the various wood pieces, which can be heavy and awkward when maneuvered from a ladder or scaffold. The end result of the project is a new, defined outdoor space that can dress up an uninspired patio.

Patio Roof

Description	Quantity/ Unit	Labor- Hours	Material
Pier form, round, fiber tube, 8" diameter	12 L.F.	2.5	26.06
Post base, 4 x 4	4 Ea.	0.3	22.80
Post, 4 x 4, pressure treated	48 L.F.	2.0	75.46
Lumber, 2 x 8, pressure treated	148 L.F.	2.5	154.51
Exterior trim, 1 x 2 cedar	74 L.F.	2.2	22.20
Exterior trim, 2 x 2 cedar	230 L.F.	8.0	121.44
Exterior trim, 1 x 6 cedar	8 L.F.	0.3	7.01
Machine bolts, 7-1/2" long	36 Ea.	2.2	41.04
Nails, 6d, galvanized	10 Lb.		16.08
Concrete, field mix, 1 C.F. per bag	2 Bag		14.64
Totals		20.0	$501.24

Project Size	12' x 12'	Contractor's Fee Including Materials	$1,698

Key to Abbreviations
C.Y.– cubic yard   Ea.– each   L.F.– linear foot   Pr.– pair   Sq.– square (100 square feet of area)
S.F.– square foot   S.Y.– square yard   V.L.F.– vertical linear foot   M.S.F.– thousand square feet

POST LIGHT & GFI RECEPTACLE

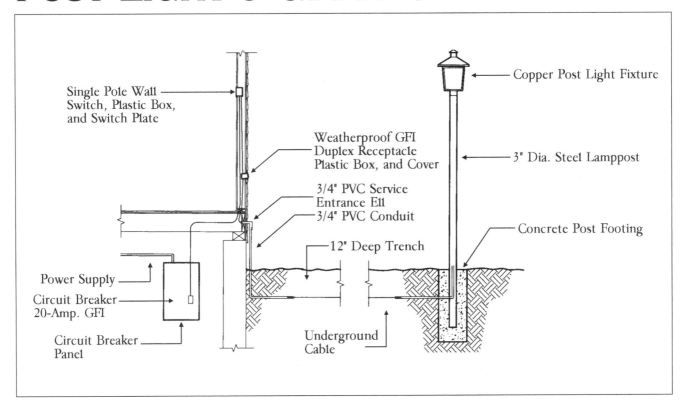

Single Pole Wall Switch, Plastic Box, and Switch Plate

Weatherproof GFI Duplex Receptacle Plastic Box, and Cover

3/4" PVC Service Entrance Ell

3/4" PVC Conduit

12" Deep Trench

Power Supply

Circuit Breaker 20-Amp. GFI

Circuit Breaker Panel

Underground Cable

Copper Post Light Fixture

3" Dia. Steel Lamppost

Concrete Post Footing

Virtually no American home today is without electric power. In fact, the standard of living we have come to consider normal is largely because of the application of affordable electrical energy. Yet, surprisingly, many homes, both new and old, lack the convenience and utility of outdoor electricity to illuminate for beauty, safety, and security and to power machines and tools for work and recreation. In the summer, the typical American backyard, with its deck or patio, barbecue, and, perhaps, pool or playset, becomes an extension of the house's living space, an outdoor "room" that becomes a center for family activities and social gatherings. All such uses can be greatly enhanced by electricity in the form of lights and receptacles. The power is present and available just inside the house wall; it needs only to be brought through to the space outside.

MATERIALS

This is a basic underground wiring project for a 120-volt receptacle and post light installation. It can easily be adapted for different types of fixtures, or expanded to include more outlets to service other

fixtures or equipment, such as walkway, security, or spot lighting.

It may be possible to get at your home's electrical system by tapping into an underused circuit. The post lamp will probably contain a 75- or 100-watt bulb, and most tools and appliances that you would be likely to plug into the receptacle do not draw much, so the demand on the circuit's current should not be very great. In most houses, branch circuits to rooms other than utility rooms, baths, and kitchens are underused and are good choices for expansion. Do not, however, make this assumption; evaluate the proposed circuit to determine if it can handle the extra load.

If you don't have an underused circuit to tap into, or if you plan on more extensive lighting or additional outlets, you'll have to add a new circuit, either at the main service panel or at a sub-panel. In many cases, this approach is actually quicker and easier than tapping an existing circuit, and it has the advantage of allowing you the option of future expansion without the need to upgrade its capacity.

The National Electrical Code (NEC) requires that all outside receptacles be equipped with a safety device called a

ground fault interrupter (GFI). GFIs are used with receptacles in areas, indoors or out, that may get wet. A GFI senses a leak of current or a false ground, such as your wet hand, and immediately cuts the power to that circuit, thereby possibly saving your life, which is a very nice feature indeed. A GFI breaker is installed in the service panel in the same way as an ordinary breaker.

From the new breaker, the wire is run to and through a hole drilled in the sole plate to the switch location, which is usually near an outside entry door. A wire from the switch is fed to the exterior receptacle box, a hole for which is cut in the outside wall 18" or so above the foundation in a convenient but unobtrusive spot. From there the wire continues through another hole in the sole plate to exit the house via 3/4" PVC conduit through a hole in the rim joist.

The service entrance ell has a watertight cover plate that can be removed to allow you to pull the necessary length of wire to the outside. The wire continues down through the conduit to an elbow, where it enters the bottom of a 12"-deep trench. Type UF (underground feed) cable has a solid plastic, impermeable covering. For residential branch circuits, the NEC allows burial of UF cable a minimum of 12" deep. It must, however, be protected by rigid conduit wherever it is above ground.

Lamp posts are usually purchased separately from lamp fixtures. Most exterior lamps come with a 3" collar that fits over the top of a standard 3" diameter post. If the lamp you choose is of another design, a post of cedar or pressure-treated wood may be easier to use. Dig a post hole about 2' deep at the selected location. Drill a hole in the post on a level with the bottom of the trench, and insert a PVC elbow through the hole into the post. Pour a concrete anchor to hold both post and conduit in place, then feed the wire to the top of the post and connect it to the lamp.

The most durable, and to many the most attractive, exterior lamps are made of copper. They are also among the most expensive. Less expensive are fixtures made of corrosion-resistant aluminum and other alloys. Beware of plastic fixtures; except for their cheap price, they have nothing to recommend them.

LEVEL OF DIFFICULTY

A wiring project such as this should be undertaken only by an expert or by a professional contractor. Knowledge, experience, and skill are needed both to assess the suitability of an existing circuit for this extension, and to tap into it efficiently so as to minimize the mess and the amount of patching needed. A tie-in to the main service panel is equally demanding. In matters electrical, ignorance is dangerous and puts both

you and your home in jeopardy. Hiring a licensed electrician eliminates the risk of hazardous errors and assures you that the correct components are properly installed. Most electricians will gladly adjust their price downward if you agree to dig the trench and post hole, mix and pour the concrete, and do the backfilling.

Both beginners and intermediates should hire professional help for the wiring, and should increase the time estimates for any tasks within their competence by 75% and 25% respectively. An expert should add 15% to the time.

WHAT TO WATCH OUT FOR

If you choose to do the electrical work yourself, remember that outdoors you are in contact with the ground and, therefore, any shock could be fatal. Never do any work on wire or equipment unless you are certain that they are completely disconnected from the main service panel. Check wires with a circuit-tester (called a "wiggy" by electricians) to be doubly certain they are dead before you handle them.

A rented trench-digging machine will make this part of the project much less arduous, especially if the post light is a great distance from the house. Small trenchers can be operated by one person, but the job is a lot easier with two, one on each side of the pull handle. After you run the cable, backfill the first 6" or so with clean fill (no stones), then lay 1' x 4' pressure-treated board or a bright-

colored nylon tape. This will eliminate the future possibility of someone's digging accidentally damaging the cable.

The PVC male adaptor fittings are used where the wire enters or exits the conduit, to prevent the wire being damaged by the cut ends of the conduit. Also, be sure to use gray conduit cement, not water pipe cement. The cement can degrade the outside coating of the wire, so complete the joints before you pull the wire through.

SUMMARY

Bringing electrical power outside your house adds a great deal of convenience and makes possible a wide range of applications, both practical and recreational. Well-planned exterior illumination can provide safety to steps and walkways, security to hidden corners and entrances, and beauty to landscaping and architectural features. An outside receptacle can provide power to any number of tools and appliances. By running the new wiring from a new 20-amp GFI breaker in the main service panel, this project becomes a relatively straightforward installation, and one that can be added to in the future without the need for upgrading.

Tapping into your home's electrical system may require the services of a contractor, but it is an investment that will pay dividends in terms of enhancing outdoor activities, thereby helping to turn your yard into a more useful and enjoyable living space.

Post Light & GFI Receptacle

Description	Quantity/ Unit	Labor- Hours	Material
Conduit in trench	8 L.F.	0.3	4.90
Outlet boxes, plastic, square, with mounting nails	1 Ea.	0.4	3.29
Switch box, 1 gang	1 Ea.	0.3	1.33
Switch device	1 Ea.	0.5	7.80
GFI receptacle	1 Ea.	0.7	39.00
Waterproof cover	1 Ea.	0.3	5.28
Outdoor post lamp incl post, fixture, 35' #14/R-type MNC cable	1 Ea.	2.3	216.00
Concrete for post footing	1 C.Y.		78.00
Placing concrete, footings, spread, under 1 C.Y.	1 C.Y.	0.9	
Totals		5.7	$355.60

Contractor's Fee Including Materials	$794

Key to Abbreviations
C.Y.– cubic yard   Ea.– each   L.F.– linear foot   Pr.– pair   Sq.– square (100 square feet of area)
S.F.– square foot   S.Y.– square yard   V.L.F.– vertical linear foot   M.S.F.– thousand square feet

RETAINING WALL

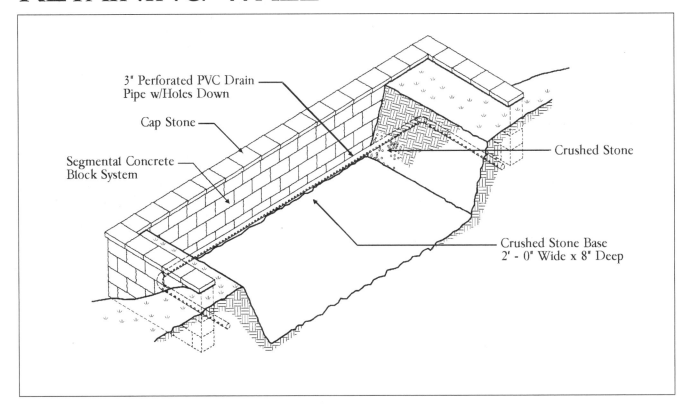

3" Perforated PVC Drain Pipe w/Holes Down

Cap Stone

Segmental Concrete Block System

Crushed Stone

Crushed Stone Base 2' - 0" Wide x 8" Deep

Sloping areas and low hillsides in a yard are often difficult to landscape or improve. The combination of poor drainage and surface water run-off often causes erosion, making the planting of grass, shrubs, or flowers a problem. By cutting across the base of the slope and building a retaining wall, you create a level area behind the wall that can be cultivated as either lawn or garden. The wall itself can add a note of visual interest to the yard by creating vertical and horizontal planes and contrasting surface textures. Imaginative and well-executed landscaping can add a great deal of beauty and value to a house.

MATERIALS

Harmonizing the terrain with the construction material is the challenge you face when planning and laying out your wall. Stone or block can be used to create curved contours, while pressure-treated landscape ties are intended to produce straight and level walls. You can accommodate very gentle curves in a hillside by cutting the ties into shorter lengths and increasing the number of joints, but this approach taken too far can result in compromising the strength

and effectiveness of the wall. If, for any reason, you must follow the varied curves and undulations of your backyard's terrain, consider building with stone or block.

A wall, dry-laid (without mortar) and made of natural fieldstone (not cut and shaped), is a very durable and attractive retainer that can be built to follow the most irregular grades and contours. Fieldstone walls are visible along almost any road in New England.

Building with fieldstone or "rubble" is simply a matter of knowing a few basic principles and techniques, and then simply getting the hang of it by going out and doing the work. In certain parts of the country, stones are plentiful and can often be acquired for only the cost of the sweat and labor involved in hauling them away. If you have to buy stones from a quarry and have them delivered to your site, a stone wall can be quite expensive, depending on the size of the wall, the type of stone you choose, and the cost of transporting it.

A third, less-expensive option is to use an interlocking concrete block system. These blocks are available in a variety of colors and textures, and offer the

advantages of durability and structural integrity at reasonable cost.

All walls begin with some sort of foundation. Once you have laid out the line of your wall with stakes and string, dig a trench about 2' wide and about 8" deep, keeping the bottom as level as possible as a base for the first course of ties or blocks. If your soil is hard-packed clay, or otherwise impermeable, it will improve the drainage to dig even deeper and increase the depth of the crushed stone to about 6". Butt the foundation ties or blocks together and level them end to end. From front to back, they should be pitched back about 1°, or approximately 1/4" off level. This is to make the wall lean slightly back to resist the pressure of the soil and water behind it, without compromising its vertical stability.

The second course should be stepped back about 1/4" from the front edge of the first course, leaving small drainage gaps between the blocks. Be sure to stagger the vertical joints between the courses. For ties, spike the upper tier into the lower ones. Once the second course is in place, connect the PVC drainpipe and lay it, holes downward, on the crushed stone base so it is a few inches behind and about level with the upper ties. Then cover the pipe with the remaining crushed stone. Hydraulic pressure in the soil can be quite powerful and potentially damaging to a wall such as this, especially in cold climates where winter frost may expand the soil and cause the wall to "heave" or become misaligned. A simple drainage system like this helps to relieve that pressure and to ensure the integrity of the wall. In warm, dry climates, or where soil conditions and terrain provide sufficient natural drainage, this system may not be needed. A local landscaper or garden center should be able to advise you on drainage requirements. Follow manufacturers' recommendations about slope stabilization if your wall is greater than 2' high.

LEVEL OF DIFFICULTY

Most landscaping work is physically demanding, but only the most sedentary soul would argue the fact that outdoor projects like this can be invigorating to perform and satisfying to complete. A block or stone retaining wall is more labor-intensive than landscape tie construction. And stones have to be individually selected and carefully fitted to ensure a strong and attractive wall. It is slow work with hard, heavy material, but in a strange way it can be very enjoyable, and it allows you the luxury of mistakes. These can be corrected by simply tearing down a section and starting over.

Landscape ties are much easier to work with. Once the foundation course is set, constructing the wall goes quickly, being a simple matter of measuring, cutting, and spiking. The design of a retaining wall like this is so basic, and the materials so limited, that, with minimal care taken in layout and measuring, serious mistakes are unlikely to be made, even by a beginner.

Considering the visual impact a retaining wall can have on a yard, beginners and intermediates would do well to get some advice regarding the design and layout of the wall, especially where complex contours or other problematic landscape features are involved. Beginners should add 100% to the time estimates; intermediates, 50%; and experts, 20%.

WHAT TO WATCH OUT FOR

Stacking stones to make a wall requires some patience and creativity. Each stone is different in shape, size, color, and texture, and almost every stone has a "face" or surface that seems to be the one that should show in the wall. Examine the space that needs to be filled, then look for the best stone or combination of stones to fill it, in much the same way that you would approach solving a jigsaw puzzle. Take time to spread out a load of stone adjacent to the wall so that you can look over and select the best stones to establish a pattern. Above all, keep in mind the most basic principle of all

masonry: one over two; two over one. This is the way a brick wall is laid, and it ensures, in both brick and stone work, that there are no contiguous vertical joints to weaken the wall or threaten to collapse it.

Pressure-treated wood is much harder than regular kiln-dried, construction-grade lumber. Cutting it requires a carbide-tipped saw blade. A carbon steel blade will be quickly dulled and possibly cause your power saw to burn out. A typical 7-1/4" circular saw blade has a depth of cut of approximately 2-1/4". In order to cut a 6" x 6" timber, you will have to square a line around all four sides, cut to the lines, and finish off the uncut 1" center square with another kind of saw: hand, bow, or reciprocating. When driving the spikes into the treated ties, you can save yourself a lot of aggravation and bent spikes by drilling pilot holes in the ties with an extra-long high-speed drill bit, and using a framing hammer or small sledge to drive the spikes in.

Backfilling can be done after the wall is complete, but it might be easier and produce better results to backfill in layers as you add each course of ties. That way you can tamp each layer of soil to compact it and eliminate the looseness and voids that can cause settling and surface depressions later.

SUMMARY

Retaining walls are often built to prevent or correct drainage and erosion problems, or to create a level-graded area for lawn plantings. A well-designed and well-constructed wall can add to the beauty of a yard by providing the visual interest of forms and textures to complement the yard's existing topographic and botanical features. Interlocking blocks are designed to provide a simple, error-free installation for straight or curved walls. Fieldstone is hard and heavy to work with, but lends itself to situations where irregular grades and contours demand curves and undulations. Pressure-treated landscape ties are easy to handle and work with, but are best used where a straight wall can be built. All involve the pleasantly strenuous kind of work associated with any landscaping project. Homeowners willing to pay the price in labor can save a good deal of money while enhancing the value and appearance of their homes.

Retaining Wall: Segmental Stone

Description	Quantity/ Unit	Labor-Hours	Material
Segmental concrete retaining wall system, 8" x 8" x 12.5" blocks	128 S.F.	7.7	1,136.64
Cap stones	32 L.F.	2.6	245.76
4" PVC perforated drainage pipe	20 L.F.	1.3	34.32
Drain trench for pipe, crushed stone bedding	1 C.Y.	0.1	18.72
Crushed stone base, 8" deep x 2' wide, for wall	1 C.Y.	0.1	18.72
Totals		11.8	$1,454.16

Project Size: 4' high x 20' long

Contractor's Fee Including Materials	$2,630

Key to Abbreviations
C.Y.– cubic yard   Ea.– each   L.F.– linear foot   Pr.– pair   Sq.– square (100 square feet of area)
S.F.– square foot   S.Y.– square yard   V.L.F.– vertical linear foot   M.S.F.– thousand square feet

SEED OR SOD

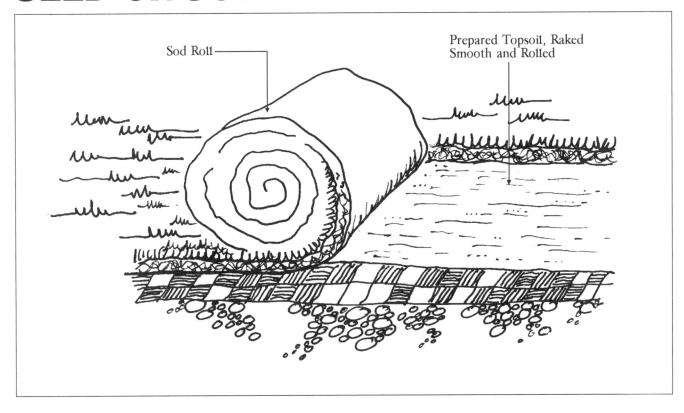

Sod Roll

Prepared Topsoil, Raked Smooth and Rolled

Visitors to England are often amazed by the carpetlike texture and look of the bright green lawns on great baronial estates, and sometimes wonder how the groundskeeper manages to achieve such beautiful results. The reply to any question usually is, "It's quite simple, really." Your lawn can hardly hope to compete on those terms, but we Americans love our lawns nonetheless, lavishing upon them great amounts of labor and money, and justly so. A lush lawn is, in itself, a delight to the eye, and can make a lovely home even lovelier, like a gem on a velvet cloth. Conversely, a dry, brown patch of weeds in front of any house greatly detracts from its appearance and usually inspires observers with unkind and negative thoughts about the pride and work habits of the owner. It may be a petty form of snobbery, but few of us can honestly plead innocent of such judgments. Whether your house is newly built and still unlandscaped, or is an older home surrounded by terminally ill grass, a new green "carpet" of lawn can wonderfully improve its appearance, enhance its value, and silence critics.

MATERIALS

To seed or to sod, that is the question. Both methods can produce a beautiful lawn, if you follow the correct procedures. In very general terms, seeding is cheaper, especially if you do the work yourself, and it gives you more flexibility to choose different grass varieties and mixes. Sod, on the other hand, produces an instant lawn, thereby satisfying the more impatient among us. It can be laid from spring through fall, it obviates erosion and splash problems, and it is less likely to fail.

Site preparation is virtually identical for both seeding and sodding. Assuming the area has a sufficient depth of topsoil, spread 2"-3" of peat moss, some starter fertilizer, and, if necessary, a dusting of ground limestone over the lawn area. Dig all these additives into the top 6" of soil so that all ingredients are thoroughly mixed. The moss will help aerate the topsoil and will keep it loose, thereby making it easier for water to permeate and for the roots to take hold. Clean, fine sand can also be mixed into soil that is especially hard-packed and dense. The amounts of all these materials depend on

both the area of the lawn and the condition of the existing topsoil. Rake the site to clear it of stones and twigs, then roll it smooth with a roller. Fill depressions, knock down high spots, and continue rolling until you have achieved a reasonably smooth and even soil surface. Consult with your local nursery or garden center to learn the characteristics of the soil in your region, and how best to chemically prepare it for grass. For example, acidic soil, typical of much of the northern United States, must be "sweetened" with lime. Some garden centers will be able to supply all your needs from soil to equipment.

Seeding is best done in the early fall or spring. The seed should be sown on the dampened soil with a spreader at a rate of up to two pounds per 1,000 square feet. Your local supplier can recommend the best mix of seed for your lawn site as well as the proper rate of application. Scratch the seed into the soil with a spring-toothed rake. Do this lightly and carefully over the lawn. Lightly roll the area to press the seed into the soil. Spread a thin layer of mulch, about 1/8", over the lawn. Plain sawdust works well for this. Use a spreader to produce an even application of mulch that does not bury the seed too deep. The mulch helps keep the seeds damp over the few weeks it takes for them to germinate and send out roots. Once most of the lawn has sprouted, let the surface dry out between waterings so the seedlings do not develop fungal diseases. When the young lawn needs mowing, use a sharp reel mower for the first several cuttings to avoid tearing out the new plants.

Sod is available in strips about 18″ wide and 6′ long, with about 3/4″ of soil attached to the roots, so prepare the surface that much lower than you want the level of the finished lawn. Lay the strips side-by-side with ends tightly butted together and with the joints staggered, as in brickwork. If the area slopes, lay the strips horizontally across the slope to minimize erosion along the seams. Roll the sod thoroughly to assure good contact with the soil, and water it daily until it is established; about three weeks in warm weather. Mow the grass when it needs it, but discourage other traffic. It may be called "instant lawn," but it is just loose sod until it truly becomes a lawn by taking root.

LEVEL OF DIFFICULTY

If your project includes removing what remains of an old lawn, the drudgery quotient of this project increases considerably. Rent a rototiller to break the old lawn up into manageable clumps. These are then speared with a pitchfork, shaken to dislodge the valuable loam

clinging to the roots, and tossed away. This is not fun. Try to remove every last trace of vetch, crabgrass, and weeds, or they will invade your new lawn as soon as you turn your back. Spreading the various additives, turning them into the soil, and rolling the area smooth is also difficult, tedious work, but it goes better if you work at a moderate, steady pace. The actual seeding is light work; the actual sodding is not. Each roll of heavy, damp sod has to be carried to the lawn, laid, and maneuvered into position, and possibly trimmed, all on the same day it is delivered. For a large lawn, be sure to get some able-bodied assistance.

A beginner, using either method, should plan on at least 150% more time than listed to accomplish the job. An intermediate should add 75% more time for sod, and 50% more time for seed; an expert should add 20% and 10%. All should line up enough help to complete the sod-laying in one day.

WHAT TO WATCH OUT FOR

When seeking advice at your local garden center, be sure to provide them with all the details of your lawn site. Factors can vary considerably even within a fairly restricted area. For example, the soil in your town may be somewhat acidic, but if your house is surrounded by pine trees, your soil will be even more so, and will require more lime to neutralize it. A lawn in constant sunlight will call for a

different type of grass seed or sod than one in constant shade. Slopes can present water run-off problems, which could lead to erosion, and low areas with slow percolation rates may require the installation of a drainage system before any type of lawn is attempted.

To prevent seed on a slope from being washed away by water and rain, cover it with tobacco netting – a lightweight gauze – available at most garden centers. This can be held down with "staples" of bent wire. The grass will sprout through the netting, which along with the wire will eventually decompose into the soil.

Special fertilizers are available that are geared to fast lawn startups. Within two to three weeks, a lawn may look lush and established, but keep in mind that a new lawn needs your care and watering to survive.

The chemical lawn care service industry has recently been subjected to a great deal of scrutiny by environmental organizations. Questions have been raised concerning the use of herbicides and pesticides, their methods of application, and the nature and extent of the health hazards they pose to humans and animals. Most of these issues are still being debated, but it is only prudent and reasonable to be aware of them, and judge for yourself the pros and cons before contracting such a service.

SUMMARY

A lush green lawn is indeed a thing of beauty, and one that can make the most ordinary house look better. Creating a lawn is a project any homeowner can successfully complete, given a little advice from a local garden center, proper materials, and a willingness to work up a sweat and raise a few blisters. Site preparation is the same for both types of lawn, seed and sod, and both have their advantages. The specific conditions of your property (soil, sunlight, drainage, etc.) may determine which method is more likely to produce the better lawn. Seeding is more cost- and labor-efficient, but sodding offers a generally higher rate of success, along with the instant gratification of a complete lawn in one day. Whichever method you choose, a healthy lawn is the main component of most home landscapes, and it returns dividends of attractiveness for whatever time and money you invest in it.

For other options or further details regarding options shown, see

Trees and shrubs

Seed or Sod

Description	Quantity/ Unit	Labor-Hours	Material
Scarify subsoil, residential, skid steer loader	1,000 S.F.	0.3	
Root raking and loading, residential, no boulders	1,000 S.F.	0.5	
Spread topsoil, skid loader	18.50 C.Y.	1.6	248.64
Spread limestone	110 S.Y.	0.1	10.56
Fertilizer, 0.2 lb./S.Y., push spreader	110 S.Y.	0.1	7.92
Till topsoil, 26″ rototiller	110 S.F.	0.9	
Rake topsoil, screened loam	1,000 S.F.	1.0	
Roll topsoil, hand-push roller	18.50 C.Y.	0.1	
Seeding, turf mix, 4 lb./M.S.F., push spreader	1,000 S.F.	1.0	7.56
Mulch, oat straw, 1″ deep, hand spread	110 S.Y.	1.9	39.60
Add to contractor's fee for sodding......$ 495.00			
Totals		7.5	$314.28

Project Size	1,000 S.F.	Contractor's Fee Including Materials	$774

Key to Abbreviations
C.Y.– cubic yard   Ea.– each   L.F.– linear foot   Pr.– pair   Sq.– square (100 square feet of area)
S.F.– square foot   S.Y.– square yard   V.L.F.– vertical linear foot   M.S.F.– thousand square feet

TREES AND SHRUBS

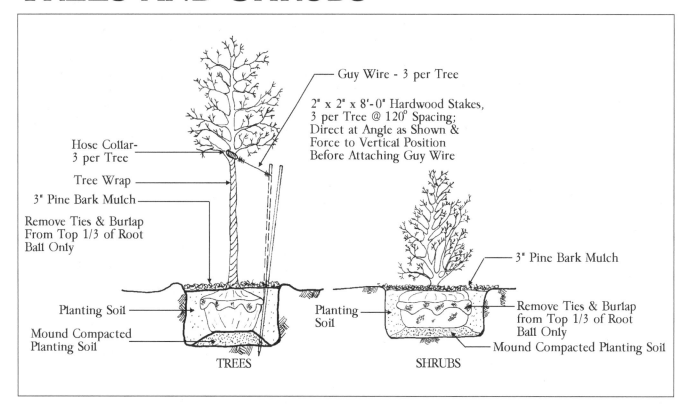

Guy Wire - 3 per Tree

2" x 2" x 8'-0" Hardwood Stakes, 3 per Tree @ 120° Spacing; Direct at Angle as Shown & Force to Vertical Position Before Attaching Guy Wire

Hose Collar- 3 per Tree

Tree Wrap

3" Pine Bark Mulch

Remove Ties & Burlap From Top 1/3 of Root Ball Only

Planting Soil

Mound Compacted Planting Soil

TREES

3" Pine Bark Mulch

Planting Soil

Remove Ties & Burlap from Top 1/3 of Root Ball Only

Mound Compacted Planting Soil

SHRUBS

A home's exterior look, sometimes called its "curbside appeal," is a combination of its architectural design and detailing, and its landscaping. The latter includes ledges, ponds, walls, fences, walkways, and the like, as well as the living, botanical elements of grass, flowers, trees, and shrubs. Whether you are landscaping from scratch around a newly built house or improving the established landscape of an older home, the careful selection and proper placement of trees and shrubs can enhance your property in terms of both beauty and value.

MATERIALS

If your property is in need of a complete makeover, you might consider engaging the services of a landscape architect to design a master plan. This will not be cheap, but it may save you money in the long run by ensuring that nothing is overlooked. Proper attention now will prevent the need to rework in ten years or so when the trees that looked so good on paper don't look so good growing on your front lawn. If your project is not so all-encompassing, the risk of less than perfect results is much less, and you can save money by laying it out yourself.

Sketch your plans on paper. Even if your artistic talents are not of the highest order, such an exercise can help you to focus on the project and perhaps anticipate problems or come up with ideas that may not otherwise have occurred to you.

It is important that you purchase trees and shrubs that are healthy. Check for injuries on the stem or trunk; look for sunscald-bark that looks discolored or appears to have bubbled up and broken away; check for insects; and choose plants that are packaged to meet your needs. Trees and shrubs come packaged in one of four ways: balled and burlapped (B & B), potted, container-grown, or bare root. The first three types are those you are most likely to find for sale at your local nursery. Bare root trees and shrubs are generally sold through mail-order catalogues. The B & B packaging is the most common of the four, and offers the widest selection in size. B & B trees and shrubs are best planted in the spring or fall, shortly after they have been dug up from the farms where they were grown.

There is an old saying: "Don't dig a $5 hole for a $10 plant." The healthiest tree or shrub will struggle to survive a poor job of planting. Dig the hole at least 12"

wider and 6" deeper than the size of the root ball. Throw some topsoil into the hole, mixed with adequate peat moss as recommended by the nursery. After pouring some water in the hole, position the plant in it at the same soil level as that of the soil contained in the burlap sack. Cut and remove as much of the burlap as possible without disturbing the root ball. Backfill the hole 3/4-full with topsoil, tamp it down, and fill with water. After the water has soaked in, finish backfilling to grade level, and tamp firmly around the plant to eliminate air pockets from the soil. Form a "water well" by building up a low dike around the perimeter of the hole; fill this with water to give the roots a thorough soaking. Ensure the survival of your new plant by watering it frequently during the first two months or so that it takes to become established. Watch for pests such as spider mites, aphids, and chewing insects. If you see any of these, contact your local garden center for advice on treatment. Forestall the growth of weeds by putting down a layer of organic mulch around the base of the plant. Do not fertilize the plant until it is well established, which may take a couple of years. Again, your local nursery expert is the best source of advice on how and when to fertilize. Trees generally do not need to be staked unless you live in a profoundly windswept region. If you do have to stake a young tree, protect the bark by cushioning the wire with a section of rubber hose where it wraps around the trunk.

LEVEL OF DIFFICULTY

Landscaping your entire property can be an enormous undertaking, possibly calling for the use of heavy equipment and large amounts of material for site preparation, drainage, and construction. If you are determined to do it yourself, attack a small section at a time, rent specialized machinery and tools, and have enough patience and vision to avoid discouragement when contemplating the number of weekends it will take to complete the job. Most homeowners who can afford it would probably do well to hire a licensed landscape contractor to come in with a crew and get the whole thing, or at least the major construction and large tree planting, done in a few days. A project more modest in scope, such as the planting of just a few shrubs and trees, is certainly one that the average homeowner can accomplish with

appropriate advice and guidance from local gardening professionals. Digging holes and muscling heavy root balls into place calls for energy, a strong back, and a willingness to get dirty, but most people find the work pleasant in the rewards that it quickly brings by improving the looks of the yard.

Beginners should add about 100% to the professional's time; intermediates should add 50%, and experts, 20%. All should consult with nursery professionals in choosing the best varieties of trees and shrubs, and in getting advice on their planting and maintenance.

WHAT TO WATCH OUT FOR

Shrubs and trees are often used for foundation plantings around the immediate perimeter of the house.

Improperly installed landscaping can cause moisture problems where none previously existed. Because new plantings require large quantities of water initially, there may be a greater chance of dampness in a basement living area. Before planting shrubs and trees or creating flower beds, check your foundation for any cracks or signs of

weakness that could leak if exposed to increased water pressure. Most drainage problems can be solved by grading away from the foundation. This can easily be accomplished by creating a low spot or swail some distance from the foundation.

Professional landscape designers follow additional guidelines and procedures when dealing with foundation plantings:

- Test the soil and, if necessary, improve it. Excessive alkalinity, acidity, or the lack of important soil nutrients such as nitrogen, phosphorous, and potassium can be determined only by testing. Mineral and chemical additives are available to restore or improve foundation soil so that it will support healthy plants. Some universities and county extension services will test soil for you.

- Don't use too many shrubs and trees. Leave enough space around them so they can develop without looking crowded.

- Don't plant too many species. Planting more than three or four types in a limited area creates a confused and jumbled effect.

- Select suitable plants. Know how tall and wide the plants will grow, what their mature appearance will be, whether they thrive in sun or shade, and whether their roots will create problems around the foundation.

- Don't plant too close to the house. Position all trees and shrubs so that when they are fully grown there will be a minimum of 3' between them and the house, lest they crowd the walls, making painting and window washing difficult and hazardous and, by the dampness they create, encouraging mildew, mold, and rot.

SUMMARY

A well-kept lawn, neat walkways, walls and fences, full flower beds, and healthy trees and shrubs all enhance the facade of your house in much the same way that a beautiful frame enhances the picture it surrounds.

For other options or further details regarding options shown, see

  Seed or sod

Trees and Shrubs

Description	Quantity/ Unit	Labor- Hours	Material
Stake out tree and plant locations	6 Ea.	0.4	
Tree pit, excavate planting pit	0.26 C.Y.	0.3	
Mix planting soil, by hand, including loam, peat, and manure	0.30 C.Y.	0.1	7.04
Backfill planting pit, by hand, prepared mix	0.30 C.Y.	0.3	
Bark mulch, hand spread 3" thick	0.35 C.Y.	0.1	0.98
Cornus Florida (white flowering dogwood) B & B, 5'-6'	1 Ea.		72.60
Picea Abies (Norway Spruce) B & B, 4'-5'	1 Ea.		57.60
Thuja occidentalis (American arborvitae) B & B, 4'-5'	1 Ea.		44.40
Syringa vulgaris (common lilac), 5-gal.	1 Ea.		19.08
Forsythia ovuta robusta (Korean forsythia) B & B, 3'-4'	1 Ea.		19.44
Rhododendron catawbeense hybrid, 5-gal.	1 Ea.		29.40
Totals		1.2	$250.54

Contractor's Fee Including Materials	$395

Key to Abbreviations
C.Y.– cubic yard   Ea.– each   L.F.– linear foot   Pr.– pair   Sq.– square (100 square feet of area)
S.F.– square foot   S.Y.– square yard   V.L.F.– vertical linear foot   M.S.F.– thousand square feet

Section Six

OUTBUILDINGS AND RECREATION EQUIPMENT

Outdoor buildings and recreation equipment can make the time you spend at home more rewarding by adding either convenience (in the case of a shed) or enjoyment (in the case of a play structure). Following are some tips for a successful project.

- Gazebos can be difficult to build because of the moderate to advanced carpentry skills and the number of odd angle saw cuts and fastening techniques that are required. Nevertheless, these projects can serve as a training ground for an intermediate do-it-yourselfer interested in improving their carpentry skills.

- When fabricating recreational equipment you should take care to make the cleanest cut possible. Your lumber dealer should be able to recommend the correct blades for sawing. Carbide blades have the advantage of not becoming dull as fast as typical saw blades. Working on the ground exposes blades to dirt and grit that can easily dull a blade. If you use pressure-treated lumber for these outdoor structures, pressure-treated wood dulls them faster than ordinary lumber. When doing any sawing, wear appropriate eye protection. This is particularly important when working with pressure-treated wood that contains poisonous substances.

- If you are contemplating building an outdoor storage shed, plan carefully what items it will contain. Be sure the door is wide enough and properly placed to allow you to move these items in and out easily.

- An accurate 2' or 4' level, calipers, and an adjustable square will all help to make the layouts and angle cutting for the gazebo project much easier and more precise.

- Most cities and towns have set-back requirements that limit construction to within a certain distance from property lines, front, side, and back. If the site chosen for your outbuilding doesn't meet these requirements, you must file an appeal for a variance.

- Be sure to lay out the posts that support the play structure platform perfectly square and plumb, using braces whenever necessary. If even one post is out of line, it will throw off almost every measurement.

- There are many types of composite materials, such as those made of recycled plastics or sawdust. They are available in a variety of colors, and offer other advantages such as nontoxic, nonsplintering surfaces. Consult your building material dealer for advice about availability and selection of these materials — particularly where children will be in close and constant contact with the materials.

- If you or your contractor are building projects that involve excavation, it's important to contact your utility companies to identify the location of (and avoid damaging) underground electrical, gas, water, and septic lines.

OCTAGONAL GAZEBO

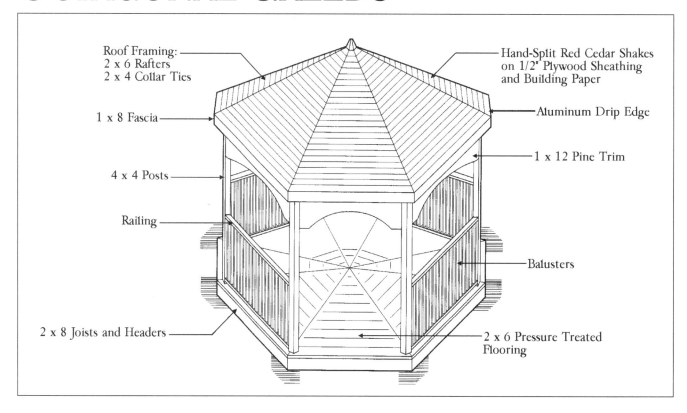

Roof Framing:
2 x 6 Rafters
2 x 4 Collar Ties

1 x 8 Fascia

4 x 4 Posts

Railing

2 x 8 Joists and Headers

Hand-Split Red Cedar Shakes
on 1/2" Plywood Sheathing
and Building Paper

Aluminum Drip Edge

1 x 12 Pine Trim

Balusters

2 x 6 Pressure Treated
Flooring

Gazebos can range from plain and rustic to very decorative and can be constructed from a wide variety of materials, and from basic to deluxe. In many ways, a gazebo is like the shell of a small house, as it consists of a foundation or support system, framed and covered walls, and a roof.

MATERIALS

Most of the materials used in this gazebo are basic products found in all exterior structures. The products are generally available at building supply outlets, regardless of geographical location. When purchasing the materials, be sure to arrange delivery, as it is a large enough order to require trucking.

The most important procedure in the early stages of the project involves establishing the layout for the gazebo and accurately placing 3' concrete pedestals, which serve as its basic support. Because of the octagonal shape and the number of pedestals required, the operation is tricky; and careful planning and workmanship are crucial if the rest of the project is to go smoothly. Be sure to double-check all measurements

before placing the concrete. The use of tube forms will add to the cost of this procedure, but they will make it easier to place and level the pedestals accurately.

The next phase of the project involves the placement of the platform and support posts. One method is to build the frame in place and then mortise it into the posts so that short legs extend below the frame onto the tops of the pedestals. Another way is to place the frame directly on the pedestals and set the posts beside the perimeter of the frame at each point. Still another is to plant the legs in the concrete at rough height, level them by sawing after the concrete has cured, and then build the frame of the platform around them. Later, the roof supports should transmit the load of the roof directly to the pedestals. The costs vary only slightly between these three methods.

Various options in both materials and methods are open to the do-it-yourselfer throughout the construction of the platform and sides. Decorative floorboard patterns and angled wainscoting are among the possibilities, and an ornate cap with trim could replace a standard railing. These are just a few of the possible variations from this plan. Generally,

extra cost and installation time should be added for all decorative work, as more material and intricate angle-cutting are usually involved.

The roof rafters are difficult to place, as they must be precisely angled on both ends if they are to fit correctly. Ceiling joists are not called for in this plan, but they can be used in place of the collar ties if the new gazebo is to have a finished ceiling. Sheathing and finished roofing are applied after the roof frame is completed. Any type of finished roofing can be used, including tile, slate, and various grades of asphalt and cedar. Although the roof is relatively small, its installation can be challenging because of the hips and the triangular shape of the roof sections. Seek some assistance if you are an inexperienced roofer, and plan on allowing extra time for cutting and scribing the roofing material. Also be aware that considerable waste is expected on roofs of this design.

The railing and steps of the gazebo also require skillful carpentry if you choose to follow the design of this plan. There are many options for the railing design. Fancy balusters, lattice, and wainscoting are some of the possibilities. Lattice and various trim styles can also be used for the opening above the rail and the fascia. Ornate trim designs tend to be more expensive than simpler styles, so compromises might be made to stay within a budget. Many stairway designs are possible, from basic wood steps to decorative styles made of concrete, brick, and stone. Although they are convenient, built-in benches do take up platform space that could be used by more patio furniture.

LEVEL OF DIFFICULTY

Gazebos are challenging projects even for expert do-it-yourselfers. Many do-it-yourselfers can complete the gazebo with patience, some preliminary reading on the subject, and help and guidance along the way. Beginners and intermediates should consider hiring a carpenter to complete the roof and some other aspects of the job. Mistakes can always be costly and time-consuming. The most challenging tasks in the project are the initial layout and the roof frame. The rental of staging may be an extra cost for this phase of the project. The detailed work of the finished roof may also be troublesome for many do-it-yourselfers. Generally, intermediates and beginners should not attempt any of the roof work. They should add 50% and 100%, respectively, to the estimated time for all other tasks, and slightly more for fancy trim work that involves expensive materials. Experts should add 20% to the professional time for all jobs, and more for all aspects of the roof work.

WHAT TO WATCH OUT FOR

A good set of carpentry tools will help facilitate work on most home improvements and the gazebo project is no exception. Having the right hand and power tools makes many of the challenging carpentry operations less difficult. Several different nailsets of various sizes will also help in the toenailing of angled pieces and in the finish nailing of the floor, balusters, fascia, and other trim pieces.

This gazebo is octagonal in shape, but five- and six-sided configurations are also authentic Victorian gazebo designs. Eight-sided structures provide more platform space, but they also require more materials and time to construct. Hexagonal and pentagonal types are slightly easier to build, because fewer pedestals are required; but the same kinds of complications still exist in angling the roof frame and applying its covering. However they may differ in size and shape, gazebos can be decoratively trimmed and authentically styled from the same kinds of materials.

SUMMARY

A gazebo is a home improvement that will enhance the atmosphere of your backyard landscape and expand your outdoor living space. While some aspects of the project may be out of reach for inexperienced do-it-yourselfers, it can still be your own personal creation, taking shape from the decorative features you choose to incorporate.

Octagonal Gazebo, 4' Sides

Description	Quantity/ Unit	Labor-Hours	Material
Excavate post holes, by hand, incl. layout	0.50 C.Y.	0.5	
Forms, round fiber tube, 8" diameter, for posts	24 L.F.	5.0	52.13
Concrete, field mix, 1 C.F. per bag, for posts	8 Bags		59.52
Placing concrete, footings, under 1 C.Y.	0.50 C.Y.	0.4	
Framing materials, pressure-treated lumber posts, 4 x 4 x 12'	96 L.F.	3.4	167.04
Headers, 2 x 8 x 12'	36 L.F.	1.7	37.58
Joists, 2 x 8 x 10'	70 L.F.	2.5	121.80
Flooring, 2 x 6 x 10'	200 L.F.	6.4	268.80
Railing with balusters	32 L.F.	11.6	345.60
Trim, 1 x 12 pine	32 L.F.	1.4	51.84
Roofing material, headers, 2 x 6 stock	36 L.F.	1.6	24.62
Rafters, 2 x 6 x 8'	128 L.F.	2.1	87.55
Collar ties, 2 x 4 stock	16 L.F.	0.3	6.91
Fascia board, 1 x 8 stock	36 L.F.	1.3	58.75
Sheathing, plywood, 1/2", 4' x 8' sheets	256 S.F.	2.9	150.53
Building paper, asphalt felt paper, 15 lb.	140 S.F.	0.3	3.36
Drip edge, aluminum, 5" girth	32 L.F.	0.6	7.68
Shakes, hand-split red cedar, 18" long, 8-1/2" exposure	2 Sq.	8.0	234.00
Stair material, stringers, 2 x 10	16 L.F.	0.6	35.52
Treads, 2 x 4 x 3'-6", 3 per tread	24 L.F.	0.8	32.26
Paint, seal and varnish, latex, sprayed	700 S.F.	11.8	58.80
Totals		**63.2**	**$1,804.29**

Project Size	10'-6" x 10'-6"		Contractor's Fee Including Materials	**$5,663**

Key to Abbreviations
C.Y.– cubic yard   Ea.– each   L.F.– linear foot   Pr.– pair   Sq.– square (100 square feet of area)
S.F.– square foot   S.Y.– square yard   V.L.F.– vertical linear foot   M.S.F.– thousand square feet

10' x 12' UTILITY SHED

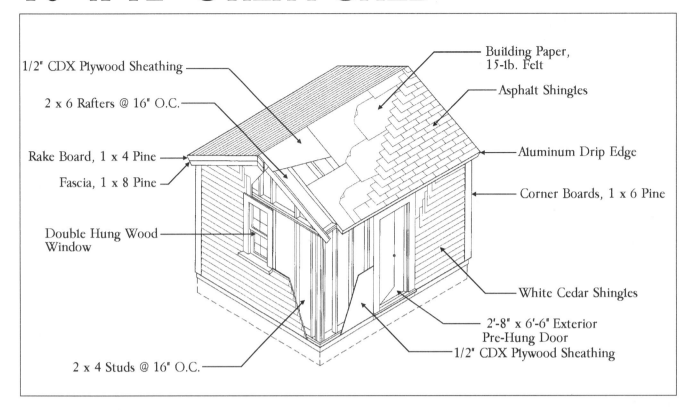

1/2" CDX Plywood Sheathing

2 x 6 Rafters @ 16" O.C.

Rake Board, 1 x 4 Pine

Fascia, 1 x 8 Pine

Double Hung Wood Window

2 x 4 Studs @ 16" O.C.

Building Paper, 15-lb. Felt

Asphalt Shingles

Aluminum Drip Edge

Corner Boards, 1 x 6 Pine

White Cedar Shingles

2'-8" x 6'-6" Exterior Pre-Hung Door

1/2" CDX Plywood Sheathing

This 10' x 12' utility shed provides 120 square feet of space for the various tools, machines, gardening supplies, and equipment that now occupy sections of your garage, basement, and back porch. The completion of a project such as this will free up that space, perhaps for the first time in years – at least temporarily.

A number of factors must be considered when selecting a location for the shed. In choosing a site for the shed, try to locate it so that it looks as if it has always been there, rather than a poorly planned afterthought. Be sure it is far enough from the house so as not to interfere with any possible additions that you or your successors might consider building in the future. Also, its location shouldn't give offense to neighbors or impinge upon their property. (Set-back codes are designed to prevent this.) Finally, try to site the shed where it will be accessible – both while you are building it and when you are using it for storage.

MATERIALS

This utility shed is a simple pitched roof design set on a slab floor/foundation.

A contractor could be hired to excavate, grade, build a form, and pour and level the concrete. But if the site is fairly level to begin with, this work could be accomplished by an experienced intermediate.

It is very important that applicable building codes be followed and that the forms be properly built and braced to produce a square and level foundation. Mixers can be rented so the concrete can be prepared on site. Some supply houses sell mixed concrete in dump trailers that you can haul to the job site on your vehicle's trailer hitch. Ordering a load from a contractor is another option, but most have a minimum requirement, which may not make this the most economical option for this project.

If a section of lawn has to be driven over by any heavy equipment, be sure there is no septic system or dry well that could collapse under the weight of a large vehicle.

A poured slab is at once a solid foundation and a durable floor. It should be set slightly above grade. If preferred or necessitated by siting considerations, a wood framed floor set on concrete piers can be substituted.

The sill is of pressure-treated lumber, attached to the slab with anchor bolts set in the concrete. As always, be sure that you build square, plumb, and level from the sill to the ridgeboard.

Wiring for lights, switches, and GFI outlets should be run as soon as the framing is complete. A narrow trench, one foot deep, will probably be required for an underground wire to tie into the house's electrical system. Depending on the nature and extent of the wiring, a subpanel in the shed may be recommended. Consult an electrical contractor.

The walls and roof are sheathed with 1/2" CDX plywood. As soon as the sheathing is on, attach the fascia and rakeboards and shingle the roof to make the structure weathertight. Roofing, siding, and trim materials can be selected to match the house or other existing outbuildings. If a different look is desired, exercise good judgment so that the finished shed is in keeping with the house and its environs.

The door need not be fancy – a homemade one would work fine – but if purchased, be sure it is an exterior model and equip it with exterior hardware.

Only one small window is planned. This allows for light and ventilation while leaving most of the interior wall area free for storage use. The interior could be fitted with shelving, pegboards, racks, etc., but it might be wise to live with it plain for a while until you decide exactly where and what built-in accessories would be most useful.

LEVEL OF DIFFICULTY

This project is an excellent opportunity for a beginner to learn construction skills. All the main phases of house carpentry are present here in miniature: site planning, excavating, laying a foundation, framing, sheathing, roofing, siding, installing a window and door, and applying finish trim.

A true beginner would be advised to work under the guidance of a more experienced hand.

An intermediate could tackle this project with enthusiasm as an opportunity to increase her or his abilities and thereby develop the confidence to undertake more ambitious tasks in the future. It is said, and it is true, that we learn from our mistakes. This little shed will forgive most of those mistakes while you learn from them.

A beginner should add at least 100% to the times required for completion; the intermediate, 50%. A person experienced in all phases of construction might come close to matching the professional's time, but it would be wise to add 10-20% to be on the safe side.

10' x 12' Utility Shed

Description	Quantity/ Unit	Labor- Hours	Material
Excavation, by hand, for slab area	1.50 C.Y.	3.0	
Fine grading, for slab on grade, confined area	13 S.Y.	0.5	
Forms in place, slab on grade, edge forms, to 6" high	88 L.F.	4.7	43.30
Welded wire fabric rolls, 6 x 6, for slab on grade reinforcing	120 S.F.	0.6	10.08
Slab, 4" thick with vapor barrier	120 S.F.	2.3	129.60
1/2" anchor bolts, 6" long	10 Ea.	0.9	11.04
Sills, 2 x 6 x 8', pressure-treated	44 L.F.	1.4	59.14
Plates, top and bottom, 2 x 4 x 8'	132 L.F.	2.6	57.02
Studs, 2 x 4 x 8'	416 L.F.	6.1	179.71
Gable framing, 2 x 4 x 8'	128 L.F.	1.9	55.30
Ridgeboard, 2 x 8 x 6'	12 L.F.	0.4	12.53
Rafters, 2 x 6 x 10', to 4 in 12 pitch	220 L.F.	3.5	150.48
Wall sheathing, 1/2" CDX plywood, 4' x 8'	400 S.F.	4.6	235.20
Roof sheathing, 1/2" CDX plywood, 4' x 8'	230 S.F.	3.3	135.24
Rakeboard, 1 x 4, pine	60 L.F.	1.1	43.20
Fascia, 1 x 8, pine	60 L.F.	1.3	86.40
Aluminum gable vent, louvers, with screen, 8" x 8"	2 Ea.	0.4	15.48
Asphalt felt sheathing paper, 15 lb.	615 S.F.	1.3	14.76
Aluminum drip edge, 0.016" x 5", mill finish	24 L.F.	0.5	5.76
Drip edge, galvanized	5 L.F.	0.1	1.32
Roof shingles, asphalt strip	2.30 Sq.	5.3	136.62
Cornerboards, 1 x 6 pine	56 L.F.	1.1	60.48
Siding, white cedar shingles	4 Sq.	13.3	518.40
Double-hung wood window, 2'-0" x 3'-0"	1 Ea.	0.8	253.20
Pre-hung exterior wood door, 2'-8" x 6'-9"	1 Ea.	1.1	298.80
Door lockset, standard	1 Ea.	0.7	45.60
Door casing, 1 x 6 pine	161 L.F.	5.2	135.24
Paint, stain shingles, 2 coats, oil base, brushwork	400 S.F.	6.7	43.20
Door and window trim, primer, oil base, brushwork	50 S.F.	0.6	1.20
Trim, primer, oil base, brushwork	116 L.F.	1.4	2.78
Door and window trim, 2 coats, oil base, brushwork	50 S.F.	1.0	3.00
Trim, 2 coats, oil base, brushwork	116 L.F.	1.4	2.78
Totals		79.1	$2,746.86

Project Size	10' x 12'	Contractor's Fee Including Materials	$7,736

Key to Abbreviations
C.Y.– cubic yard   Ea.– each   L.F.– linear foot   Pr.– pair   Sq.– square (100 square feet of area)
S.F.– square foot   S.Y.– square yard   V.L.F.– vertical linear foot   M.S.F.– thousand square feet

WHAT TO WATCH OUT FOR

The two areas where the beginner/intermediate will need professional guidance and advice are laying the sill and framing the roof. Whether a slab is poured or a deck is built on piers, the sills must be solidly anchored, square, and level.

Framing a roof is the most difficult task involved in rough carpentry. A pitched roof is one of the simplest to accomplish, but the beginner would be well advised to hire professional help. Work the numbers out clearly and in an orderly fashion on paper, taking care not to overlook anything. If necessary, "loft" the roof frame on a suitably large and clean surface. This simply means drawing a full-size diagram in cross section on the entire frame – ridge, rafters, and plates – so you can check the accuracy of your calculations. Finally, make test cuts on scrap lumber of all rafter cuts – ridge, bird's mouth, tail – before going to work on the actual (expensive) rafter lumber.

SUMMARY

There is no interior finish work required in this shed, nor is there plumbing, heating, or insulation, and the electrical wiring is minimal. A helper would be handy throughout the project, but most of the tasks can be handled by one person.

For other options or further details regarding options shown, see

Residing with wood clapboard
Standard window installation
Trash can shed with recycling bins

TRASH CAN SHED WITH RECYCLING BINS

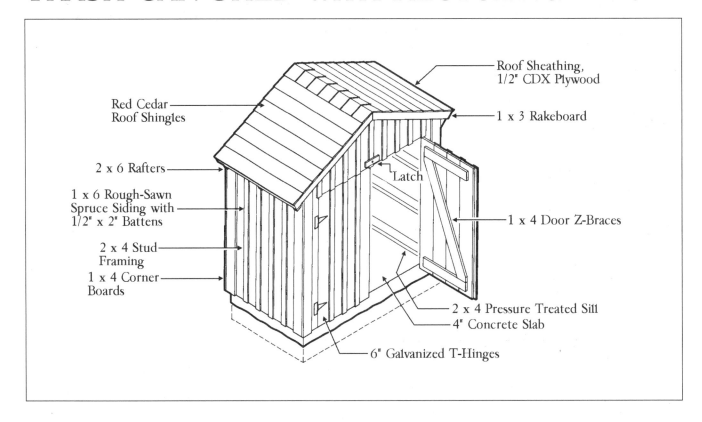

Red Cedar Roof Shingles

2 x 6 Rafters

1 x 6 Rough-Sawn Spruce Siding with 1/2" x 2" Battens

2 x 4 Stud Framing

1 x 4 Corner Boards

Roof Sheathing, 1/2" CDX Plywood

1 x 3 Rakeboard

Latch

1 x 4 Door Z-Braces

2 x 4 Pressure Treated Sill

4" Concrete Slab

6" Galvanized T-Hinges

The down side of a generally high standard of living is the incredible amount of garbage and pollution produced by the lifestyles of a consumer society. We are finally coming to realize that the price we must pay for the convenience of cheap, disposable goods is, in the long run, unacceptably high – overflowing landfills, polluted groundwater, depleted natural resources, and unhealthy air. Our collectively raised consciousness has resulted in much new legislation encouraging or requiring the recycling of many products and materials. Until now, a couple of trash cans tucked into a corner of a garage, basement, or tool shed sufficed for the household refuse, but today a homeowner must find a place to separate and store recyclables until they are picked up or taken to the local disposal site. A small backyard shed such as this one is designed to serve as both a trash can hideaway and a home recycling center.

MATERIALS

Prefabricated sheds in kit form can be purchased from many building supply houses and garden centers, and most providers can arrange to have them built

for you. The convenience of this solution might appeal to you; but it does limit your choice of designs and sizes, and the final cost will be much higher than if you do it yourself. You are, after all, the cheapest contractor you can hire. This shed is wide enough to hold two 32-gallon (standard-sized) trash cans, leaving about 24" for a stack of plastic recycling bins. Its rough-sawn board-and-batten design is traditional, and can be stained to match your house and other outbuildings, or to blend into a background of trees and shrubs. Its wood-shingled pitched roof and double doors combine to give it the look of a tiny barn, charming enough to enhance its surroundings.

The site chosen for the shed should be a level area of about 15 square feet, located not far from the back or side entrance to your house. (You will appreciate this proximity when you have to take out the trash in foul or freezing weather.) Prepare the area and build a form, properly squared and braced, for the poured concrete slab. Insert six anchor bolts in the wet concrete for the 2 x 4 sills. Once the concrete has cured, the three walls can be framed and nailed together, and the roof framed. The siding is of

rough-sawn 1 x 6 spruce with 2" battens, attached to the framing with 6d hot-dipped galvanized nails. Once the front gable is boarded up (horizontally, without battens) the 1/2" plywood roof sheathing can be cut and nailed, allowing a 3" overhang front and back, and a 1" overhang on the rafter ends. Install red cedar shingles, in courses about 6" to the weather, over a layer of roofing felt, using 3d galvanized box nails. Leaving the roof overhang unenclosed by a soffit saves materials and labor, and has the beneficial effect of helping to ventilate the shed. The doors are built by nailing boards to a Z-frame of 1 x 4s, and are hung by being attached to the corner framing with galvanized T-hinges. A combination of a wooden leaf latch and a metal barrel bolt will keep the doors securely shut. The siding should be stained with a wood preservative; the roof shingles can be left to weather naturally.

For sorting and storing recyclables, sturdy, ribbed plastic bins with interlocking feet can be stacked up to six high. Communities have different recycling requirements, but four categories – paper, glass, tin/aluminum, and plastic – are typical. These bins have an 11-gallon capacity, which should suffice for an average family with weekly trash pickup. Large families and/or less frequent pickups might require more of these bins, or even larger ones. The bins themselves can be used to transport their contents to the town recycling facility, and being plastic, they are both lightweight and easy to clean. If more space is needed in the shed itself, the overall dimensions can, of course, be increased, or storage-enhancing alterations made, such as installing an "attic floor" shelf that could be reached through a small door in the front gable.

LEVEL OF DIFFICULTY

Like the utility shed project, this little building is perfect for the beginner to learn a number of construction skills that can later be applied to larger, more ambitious home improvements. The walls, which use horizontal nailers, are not the usual stud framing. Nevertheless, the ability to measure carefully, cut accurately, and build walls of any type, plumb, square, and level are essential elements of the carpenter's craft, and are as necessary to good results in this tiny shed as they are in the largest house. Roof framing is a challenge for a novice, but the use of 45° angles makes cutting the rafters somewhat easier than it otherwise might be, and, of course, there are only six of them. You could hardly expect to find an easier roof to shingle, and although cedar shingles are more expensive and take longer to install than asphalt strip shingles, they are much more attractive, especially when seen close up. If the side walls are perfectly plumb and in line with one another, hanging the door should also be quite easy. This is a low-risk, very forgiving project for a beginner to undertake. Not much can really go wrong that can't be corrected, and the consequences of any mistakes will certainly not be dire. With the possible exceptions of the digging and the pouring of the concrete slab, there is no heavy, messy, drudge work. As is the case with most new construction, the tasks involved are actually quite pleasant and rewarding. A beginner working alone should add 200% to the time estimated for completion. An intermediate should add 50%, and an expert, 20%.

WHAT TO WATCH OUT FOR

A good building starts with a good foundation. Be sure the form you build for the concrete slab is perfectly rectangular and level. Use braces and packed earth to ensure that it will not move, bulge, or rack out of square. If you plan and build the form carefully, its top edges can serve as your screeding guide. "Screeding" concrete means to pack and level it by moving a straightedge board across the surface, with a back-and-forth sawing motion, while at the same time pushing the board forward. In this way, the concrete is lightly agitated to make it settle and fill any air pockets that may have formed, and the excess concrete is scraped off, leaving a smooth surface level with the top of the form. Insert the sill anchors in the appropriate locations, and cover the slab with a sheet of plastic, leaving it for a few days to slowly cure. Once the concrete has set, the forms can be removed and cleaned for future use, and the area around the slab backfilled and tamped.

Before cutting the lumber for the rafters, measure, mark, and make cuts for the ridge, tail, and bird's mouth on a piece of 1 x 6 spruce. If you are going to make mistakes, make them on the less expensive lumber, so you can correct them before making the final cuts on the expensive 2 x 6s.

Trash Can Shed with Recycling Bins

Description	Quantity/Unit	Labor-Hours	Material
Excavation, by hand, for slab area	0.25 C.Y.	0.5	
Fine grading, for slab on grade, confined area	1.50 S.Y.	0.1	
Forms in place, slab on grade, edge forms, to 6" high	32 L.F.	1.7	15.74
Concrete slab on grade, 4" thick	0.20 C.Y.	0.3	20.28
Sill anchor bolt, 1/2" diam., 6" long, including nut and washer	6 Ea.	0.5	6.62
Sill, 2 x 4 x 8', pressure-treated	16 L.F.	0.5	9.60
Framing, 2 x 4 x 8', KD studs	15 Ea.	1.5	51.84
Rafters, 2 x 6 x 8'	24 L.F.	0.5	16.42
Roof sheathing, 1/2" CDX plywood, 4' x 8' sheets	32 S.F.	0.5	18.82
Asphalt felt sheathing paper, 15 lb.	15 S.F.	0.1	0.36
Red cedar wood roofing shingles, no. 1 grade	1 Sq.	3.2	181.20
Siding, spruce, rough-sawn, 1 x 8, natural	330 L.F.	9.6	265.32
Bins, plastic, 11-gallon, 18-1/2" W x 20" D x 12" H	4 Ea.	0.1	44.64
Door hardware, 6" galvanized T-hinge	2 Pr.		55.20
4" galvanized barrel bolt	1 Ea.	0.1	3.55
Galvanized handle	1 Ea.	0.3	2.15
Nails, 6d H.D. galvanized	5 lbs.		7.08
Nails, 3d galvanized box	2 lbs.		2.83
Drywall screws, 2" galvanized	1,000 Ea.		10.38
Drywall screws, 1-1/4" galvanized	1,000 Ea.		7.98
Stain, exterior, 2 coats, oil base, brushwork	120 S.F.	2.0	12.96
Totals		21.5	$732.97

Project Size	2'-10" x 5'	Contractor's Fee Including Materials	$2,100

Key to Abbreviations
C.Y.– cubic yard   Ea.– each   L.F.– linear foot   Pr.– pair   Sq.– square (100 square feet of area)
S.F.– square foot   S.Y.– square yard   V.L.F.– vertical linear foot   M.S.F.– thousand square feet

SUMMARY

A small shed that can be used as a trash can hideaway as well as storage for sorted recyclables is a very useful and convenient amenity, and one that, for a large family with limited space and a less-than-weekly disposal schedule, can border on being a necessity.

For other options or further details regarding options shown, see

10' x 12' utility shed

PLAY STRUCTURE

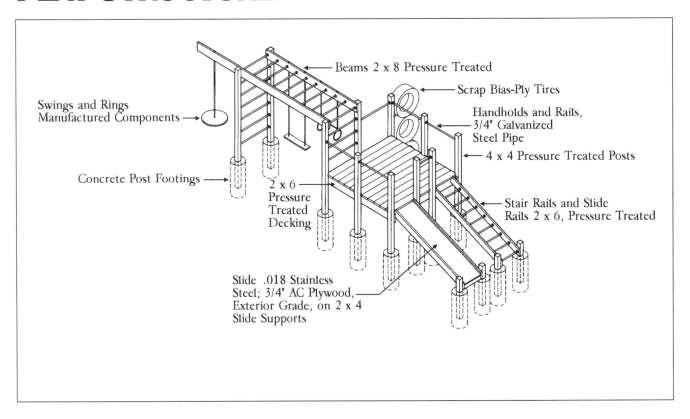

Beams 2 x 8 Pressure Treated

Scrap Bias-Ply Tires

Swings and Rings
Manufactured Components →

Handholds and Rails,
3/4" Galvanized
Steel Pipe

← 4 x 4 Pressure Treated Posts

Concrete Post Footings →

2 x 6
Pressure
Treated
Decking

Stair Rails and Slide
Rails 2 x 6, Pressure Treated

Slide .018 Stainless
Steel; 3/4" AC Plywood,
Exterior Grade, on 2 x 4
Slide Supports

If you have active young children and a spacious level yard, you already have two of the primary components of a backyard play structure. A carefully designed, well built play structure will provide years of safe, healthy fun.

MATERIALS

This design is representative of play structures, which typically include swings, slide, ladders, monkey bars, and a raised platform. How-to books dealing with outdoor building projects will show you other layouts and variations that you can freely adapt to your child's preferences and to the dimensions of your backyard site.

Play structures are available in kits that come complete with instructions, fasteners, hangers, ladder rungs, swing seats and chains, and precut, predrilled lumber. Others, called "hardware only" kits, provide everything listed above except the lumber. Purchasing a high-quality kit can be a good investment for a homeowner who lacks the tools, skills, and/or time to build a play structure from scratch. Complete kits involve simply assembling parts by following detailed

written and illustrated instructions. Hardware-only kits provide instructions for the cutting and assembly of the lumber, which you purchase separately. Many kit-built play structures can also be disassembled and moved—an attractive feature for those who might move to another house before their children outgrow the play set. The best kit designs have been field tested and approved for strength and safety by independent rating organizations; look for an indication of this on any kit you consider buying.

If you build your own play structure or work from a hardware- only kit, you must decide on the type of lumber to use. Standard construction-grade stock is not recommended because, even if coated with preservative, it is prone to deterioration from ground contact and constant weather exposure. This deterioration might not always be visible, and could go so far as to weaken the wood sufficiently to cause collapse or breaking, which might well result in serious injury to a child. The best types of weather resistant wood are cedar, redwood, and pressure-treated lumber, usually yellow pine. Treated lumber will last longest in damp ground or concrete, and thus is the best choice for posts.

Unfortunately, it tends to be splintery—more so as it weathers – and the chemicals used in the treatment (chromated copper arsenate) can cause health problems. Make sure to round over all corners and edges to a 1/4" radius, and regularly monitor the condition of the lumber, touching it up whenever necessary. (This applies to all types of lumber.) Treated lumber is also the most economical and readily available. Any well-stocked lumber yard will offer a wide selection of lengths and sizes. When purchasing your lumber, choose pieces that are relatively straight, not warped, and, most importantly, free of large knots, which cause the wood to be structurally weak.

For safety's sake, all sharp hardware, such as bolt ends and nuts, should be countersunk so they don't protrude above the surface of the wood. You should also spread a 10" layer of cushioning material under the play set—fine sand or shredded bark are good choices. Grass in this area will quickly be worn away and the soil compacted to the consistency and hardness of asphalt, not to mention the mud and puddles that will develop on bare ground in wet weather. Any chains used for swings should be encased in a length of rubber or plastic hose to obviate the possibility of pinched fingers and the chance of strangulation. The 3/4" galvanized pipe used for the monkey bars should be pinned into the 2 x 8 rails by drilling a pilot hole through the top of the rail, into the pipe, and then driving a 3" galvanized screw. This will keep them from turning in a child's grasp and causing a fall.

Once you have completed the basic structure, you can equip it with manufactured components: swings, gliders, rings, trapeze, cargo net, rope ladder, tire swing, and whatever else can be acquired through catalogs or at playground supply centers.

LEVEL OF DIFFICULTY

Building a play structure from a complete kit is the easiest route to take. All you need do is follow the directions and assemble the parts in sequence, using common household tools. Some kits are designed with bases and angled supports that require no digging and need no concrete, a feature that saves a lot of labor, although such a kit would naturally lack the rock solid stability of posts anchored in the ground. Working with a hardware-only kit means you have to purchase the lumber and then measure, cut, drill, and assemble it according to the plans supplied. With both types of kits you are relieved of the tasks of design and layout and, assuming the plans are accurate and your work is faithful to them, your chances of making a major error are almost completely eliminated.

Designing and building your own custom play structure is a complex undertaking and demands a fairly substantial investment of time to complete. Site preparation—excavating the area to a depth of 10" and filling it with soft landscaping material—is necessary for any play set, the size of the area being the only variable that will determine the amount of time you'll need to spend with a pick and shovel.

A beginner could do site preparation and could assemble either type of kit, but should work under the guidance of an experienced hand in constructing a custom play structure, adding about 200% to the times estimated. An intermediate should add about 100%, and an expert about 20%.

WHAT TO WATCH OUT FOR

Make sure the site you choose is level. Irregular terrain under the play set will make it both difficult to build and hazardous to use. Build the structure after excavating and before laying down the cushioning material. Remember to take the 10" of fill into account when figuring the heights above grade. Take measurements on the structure itself as you progress, rather than only from the plans. This will allow you to adjust for slight variations in the dimensions of the lumber and will result in much better joints and fits.

The climbing tires are bolted to the post, and each should have a drainage hole cut in the bottom with a hole saw. (Use bias-ply tires, not steel-belted.) Also, be sure to drill pilot holes in the ends of the deck boards; nailing directly into the wood can cause it to split.

From time to time, and especially during the first several weeks of use, it is a good idea to check all bolts and fasteners for tightness. The wood tends to shrink and the regular use of the structure can cause loosening of bolts as well.

SUMMARY

A kit-built play structure will surely delight your children, and probably those of your neighbors as well. It can also usually be taken apart if you move and want to take it with you. A custom-designed play structure is a good deal more challenging and time-consuming, and thus is recommended only for the advanced inttermediate or expert.

Play Structure

Description	Quantity/Unit	Labor-Hours	Material
Layout, clearing and excavating, by hand	5 C.Y.	10.0	
Compaction, vibratory, 8" lifts, common fill	5 C.Y.	0.2	
Fine sand	5 C.Y.		108.00
Backfill sand, by hand, light soil	5 C.Y.	2.9	
Forms in place, columns, round fiber tube, 8" diameter	60 L.F.	12.4	130.32
Concrete, redi-mix, 2000 psi	3.25 C.Y.		253.50
Footings, spread type, placing of concrete	3.25 C.Y.	2.8	
Posts, 4 x 4 x 10', treated	140 L.F.	5.0	243.60
Rim joists, 2 x 8 x 12', treated	24 L.F.	0.9	41.76
Deck joists and beam supports, 2 x 6 x 12', treated	36 L.F.	1.3	62.64
Decking, 2 x 6 x 8', treated	120 L.F.	4.3	208.80
Beams, 2 x 8 x 10', treated	20 L.F.	0.7	34.80
Slide rails, 2 x 6 x 10', treated	40 L.F.	1.4	69.60
Ladder rungs, slide supports, 2 x 4, treated	20 L.F.	0.6	12.00
Joist hangers, galvanized, 2 x 6 to 2 x 10 joists	8 Ea.	0.4	4.13
Bolts, square head, nut and washers incl., 3/8" x 4", galvanized	75 Ea.		18.00
Nails, 12d common, galvanized	5 lbs.		7.08
3/4" galvanized steel pipe	90 L.F.	11.8	158.76
3/4" galvanized elbows	4 Ea.	0.6	59.28
3/4" plywood, 4' x 8' sheets	32 S.F.	0.4	25.34
0.018 stainless steel, 2' x 10' sheets	20 S.F.	1.0	78.00
Tires, bias-ply, scrap	4 Ea.	4.0	9.60
Swings, manufactured	1 Ea.	1.5	74.40
Trapeze with rings	1 Ea.	1.9	46.80
Tire swing, manufactured	1 Ea.	1.0	2.40
Totals		65.1	$1,648.81

Contractor's Fee Including Materials	**$5,307**

Key to Abbreviations
C.Y.– cubic yard   Ea.– each   L.F.– linear foot   Pr.– pair   Sq.– square (100 square feet of area)
S.F.– square foot   S.Y.– square yard   V.L.F.– vertical linear foot   M.S.F.– thousand square feet

REDWOOD HOT TUB

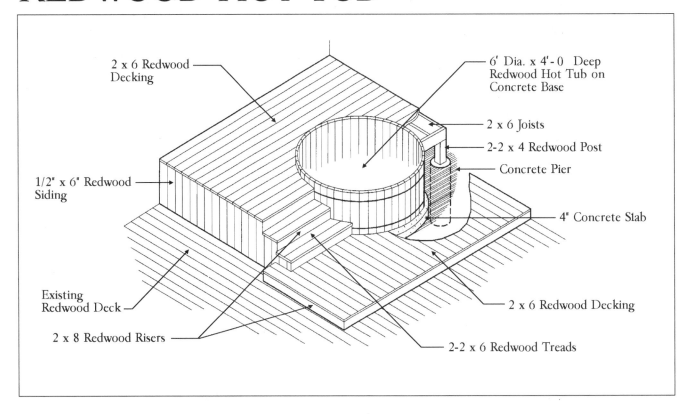

2 x 6 Redwood Decking

6' Dia. x 4'- 0 Deep Redwood Hot Tub on Concrete Base

2 x 6 Joists

2-2 x 4 Redwood Post

Concrete Pier

1/2" x 6" Redwood Siding

4" Concrete Slab

Existing Redwood Deck

2 x 6 Redwood Decking

2 x 8 Redwood Risers

2-2 x 6 Redwood Treads

Hot tubs can be installed in interior or exterior settings and often complement another facility, such as a deluxe bath, exercise room, sauna, sun deck, or swimming pool. This plan demonstrates how a 6' factory-prepared redwood hot tub can be placed on an existing redwood deck. Although many locations and designs are possible, this project covers the basic materials, procedures, and considerations for a medium-sized exterior installation.

MATERIALS

The materials list starts with redwood support members and decking. These are required to alter and prepare the existing deck to receive the hot tub. The remaining components are the precut and prefinished 6' hot tub kit and incidental finish products. The framing and decking lumber are available from most lumber yards, as are some of the finish materials. Redwood may have to be special-ordered, however, in some areas of the country. The redwood tub kit can be purchased from a specialty retailer, but some general building suppliers may also carry hot tubs and spas on a special order basis. Because the

redwood deck materials and hot tub are expensive products, do-it-yourselfers should thoroughly research their options before finalizing the design of the facility and the materials order.

The 6' hot tub suggested in this plan weighs well over three tons when filled and in use. Consequently, a considerable amount of support must be added to the existing deck if it is to bear the load. Each installation will present a different set of conditions, and the requirements for reinforcing your deck's support system may vary from what is recommended in this plan. Consult local building codes, follow the manufacturer's guidelines, and get a professional opinion before you determine the arrangement and size of the deck joists and supports. In some cases, extra joists and bridging may be enough; in others, additional footings, posts, and supports may be required. Most manufacturers of hot tubs recommend an independent ground support system for the unit, separate from the deck. Whenever possible, this method of installation should be employed, even though it may involve considerable extra work and expense. When placing the knee wall and partial perimeter deck structure around the hot tub, be careful

not to drive the fasteners into the side of the unit. Also, leave a small space for expansion and drainage between the tub and the deck frame. Use quality hot-dipped galvanized nails and lag screws for the decking and supports to prevent corrosion and rusting.

Many different sizes, shapes, and material options are open to you when selecting the hot tub kit. The unit in this plan features standard-grade bottom pieces and redwood staves collared with adjustable stainless steel hoops. The kit is precision cut and fitted at the factory, and delivered unassembled to the consumer. Also included in the price of the kit are the basic plumbing and electrical support packages for the operation of the facility. Many specialty retailers will provide, at extra charge, an on-site assembly service for the tub itself. Intermediates and beginners might consider this service, as the hot tub is fairly complicated and must be accurately fitted together so that it will be sturdy and watertight. The cost of this service covers only the basic tub and its attached components, however, and does not include plumbing and electrical work. As noted earlier, careful planning and research will help you determine the size and design features that meet your practical and aesthetic requirements. In addition to redwood, hot tubs are available in teak, cedar, mahogany, and cyprus. Coated fiberglass, stainless steel, and other materials are also used. Whatever they are made of, hot tubs come in different sizes and are designed as freestanding, exposed units or as inserts for redwood or other raised perimeter decks. Still others can be recessed into the raised section of an interior floor. The cost of these alternative sizes, designs, and settings will, of course, significantly affect the cost of the project.

After the support system, perimeter decking, and hot tub are in place, the installation can be left as is or finished with stain or sealer. If left untreated, the redwood will eventually weather to a tanned leather color, but it will not rot or deteriorate. However, because of the dampness, a sealer with a mildewcide additive is recommended for the exposed exterior surfaces of the tub and the surrounding deck. If the deck and tub are made from different materials, one or both may be stained to blend the new and old installations. In most cases, a qualified plumber and electrician should be hired to tie in the support packages to the house utilities. Because the water pump, heater, filters, and other tub support facilities are often remotely located, extra expense may be incurred for additional piping and underground placement. Increased cost is to be expected for convenience options such as air switches, therapy jets, chemical feeders, timers, lighting, and solar heat packages. Shelves, steps, and ladders are other options that can be obtained at additional cost.

LEVEL OF DIFFICULTY

This project calls for some expensive materials – both the redwood decking and the tub kit. This factor, together with precise installation requirements, makes it a challenge for all do-it-yourselfers. Beginners and most intermediates might want to stick to the more basic tasks, like the preliminary site preparation and the surface finishing. Experts who opt to assemble the kit themselves should follow the manufacturer's instructions precisely and consult a professional before tackling the operation. Experts should add 25% to the labor-hours for all carpentry work. Beginners and intermediates should add 100% and 50%, respectively, to the time estimate for the basic tasks that they attempt. All do-it-yourselfers should hire qualified tradespeople to accomplish the electrical and plumbing work.

WHAT TO WATCH OUT FOR

Provisions should be made in the planning stages of this project for features such as proper drainage piping and protection against the effects of freezing temperatures. The installation should allow for complete drainage during the period when the facility is not in use. In cold climates where the tub is to be used year-round, an optional freeze protection package can be purchased at additional cost. Floating or attached covers made of wood and plastic are other extras that will reduce heat loss and protect the tub. If the tub is to be used during cooler weather, you might consider installing a wind screen adjacent to the facility for greater comfort and reduced heat loss. All of these components will increase the cost of the project, but they will also extend the length of the season during which the facility can be used.

SUMMARY

Although the installation of a hot tub is a relatively expensive undertaking, the comfort and potential benefits can add a unique luxury to your home life. When carefully designed to complement an existing deck, exercise area, or swimming pool, it can also enhance your home's exterior appearance.

For other options or further details regarding options shown, see

> *Deluxe bath/spa**
>
> *Elevated deck*
>
> *Sauna**
>
> * In Interior Home Improvement Costs.

Redwood Hot Tub

Description	Quantity/ Unit	Labor- Hours	Material
Remove portion of existing deck, per 5 S.F.	6 Ea.	9.6	
Reinforce existing deck, 2 x 8, redwood stock	24 L.F.	0.9	41.76
Base support slab, hand grading	5.50 S.Y.	0.2	
Forms in place, edge forms for slab on grade, 6" high	24 L.F.	1.3	11.81
Concrete, ready mix, 3000 psi	1 C.Y.		82.80
Placing concrete, slab on grade, 4" thick	1 C.Y.	0.4	
Finishing concrete, broom finish	1 C.Y.	0.1	
Platform framing, redwood, posts, doubled, 2 x 4 stock	48 L.F.	1.4	28.80
Joists, 2 x 6 stock	130 L.F.	4.2	174.72
Decking, 2 x 6 x 12'	252 L.F.	8.1	338.69
Stair stringers, 2 x 8 stock	12 L.F.	0.4	20.88
Stair treads, 2 x 6 stock	18 L.F.	0.6	24.19
Stair risers, 2 x 8 stock	14 L.F.	0.5	24.36
Wood siding boards, vertical grain, clear beveled 1/2" x 6"	48 L.F.	1.5	122.11
Redwood hot tub kit, 6' dia., 4' deep w/circ. pump & drain system	1 Ea.	20.0	2,400.00
Plumbing, rough-in and supply	1 Ea.	7.7	147.60
Electrical rough-in, hook-up	1 Ea.	1.6	93.60
Totals		58.5	$3,511.32

Project Size	12' x 14'	Contractor's Fee Including Materials	$7,897

Key to Abbreviations
C.Y.– cubic yard   Ea.– each   L.F.– linear foot   Pr.– pair   Sq.– square (100 square feet of area)
S.F.– square foot   S.Y.– square yard   V.L.F.– vertical linear foot   M.S.F.– thousand square feet

Section Seven
PORCHES AND ENTRYWAYS

Because a new entryway or porch can make a dramatic change in the appearance of your home, it should be treated as a major renovation. Following are some tips for planning your project.

- Entry doors can be made of steel, wood, or fiberglass. Talk to your supplier about the door most appropriate for your situation. Steel doors are energy efficient, durable and offer the best security.

- A storm door prolongs the life of your entry door by protecting it from the elements. When choosing a storm door, look for models that feature a solid inner core, low maintenance finish, and a seamless outer shell.

- A storm door takes more abuse than the door it protects. It is therefore imperative that the storm door be properly installed and secured. This may require use of larger (e.g., 3") screws to ensure that the door is secured through the trim into the jamb.

- Check the manufacturer's recommendations and restrictions of your entry door before buying a storm door.

- Breezeways, porches, and roofed-over entryways can be converted into comfortable living spaces by most do-it-yourselfers.

- If your porch or breezeway is large enough to warrant conversion, the materials order will probably be large enough to require trucking. Be sure to work out the conditions of delivery before you finalize the purchase agreement.

- Before beginning construction, be sure to inspect the condition of the existing porch or breezeway's roof and support system. Ask an expert for their opinion if you do not have experience with these systems.

- Particularly in colder climates, the floor of a converted room should be insulated for comfort and energy savings.

- Screen panel options include conventional wood-framed panels, aluminum-framed panels, fiberglass screening, and copper screening. Consider your options carefully before deciding on a final design.

FINISHED BREEZEWAY WITH SCREENS

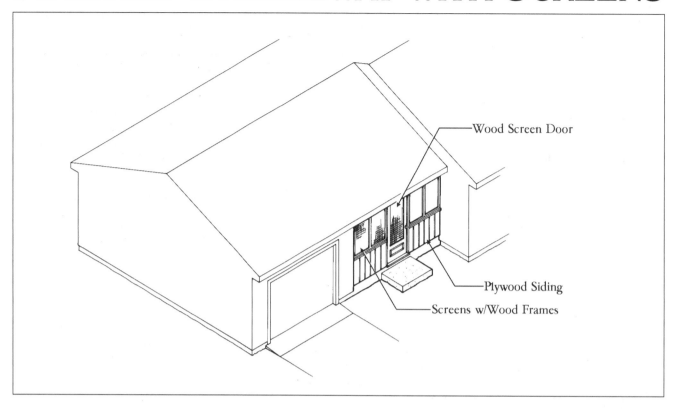

Wood Screen Door

Plywood Siding

Screens w/Wood Frames

A screened-in sitting or eating area can add pleasure and comfort to any home. Such a facility can be created without building a separate structure if an existing breezeway is used. Because the roof, side walls, and floor are already in place, the cost of remodeling a breezeway is considerably less than that of a new addition. Also, because breezeways are generally at the first-floor level, the project site is easily accessible and can usually be framed and screened with few logistical problems. Like deck projects, breezeway conversions normally cause little or no disruption to the home's routine and, therefore, can be completed at a slower pace. For these reasons, this project provides an excellent opportunity for the homeowner to cut costs by accomplishing most of the work on his or her own.

MATERIALS

The materials needed to complete this breezeway remodeling project include standard 2 x 4s to frame the knee wall and door opening, 2 x 6s for the ceiling joists (if needed), appropriate coverings for the knee wall, ceiling, and floor, and the screen panels and door. All of these materials are readily available from

building supply outlets, so it pays to compare prices from several different retailers. The screen panels are specialty items that have to be custom fitted to the opening created by the knee wall frame. If you hire a contractor to build the panels, it might be a good idea to have him or her install them as well. If you intend to build the panels on your own, develop a master plan for all of the screening to determine the most cost-effective width for the individual panels, and then base the order for the screening and wood frame on that factor. Plan to include delivery as part of the purchase order for all of the materials, as the lumber needed to frame the enclosure, the knee wall, and ceiling coverings is cumbersome and will have to be trucked.

Precise framing is required for the knee wall, as it serves as the primary support for the screen panels. If it is not level and plumb, the panels will not seat correctly and openings will result along the abutting edges. If the existing wood or concrete floor of the breezeway is uneven or slanted, you will have to shim or angle the bottom of the knee wall frame to attain a level surface on the top. Do-it-yourselfers who are inexperienced in framing procedures should get some

assistance before or during the knee wall construction, as its placement must be correct for the rest of the job to go smoothly. After the knee wall is in place and the doorway has been framed, rough-sawn plywood siding is applied to the exterior.

In this enclosure, the ceiling is prepared with 2 x 6 joists, spaced 24″ on center. Although 16″ spacing is the accepted standard in other types of construction, 24″ spacing is adequate for this project. This plan includes 3/8″ rough-sawn fir plywood for the ceiling surface, but other materials can be installed at varying costs. The price of indoor/outdoor carpeting is also included, though other types of floor materials, such as weather-resistant tiling and exterior floor paint, can also be used. The inside of the knee wall is covered with fir plywood in this model but, once again, other choices are available. Be sure to include the additional expense for any electrical work, such as a ceiling light fixture, door lamp, and watertight duplex outlets. Especially if you plan to use the enclosure at night, some provision for lighting should be made in the master plan before construction begins. If you wait until the new materials are in place, the labor cost for electrical work will be higher than if you have it done beforehand.

Particular care should be exercised in the planning, building, and installation of the screen panels. A prefabricated type is included in this estimate, but do-it-yourselfers may consider making their own. However, they are tricky to make and require the proper tools and know-how. Fiberglass and aluminum screening are available in different grades and at varying costs, so shop carefully if you are making the screens yourself. Even tension of the screening is important for both the appearance and longevity of the panels, so get some instruction from a knowledgeable person on the ins and outs of screen panel manufacture before you start. You might consider installing them in such a way that they can be conveniently removed for periodic cleaning, painting, and storage during the off-season. Prefabricated aluminum screen panels can also be used as an alternative to the suggested wood-framed units, but they have to be custom made and, therefore, cost more.

LEVEL OF DIFFICULTY

This project is a manageable undertaking for most do-it-yourselfers who have a basic knowledge of carpentry and experience in the use of tools. As noted earlier, the planning, building, and installation of the screen panels should be left to professionals or expert do-it-yourselfers. The hanging of the screen door also requires advanced carpentry skills. Generally, beginners should not attempt the frame, joist, door, carpeting, and screening tasks; and they should double the professional time estimates for all other jobs. They should seek professional help on unfamiliar tasks. Intermediates should stop short of the screen panel fabrication and carpeting; but they should be able to complete the rest of the project with the addition of about 40% to the labor-hour estimates. Experts should allow an additional 10% in time for all tasks, and more for the screen panels if they have no experience in this operation.

WHAT TO WATCH OUT FOR

The details of the wood-framed screen panel installation will vary according to the desired finish appearance of the enclosure. In a basic installation, the panels can be secured at the top and bottom to the interior or exterior face of the new wall. In a more finished installation, the panels can be placed within the framed opening and secured with stops made from trim material, such as quarter-round molding or 1 x 1 pine strips.

Brass or galvanized wood screws can be used as fasteners to allow for easy removal of the panels for maintenance or storage during the off-season. This type of finished installation will require additional materials and more time to accomplish; but the extra cost and effort may prove worthwhile.

SUMMARY

An existing breezeway already has the basic requirements of a roof and two walls that make it easily convertible to a screened-in porch. It is in a convenient work location, and many of the tasks are within reach of do-it-yourselfers. The costs can be kept down because of the relatively small quantities of materials required and the potential savings on labor.

For other options or further details regarding options shown, see

> *Finished breezeway with windows*
> *New breezeway*
> *Skylights*

Finished Breezeway with Screens

Description	Quantity/ Unit	Labor- Hours	Material
Framing for knee wall, plates 2 x 4 stock	96 L.F.	1.9	41.47
Framing for knee wall, studs, 2 x 4 x 3'	60 L.F.	1.3	18.00
Framing for doorways, 2 x 4 x 8'	72 L.F.	1.1	31.10
Ceiling joists, 2 x 6 x 24', 24" O.C.	240 L.F.	3.1	164.16
Ceiling, rough-sawn cedar, plywood, 3/8" thick, pre-stained	288 S.F.	6.8	414.72
Siding, knee wall, 2 sides 5/8" rough-sawn cedar, pre-stained	64 S.F.	1.5	92.16
Wood screens, 1-1/8" frames	200 S.F.	8.5	1,368.00
Screen door, wood, 2'-8" x 6'-9"	2 Ea.	2.7	273.60
Paint, inside wall and trim, primer	200 S.F.	1.4	12.00
Paint, inside wall and trim, 2 coats	200 S.F.	2.4	21.60
Indoor/outdoor carpet, 3/8" thick	32 S.Y.	25.6	1,123.20
Totals		56.3	$3,560.01

Project Size	24' x 12'	Contractor's Fee Including Materials	$7,776

Key to Abbreviations
C.Y.– cubic yard   Ea.– each   L.F.– linear foot   Pr.– pair   Sq.– square (100 square feet of area)
S.F.– square foot   S.Y.– square yard   V.L.F.– vertical linear foot   M.S.F.– thousand square feet

FINISHED BREEZEWAY WITH WINDOWS

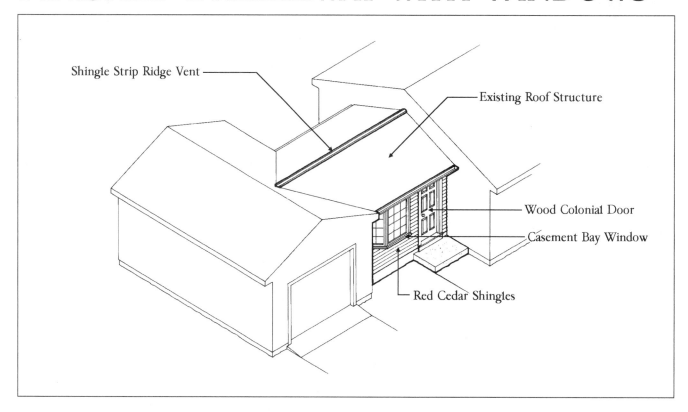

Shingle Strip Ridge Vent

Existing Roof Structure

Wood Colonial Door

Casement Bay Window

Red Cedar Shingles

Breezeways can be readily converted to interior living space because two walls, the roof, floor, and support structure are already in place. The project requires framing the two open sides and all of the tasks necessary to finish the exterior of the two new walls, including the installation of a door and windows. After the breezeway has been closed in, the necessary interior finish work is completed, as in other types of conversion projects. Because the breezeway is usually located on the first floor with immediate accessibility from the yard and garage, the logistics of materials storage and transport are minimal. This plan demonstrates how a 12' x 12' breezeway can be converted to interior living space at reasonable cost. With some or all of the work done by the homeowner, the price of the improvement can be reduced substantially. Beginners and less skilled intermediates should not attempt to complete the entire project, but they can accomplish most of it.

MATERIALS

The materials included in this plan consist of standard construction products available at building supply outlets and lumber yards. First, framing materials will be needed for the two open walls. Exterior and interior sheathing is also required, along with a door, windows, and insulation. Floor coverings, trim work, and a minimal electrical installation complete the list of supplies for the interior work.

The framing for the two open walls is constructed from 2 x 4s set 16" on center with a single bottom and a double top plate. The door and window rough openings should be built into the partition using jack studs and 2 x 8 headers. After the framing is in place, 1/2" plywood sheathing is placed on the outside. When the door and windows have been installed, the siding is put on. Try to match closely the siding on the new room to that of the rest of the house by using the same material, following existing course lines, and painting and finishing it in the same style. Additional cost can be anticipated for the matching of expensive sidings and for any special installation techniques required to install them.

The door and window units included in the plan can be altered to meet the style and design of the house. The three-lite bay unit, for example, is an attractive and practical window, but two double-hung windows may be more appropriate. As a general rule, a quality window product is a worthwhile investment. Years of energy efficiency, easy operation, and low maintenance will return the extra initial expense. The same rule holds for the selection of the exterior door unit. Before you attempt to install the door and windows, line up a helper, because these items are heavy, awkward to manipulate, and expensive to repair if they are dropped or mishandled.

The materials used in the interior of the breezeway include new 2 x 6 ceiling joists (if none are in place), strapping, insulation, and drywall for the ceiling and walls, and carpeting for the floor. As in other interior remodeling projects, many alternative materials are available at varying costs to suit your personal tastes and home decor. Remember that if portions of the exterior house and garage walls are included in the new room, they must be stripped of their exterior siding and drywalled or covered with suitable interior wall material. The garage wall may also have to be insulated and the garage doorway dressed up or relocated. All of these operations will add cost and time to the job.

Before the interior wall and ceiling coverings are installed, the rough electrical wiring and boxes have to be in place. This plan includes the basic electrical service of four wall outlets and a switch box. If you want to put in an overhead fixture, recessed or track lighting, a door lamp, or other electrical facilities, the cost for the project will increase. Hire a qualified electrician to do the work if you are unfamiliar with electrical installation procedures. Do-it-yourselfers should not attempt to tie in the new wiring to the panel, even if they are experienced enough to do all other parts of the electrical installation.

LEVEL OF DIFFICULTY

Experts, as well as intermediates with a command of basic carpentry and remodeling skills, should be able to complete most of this project. The site is convenient, the extent of remodeling is fairly limited, and the tasks are routine, with the exception of the electrical and carpeting installations. Specialty contractors should be hired to perform these operations in most cases. Even experts should weigh the savings of doing these jobs on their own against the benefits of fast, high-quality professional work. For all tasks, except the electrical and flooring work, intermediates should add 40% to the professional time, and experts 10%. Beginners should hire a contractor to do all of the exterior work, including the window and door installations. With some instruction and guidance, they can perform some of the insulating, drywalling, painting, and trim work. They should, however, leave the placement of the ceiling joists, electrical installation, and carpeting to professionals. Beginners should add 80% to the professional time estimate for any interior jobs they do take on.

WHAT TO WATCH OUT FOR

Although this breezeway project presents an economical means of converting exterior space to interior living area, it contains some potential problems that may add to its cost and increase the installation time. The doorway into the house will require attention, because it should be converted into an interior opening and trimmed with the same materials used in the new room. If the doorway is one step above the new floor level, a suitable interior step should be built. The old door from this location can be used for the new exterior installation as a cost-cutting measure.

SUMMARY

Because a breezeway already has a floor, roof and two walls, it is a practical, convenient source of new interior living space. The tasks involved in its conversion are within the realm of many do-it-yourselfers, and this aspect of the project can make it an economical undertaking as well.

For other options or further details regarding options shown, see

Bay window
Finished breezeway with screens
New breezeway
Skylights
Standard window installation

Finished Breezeway with Windows

Description	Quantity/Unit	Labor-Hours	Material
Wall framing: studs, 2 x 4, 16" O.C., 8' long	240 L.F.	7.8	357.12
Plates, single bottom, double top, 2 x 4	80 L.F.	1.6	34.56
Headers, over windows & door, 2 x 8	30 L.F.	1.4	31.32
Sheathing, plywood, 1/2" thick, 4' x 8' sheets	192 S.F.	2.7	112.90
Housewrap, spun bonded polypropylene	240 S.F.	0.5	46.08
Ceiling joists, 2 x 6, 16" O.C., 12' long	12 Ea.	0.2	8.21
Ceiling furring, 1 x 3 strapping, 16" O.C.	150 L.F.	3.4	34.20
Insulation, ceiling, 6" thick, R-19, paper-backed	150 S.F.	1.2	59.40
Insulation, walls, 3-1/2" thick, R-11, paper-backed	200 S.F.	1.4	55.20
Sheetrock, walls & ceilings, 1/2" thick, 4' x 8', taped & finished	576 S.F.	9.6	172.80
Door, exterior, 2-lite, colonial, 2'-8" x 6'-8"	1 Ea.	1.0	552.00
Window, 3-lite bay, casement type, vinyl-clad, premium	1 Ea.	2.6	1,824.00
Window, 2-lite casement window, plastic-clad, premium	2 Ea.	1.8	629.22
Siding, #1 red cedar shingles, 7-1/2" exposure	2 Sq.	7.8	242.40
Ridge vent, molded polyethylene	12 L.F.	0.6	36.72
Trim, colonial casing, windows, door, and base	120 L.F.	4.5	214.56
Paint, walls, ceiling, and trim, primer	575 S.F.	4.0	34.50
Paint, walls, ceiling, and trim, 2 coats	575 S.F.	6.8	62.10
Carpet, nylon anti-static	16 S.Y.	2.6	352.51
Pad, sponge rubber	16 S.Y.	0.9	53.57
Electrical work, 1 switch	1 Ea.	1.4	21.36
Electrical work, 4 plugs	4 Ea.	6.0	78.48
Totals		69.8	$5,013.21

Project Size: 12' x 12'

Contractor's Fee Including Materials: $10,610

Key to Abbreviations
C.Y.– cubic yard   Ea.– each   L.F.– linear foot   Pr.– pair   Sq.– square (100 square feet of area)
S.F.– square foot   S.Y.– square yard   V.L.F.– vertical linear foot   M.S.F.– thousand square feet

New Breezeway

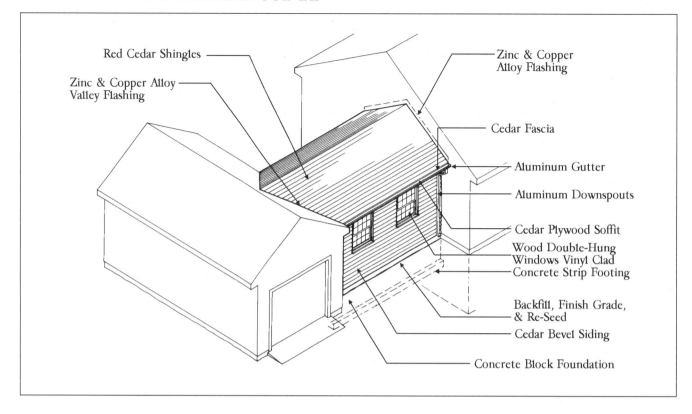

Red Cedar Shingles

Zinc & Copper Alloy
Valley Flashing

Zinc & Copper
Alloy Flashing

Cedar Fascia

Aluminum Gutter

Aluminum Downspouts

Cedar Plywood Soffit

Wood Double-Hung
Windows Vinyl Clad

Concrete Strip Footing

Backfill, Finish Grade,
& Re-Seed

Cedar Bevel Siding

Concrete Block Foundation

This plan demonstrates how the area between a detached garage and house can be converted to living space. The new structure is connected to the end wall of the house and the side wall of the garage. Depending on the configuration of the existing structures, alternative designs may be required, with accompanying variations in cost. Because the breezeway room has to be constructed within established boundaries, it is a challenging project for do-it-yourselfers, but one that can be completed by those who have a reasonable level of skill. The results will enhance your home's appearance and value, and add considerably to its living area.

MATERIALS

The exterior components of this breezeway are essentially the same as those used to construct the shell of a complete house. A foundation will support the load of the new structure, and framing, wall, and roof materials will complete the shell and make it weathertight. Materials for finishing the interior have not been listed in the plan, but you can get estimates of their cost

and installation time from the finished breezeway plans in this section.

The foundation materials consist of a concrete footing and concrete block wall. Before the footing and wall are placed, excavation of the site is required to make room for the new foundation. Because of the close quarters between the existing structures, much of the digging will probably have to be done by hand, so plan accordingly and allow plenty of time for the job. After the foundation wall is in place, 2 x 8 joists and plywood decking are placed to frame the rough floor of the new room. Before installing the joists, cover the area with a vapor barrier, then insulate between the joists before laying the subfloor. The foundation and floor must be correctly installed to ensure a firm and level starting point for the rest of the project.

The walls of the breezeway room are constructed from standard materials. Before the framing begins, allow some time to prepare the adjacent walls of the house and garage to receive the new materials. The siding must be stripped and, in some cases, a door or window may have to be removed or relocated if it will interfere with the framing process. The new partitions should be carefully laid

out, since they must be erected to a predetermined height and length as established by the existing garage and house. This plan suggests 25/32″ wood fiberboard for sheathing, but 5/8″ CDX plywood can also be used at a slight increase in cost. Make sure that the wall framing is even with the face of the concrete to allow the siding to extend over the lip of the foundation for proper weather protection.

When the rough walls have been erected, the 2 x 6 ceiling joists should be installed to keep the new walls in place before the roof is framed. The placing of the rafters is the most challenging part of the project. Advanced framing skills and know-how are required to do the job, so get help if you have not done it before. Be sure that you check the condition of the garage shingles, support, and sheathing before you begin the framing for the new roof. Do not proceed with the installation of the new roof until any deficiencies have been corrected. Additional cost and time will be required if you are careless in stripping the old roofing, so work slowly and remove only as much as you have to. After 1/2″ plywood has been placed on the rafters, the roofing material is installed to make the new structure weathertight. The cost for the roofing material and the time to install it will vary with different products. Because roofing is a specialty trade, even accomplished intermediates might consider subcontracting this part of the job. In most cases, a qualified tradesperson will do this work faster and more effectively than even expert do-it-yourselfers.

After the roof and soffit are in place, the door and windows can be installed and the exterior walls sided. This plan suggests a high-quality vinyl-clad sliding glass door unit and two matching thermopane windows. Both of these products feature the advantages of low maintenance, neat design, and energy savings. Costs will vary considerably with the choice of alternative door and window products, so shop wisely and select quality units that will fit your design standards and budget. Like the roofing, the siding material for the new room should also match or complement the material on the house and garage, and it should be appropriately finished with paint or stain to blend with the existing structures. Gutters and downspouts are an integral part of the weatherproofing of the shell, so don't cut corners by leaving them out of the exterior finishing.

LEVEL OF DIFFICULTY

Building this breezeway project can serve as an excellent opportunity to advance your building skills. All of the basic operations in exterior house construction are required to complete the job, from foundation work and framing to exterior finish work like siding and gutter installations. Because the foundation involves specialized know-how, inexperienced do-it-yourselfers should consider hiring a qualified tradesperson for this part of the project. Beginners should add 100% to the professional time for the basic tasks they take on, such as excavation, installing the sheathing, painting, and landscaping. Intermediates should add 40% to the time for the basic tasks and wall framing, and 60% for roof framing. Experts should add 10% to the professional time for all tasks, except the foundation work and roofing. The time increase for these jobs should be gauged according to the amount of experience you have in these areas.

WHAT TO WATCH OUT FOR

Hiring specialty contractors often makes sense even if you have experience and skill enough to complete the job on your own. The roofing for this project is a good example. A capable roofer can close in this structure quickly and thoroughly, even though there is a considerable amount of flashing involved. The correct installation of the valley is particularly important to prevent leaking and major water damage to the interior later on. A professional roofing job ensures that such problems won't arise.

SUMMARY

This breezeway plan demonstrates how the space between the house and garage can be converted to a comfortable interior living area. If your property is laid out in the appropriate way, much of the work can be done by do-it-yourselfers, with considerable savings in labor costs.

For other options or further details regarding options shown, see

> *Bay window*
> *Finished breezeway with screens*
> *Finished breezeway with windows*
> *Skylights*
> *Standard window installation*

New Breezeway

Description	Quantity/ Unit	Labor- Hours	Material
Excavate by hand for footing, 4' wide, 4' deep	19 C.Y.	19.0	
Footings, 8" thick x 16" wide x 32' long	2 C.Y.	5.5	218.40
Foundation, 8" concrete block, 3'-4" high	160 S.F.	14.1	301.44
Vapor barrier, 6 mil polyethylene	1.60 Sq.	0.4	5.63
Floor framing, sill plate, 2 x 6	32 L.F.	0.9	21.89
Joists, 2 x 8 x 10', 16" O.C.	160 L.F.	2.3	167.04
Ceiling and floor insulation, 6" thick, R-19	320 S.F.	1.9	157.44
Subfloor, 5/8" x 4' x 8' plywood	160 S.F.	1.9	109.44
Wall insulation, fiberglass, kraft faced, 3-1/2" thick, R-11	256 S.F.	1.3	70.66
Wall framing, 2 x 4 x 8', 16" O.C.	32 L.F.	5.1	122.88
Headers over openings, 2 x 8	48 L.F.	2.3	50.11
Sheathing for walls, wood fiberboard 25/32" thick, 4' x 8' sheets	256 S.F.	3.4	107.52
Housewrap, spun bonded polypropylene	320 S.F.	0.7	61.44
Roof framing, joists, 2 x 6 x 10', 16" O.C.	130 L.F.	1.7	88.92
Ridge board, 1 x 8 x 12'	24 L.F.	0.7	34.56
Rafters, 2 x 6 x 12', 16" O.C.	312 L.F.	5.0	213.41
Valley rafters, 2 x 6	32 L.F.	0.7	21.89
Sub-fascia, 2 x 8 x 16'	32 L.F.	2.3	33.41
Sheathing for roof, 1/2" thick, 4' x 8' sheets	240 S.F.	2.7	141.12
Felt paper, 15 lb.	240 S.F.	0.3	6.13
Shingles, #1 red cedar, 5-1/2" exposure, on roof	3 Sq.	9.6	543.60
Ridge vent, polyethylene	16 L.F.	0.8	48.96
Flashing, zinc & copper alloy, 0.020" thick	140 S.F.	7.2	547.68
Fascia, 1 x 8 rough-sawn cedar	32 L.F.	1.1	34.56
Soffit, 3/8" rough-sawn cedar plywood, stained	48 S.F.	1.1	69.12
Vented drip edge, aluminum	32 L.F.	0.6	59.14
Gutters & downspouts, aluminum	52 L.F.	3.5	75.50
Windows, wood, vinyl-clad, insulated 3' x 4'	2 Ea.	1.8	592.80
Sliding glass door, wood, vinyl-clad, 6' x 6'-10"	1 Ea.	4.0	924.00
Siding, cedar-beveled, rough-sawn, stained	220 S.F.	7.3	704.88
Backfill, finish grade and re-seed	1 C.Y.	0.6	
Totals		109.8	$5,533.57

Project Size	10' x 16'	Contractor's Fee Including Materials	$13,072

Key to Abbreviations
C.Y.– cubic yard   Ea.– each   L.F.– linear foot   Pr.– pair   Sq.– square (100 square feet of area)
S.F.– square foot   S.Y.– square yard   V.L.F.– vertical linear foot   M.S.F.– thousand square feet

ENCLOSED PORCH WITH SCREENS

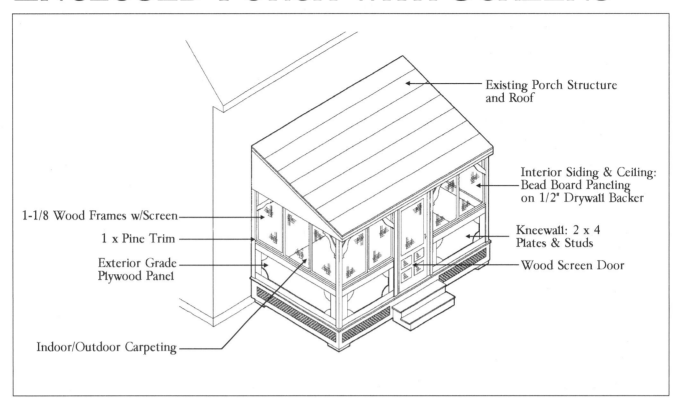

Existing Porch Structure and Roof

Interior Siding & Ceiling: Bead Board Paneling on 1/2" Drywall Backer

1-1/8 Wood Frames w/Screen

1 x Pine Trim

Exterior Grade Plywood Panel

Kneewall: 2 x 4 Plates & Studs

Wood Screen Door

Indoor/Outdoor Carpeting

A roofed-over exterior space, such as a porch, can be screened in for a modest investment of time and money. It is an ideal project for do-it-yourselfers of all ability levels, as only a limited amount of specialized work is required.

MATERIALS

As in other porch and breezeway conversions, the amount and type of materials are determined by the fixed dimensions of the structure and the style of the house. If the dimensions of your project differ from this model, the cost will have to be adjusted. All of the products are conventional construction materials that can be purchased at competitive prices from most building supply outlets.

Before the new materials are installed, the existing porch should be thoroughly inspected and its condition evaluated. If the roof or floor is in need of repair, it should be fixed. Some additional costs may be incurred during this stage of the project if repairs are needed. Any extra work, like installing a gutter and downspout, finishing the fascia and soffit, or building a new set of stairs should also be completed at this stage.

The frame for the knee wall of the porch is constructed from standard 2 x 4s between the existing roof support posts. Unless there is good reason to move them, all support posts should be kept in position to maintain the structural integrity of the roof support system. In situations where a post has to be relocated or replaced, temporary bracing must be provided. Extra cost and labor-hours will also have to be added. Inexperienced do-it-yourselfers should consult a contractor if extensive post relocation or replacement is required. Once the framing has been completed, the exterior walls are covered with a single layer of sheet siding, or sheathed and then covered with appropriate cedar, vinyl, or aluminum exterior finish. A decorative cap with trim for the knee wall is an option that can be applied at extra cost. With some ingenuity, the trim for the cap can be designed to serve as the bottom stop for the screen panels.

Inside the enclosure, 2 x 6 ceiling joists are put into place, and the finish coverings for the ceiling, walls, and floor are applied. Many options are available for the type and quality of materials used in the interior, depending on the degree of finish work desired. If you want the

effect of a rustic finish, the ceiling can be left open to the rafters, the inside of the exterior walls left uncovered, and the existing floor repainted or left as is. The cost of the project will be reduced significantly in this case, and the finishing touches could be added at a later date. The moderate finish work suggested in this plan provides covering for the ceiling, knee wall, and floor, but not the interior wall. Sheet goods, like rough-sawn cedar plywood, install quickly and provide an attractive, low-maintenance finish. Barnboard, tongue-and-groove, beadboard and plank coverings are also available. They are usually more expensive and require more time to install. Electrical work is not included in this plan, but you might consider installing outdoor-approved outlets and an overhead or wall lighting fixture. If you plan to include these electrical amenities, be sure that the rough-in is completed before the wall and ceiling coverings are put into place.

The screen panels for the new porch are the most important components in the structure. They deserve special consideration before you decide on their design and the best installation method. In most cases, removable wood-framed panels are the most effective system to use. They may cost more and take longer to fabricate than permanent models, but they have many advantages. They can be removed and stored when the porch is not in use and can be cleaned and painted easily on the ground and then put in place. Repair work, restretching, and replacement can also be performed more conveniently because of the modular design of these panels. Beginners might consider hiring a carpenter to manufacture the panels and, later, to install the screen door. Both of these operations require some know-how, and precision carpentry skills are a prerequisite. Intermediates and experts who intend to make the screen panels should weigh the benefits of the potential savings in labor costs against the expediency and professional workmanship of an experienced carpenter. Screen panels are also available in standard factory-prepared models with both wood and aluminum frames, but

they may require special ordering. If you intend to make the panels on your own, be sure to read up on the techniques involved in their manufacture and consult a knowledgeable person for tips. Priming and painting the stock for the frames will save time in the finishing stages of the project.

LEVEL OF DIFFICULTY

The carpentry skills required to complete this enclosure are within the realm of most do-it-yourselfers. The heavy exterior work is already done, and the filling in of the shell is all that is required. Professional help may be needed if extensive reconditioning is necessary for the roof, floor, or support system, but minor problems can be corrected by the do-it-yourselfer. Beginners should consider hiring a contractor or retailer who specializes in screen fabrication to manufacture the screen panels. For all

other tasks, beginners should add 100% to the labor-hour estimates. Intermediates and experts who tackle the screen panel manufacture should add extra time to their normal work rate to allow for the precise nature of the operation. For all other procedures, they should add 50% and 20%, respectively.

WHAT TO WATCH OUT FOR

Before you make a decision on the screen panel design and installation method, investigate all possibilities. Conventional wood-framed panels meet durability standards, but they require maintenance and are prone to bowing and warping if they are poorly made or installed. Aluminum-framed panels may not be as attractive as wood-framed units, but they will last longer, and periodic cleaning is the only required maintenance. Aluminum screen material is long-lasting and capable of withstanding abuse, but it discolors with corrosion after a few years. Fiberglass screening tends to stretch and bow more easily than aluminum, but it retains its original luster and attractive appearance longer. Copper screening performs and looks better than both aluminum and fiberglass, but it is very expensive. The type of screening should be carefully considered before the final design has been determined.

SUMMARY

A screened-in porch can provide years of enjoyment for you and your family. If the basic roof and support structure are already in place, the facility can be completed at a reasonable cost, as the work can be done by most do-it-yourselfers.

For other options or further details regarding options shown, see

Enclosed porch with sliding glass doors
Enclosed porch with windows
Entryway steps
Farmer's porch

Enclosed Porch with Screens

Description	Quantity/ Unit	Labor- Hours	Material
Framing for knee wall, 2 x 4 plates	96 L.F.	2.4	41.47
Studs, 2 x 4 x 3', 16" O.C.	84 L.F.	1.2	36.29
Framing, for doorway, 2 x 4 x 8'	32 L.F.	1.0	13.82
Ceiling joists, 2 x 6 x 8', 16" O.C.	88 L.F.	1.1	60.19
Ceiling & kneewall stiffener, gypsum board, 3/8"	224 S.F.	1.8	51.07
Ceiling and knee walls, bead board paneling, 1/4", prefinished	224 S.F.	8.5	301.06
Siding, exterior, for knee wall, plywood panels, 7/16", smooth	72 S.F.	1.7	63.07
Wood screens, 1-1/8" frames, pre-fabricated	140 S.F.	6.0	957.60
Screen door, wood, 2'-8" x 6'-9"	1 Ea.	1.8	277.20
Trim, 1 x 4 pine	72 L.F.	2.9	57.89
Indoor/outdoor carpet, 3/8" thick	112 S.F.	2.0	274.18
Paint, screens, primer and 2 coats, oil base, brushwork	10 Ea.	13.3	28.56
Panels, 2 coats, oil base, brushwork	72 S.F.	0.9	9.50
Door, primer coat, oil base, brushwork	1 Ea.	0.8	2.46
Door, 2 coats, oil base, brushwork	1 Ea.	5.3	15.89
Totals		50.7	$2,190.25

Project Size	8' x 14'	Contractor's Fee Including Materials	$5,557

Key to Abbreviations
C.Y.– cubic yard   Ea.– each   L.F.– linear foot   Pr.– pair   Sq.– square (100 square feet of area)
S.F.– square foot   S.Y.– square yard   V.L.F.– vertical linear foot   M.S.F.– thousand square feet

ENCLOSED PORCH WITH SLIDING GLASS DOORS

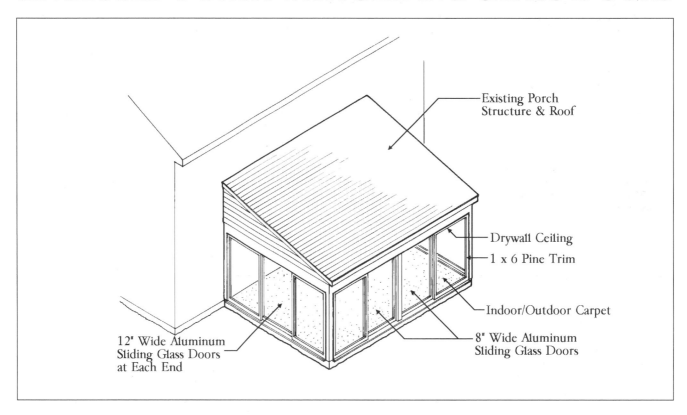

Existing Porch Structure & Roof

Drywall Ceiling

1 x 6 Pine Trim

Indoor/Outdoor Carpet

12" Wide Aluminum Sliding Glass Doors at Each End

8" Wide Aluminum Sliding Glass Doors

Open, attached porches can be remodeled with a standard approach or converted to a more deluxe interior space. This plan presents the latter type of improvement and features sliding glass doors to provide the benefits of natural light and fresh air. If you want to increase the light and ventilation still further, you might invest a bit more by adding skylights. Generally, this is a moderate-to-difficult project because of its size and the specialized skills required for several of the tasks.

MATERIALS

As in other porch plans included in this section, it is assumed that a weathertight, sturdy roof and a solid foundation are already in place. The materials list, therefore, includes only products that are required to fill in the walls, finish the interior, and otherwise enhance the porch's function and appearance. Because many variables exist, the particulars of an individual project will modify this estimate. The size, roof style, and the amount of reconditioning required for the existing structure will affect the overall cost, as will the type and extent of finish work that goes into the new room. The basic structural materials are readily

available from building supply retailers, but some sliding glass door models may require special ordering.

Before any new materials are installed, the existing structure should be thoroughly inspected and evaluated. The roof should be dry and its sheathing solid and free from rot or deterioration. Carefully check the flashing along the seam between the roof and the wall of the house, and examine the underside of the roof's subsurface and rafters. If the roof is held up by a post-and-beam support system, check the condition and alignment of the supporting members. This plan depicts a concrete slab as the floor and foundation for the porch, but a concrete block or poured-in-place foundation with a wood floor may better describe your porch's arrangement. Regardless of the type, check and evaluate its condition before continuing. If deficiencies or weaknesses are found in the existing porch's components, correct them before proceeding with the project. Extensive reconditioning may require the services of a roofing or foundation specialist.

The sliding glass doors are the primary components of the conversion, so give careful consideration to their selection

and installation. Generally, the extra expense for top-quality units is worthwhile, as they will provide years of reliable, weathertight service. The most important aspect of the door installation is the correct sizing and precise leveling, plumbing, and squaring of the rough openings. The units are available in various standard sizes, including the 6' and 8' models used in this plan. Odd sizes require special ordering and cost more because they must be custom made. The estimate given in this plan is for a high-quality slider unit. Because of its expense, it should be handled carefully, with plenty of time allowed for installation.

The doorway from the porch area to the house may require some attention after the sliders and interior finish are put into place. If the porch area is to become a new room in the house, you may want to enlarge the existing doorway, build a rough frame with appropriate jack studs and headers, and then trim the opening. A slider unit or French door can also be installed in this location. If the porch is to be winterized but will still be considered more as an exterior area, the old doorway can be left as is with no additional work required. The exterior siding and any existing windows can be left in place or removed, depending on the extent of finish desired in the new porch. Your choice of treatment for this area will have a substantial impact on the cost of the project. Plan carefully if the access is to be redesigned.

LEVEL OF DIFFICULTY

Because of the specialty work involved in the sliding glass door installations, beginners and intermediates with limited remodeling experience should consider hiring professionals to complete this task. Even experts and experienced intermediates should seek some assistance. The components are expensive, and precise workmanship is required to ensure that the results of the installation are commensurate with the investment in materials. The rest of the project should be within the reach of most do-it-yourselfers. Beginners should double the professional time for all work and seek advice when needed. Intermediates and experts should add 50% and 20%, respectively, to the labor-hour estimates for basic procedures, and 60% and 30% for the sliding door installations. All do-it-yourselfers will have to add both time and expense for extra interior finish work if the porch is to be integrated into the first-floor plan of the house.

WHAT TO WATCH OUT FOR

If the new glassed-in porch is to be used essentially as an interior facility, local building codes may require a minimum number of electrical outlets. Extra lighting fixtures and wall switches may also be desired by the homeowner. These necessities and amenities should be provided for before the ceiling and walls are framed and covered, to save time and money and to simplify the procedure. Space may be required between the rough framing members of the slider units if wiring and switch or outlet boxes are to be installed in this location. Also, supply wires for the wall switches, outlets, and overhead lighting fixtures must be run above the finished ceiling and then fed to the electrical service panel. A qualified electrician should tie in the new circuit to the electrical supply and perform any parts of the installation that are too difficult for the do-it-yourselfer. Beginners should hire a professional for all of the electrical work.

SUMMARY

This glassed-in porch enclosure will add an attractive and comfortable new living space to your home. If the conditions of the existing roofed-over structure are right, this deluxe conversion can be performed for a reasonable cost.

For other options or further details regarding options shown, see

> Electrical system: Light fixtures*
>
> Electrical system: Receptacles*
>
> Enclosed porch with screens
>
> Enclosed porch with windows
>
> Farmer's porch
>
> Interior doors*
>
> Patio & sliding glass doors
>
> Skylights
>
> * In Interior Home Improvement Costs.

Enclosed Porch with Sliding Glass Doors

Description	Quantity/Unit	Labor-Hours	Material
Demolition, existing door 3' x 7', and enlarge opening	1 Ea.	0.5	
Door frame demolition, including trim, wood	1 Ea.	0.5	
Frame opening into house, 2 x 4 studs	32 L.F.	1.0	13.82
Header over opening, 2 x 10 stock	20 L.F.	1.0	29.76
Trim at opening, 1 x 6 jamb, stock pine	24 L.F.	0.8	17.57
Casing, stock pine, 11/16" x 2-1/2"	24 L.F.	0.8	33.41
Sliding glass doors, aluminum, 5/8" insul. glass, 8' wide	2 Ea.	10.7	3,000.00
Sliding glass doors, aluminum, 5/8" insul. glass, 12' wide	2 Ea.	12.8	5,040.00
Miscellaneous wood blocking for ceiling, 2 x 4 stock	48 L.F.	1.5	20.74
Insulation, fiberglass, kraft-faced batts, 6" thick, R-19	192 S.F.	1.1	94.46
Drywall at ceiling, 1/2" thick, 4' x 8' sheets, taped and finished	192 S.F.	3.2	57.60
Trim, cove molding, 9/16" x 1-3/4"	60 L.F.	1.8	39.60
Indoor/outdoor carpet, 3/8" thick	189 S.F.	3.4	462.67
Paint, trim, wood, incl. puttying, primer, oil base, brushwork	50 L.F.	0.6	1.20
Trim, wood, incl. puttying, 1 coat, oil base, brushwork	50 L.F.	0.6	1.80
Ceiling, primer, oil base, roller	192 S.F.	0.8	9.22
Ceiling, 1 coat, oil base, roller	192 S.F.	1.2	11.52
Totals		42.3	$8,833.37

Project Size	12' x 16'	Contractor's Fee Including Materials	$14,649

Key to Abbreviations
C.Y.– cubic yard   Ea.– each   L.F.– linear foot   Pr.– pair   Sq.– square (100 square feet of area)
S.F.– square foot   S.Y.– square yard   V.L.F.– vertical linear foot   M.S.F.– thousand square feet

ENCLOSED PORCH/MUDROOM/LAUNDRY

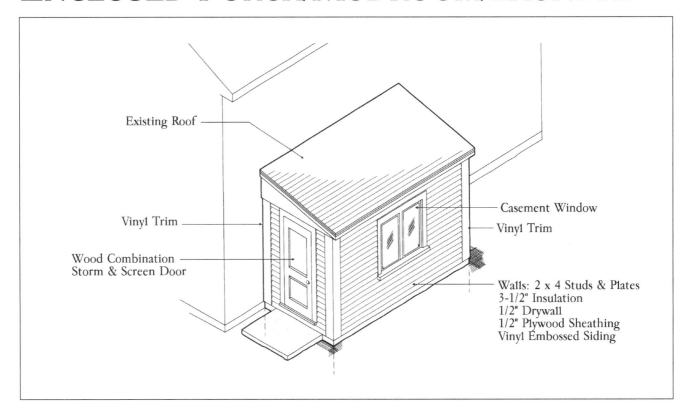

Existing Roof

Vinyl Trim

Wood Combination
Storm & Screen Door

Casement Window

Vinyl Trim

Walls: 2 x 4 Studs & Plates
3-1/2" Insulation
1/2" Drywall
1/2" Plywood Sheathing
Vinyl Embossed Siding

Like breezeways, porches and roofed-over entryways can be converted into usable interior spaces. This plan demonstrates how a 6' x 12' porch can become a small but practical mudroom/laundry. With the roof, floor, and support system already in place, the conversion can be accomplished quickly and economically by most do-it-yourselfers.

MATERIALS

The amount of required material will depend on the dimensions of the existing structure, the style and design of the house, and the extent to which the porch must be prepared and reconditioned.

Two critical areas of the existing porch should be examined before you begin construction: the roof and the support system. Especially in older structures, these two vital components may need reconditioning or replacing if they are to protect the new interior and carry the load of the new materials. Examine the flashing and general condition of the roof covering and rafters for signs of leakage and deterioration. If you can crawl under the structure, make a thorough inspection of the floor joists, decking, and

foundation. Rot, insect infestations, moisture, and weakness of the frame and flooring are problems that must be corrected before the project continues. If you cannot get under the porch, the condition of the floor and its frame can be checked by removing some of the floor boards in several places. These preliminary measures take time and may involve extra expense if reconditioning work is called for. If a problem looks serious, hire a specialist to investigate further and correct the situation. In short, be sure that the existing porch roof, floor, and foundation are in top condition before you install any new material. After the porch has been reconditioned, the 2 x 4 walls are framed and the exterior of the structure is sheathed, equipped with a window and door, and then sided. In most cases, the framing will fill areas between the existing roof support posts, so allow extra time, and possibly cost, for extra cutting and measuring. If any of the posts have to be replaced or relocated, be sure to provide temporary roof support posts during the framing operation. The 1/2" plywood sheathing and new window and door units are put into place once the framing is complete. This plan includes a

medium-sized thermopane casement window unit, but double-hung or jalousie models may be more appropriate for your home. Embossed vinyl siding is suggested in this model, but cedar, aluminum, or a different style and grade of vinyl may better suit your home. Your total cost will vary somewhat with these substitutions. Refurbishing or changing the exterior trim or adding a gutter and downspout, if required, will also add time and expense to the project.

Within the enclosure, the old exterior wall must be prepared for new materials. Ceiling joists and insulation must be installed, and finish coverings applied to the ceiling, walls, and floor. The wall preparation involves stripping the old siding and reconditioning the existing entryway for the new interior door. The exterior ends of the new ceiling joists must be tapered to fit the existing roof line and the interior ends plated and lagged into the side of the house. The ceiling and new exterior walls should be insulated to the recommended R-value ratings for your geographical area. Before drywall or paneling is applied, all rough electrical and plumbing work should be completed. You can place the underlayment for the carpeting or vinyl flooring directly on the old porch floorboards, but be sure to use threaded flooring nails and then level the surface before the finish is applied. A layer of building paper between the old floor and underlayment will reduce the chance of moisture penetration while helping to eliminate squeaks. Install an air infiltration barrier before you install the new siding.

LEVEL OF DIFFICULTY

This porch conversion can be performed by all do-it-yourselfers. Because the site is conveniently located and much of the heavy exterior work has already been done, the project is considerably less difficult and expensive than a complete room addition of the same size. Assuming that only minimal reconditioning is needed, the most challenging operations are the wall framing and ceiling joist placement. If you are unfamiliar with framing, the procedures can be learned from an experienced carpenter. As in other conversion projects, like the breezeway, attic, and garage plans, the work can usually be completed at a comfortable pace without disturbing the routine of the house. Beginners should double the professional time estimates throughout the project and get help before they begin tasks that are new to them. Intermediates should add 50% to the estimated labor-hours, and experts 20%. All do-it-yourselfers should consider hiring an electrician and a plumber for these aspects of the job.

WHAT TO WATCH OUT FOR

Installing the rough plumbing and insulating the floor can be challenging, even if the floor is accessible from underneath the structure. You may have to work from a prone position and under cramped conditions. Space problems may require that batts, rather than rolled material, be used. Once the insulation is in place, some means of permanent support should be installed. Strapping can be nailed across the joists at 16″ intervals, or chicken wire or inexpensive screening can be stapled to the supports to cover the newly insulated area. If the floor is inaccessible from below, styrofoam sheets can be placed between sleepers made from strapping or 2 x 4s on the old floor surface. With this method, however, new decking as well as underlayment must be installed before the finished floor is laid. In this case, you may also have to make adjustments for the new floor height at both entryways. Although the materials and extra work required to insulate the floor will add cost and time to the project, the benefits of a warm floor and reduced heating costs may make the investment worthwhile.

SUMMARY

Converting and winterizing an existing porch like this one can increase the convenience of your home without the large expense of a complete room addition. Because the roof and support system are already in place, many do-it-yourselfers can handle the finishing tasks at their convenience, without having to deal with the challenges of a completely new structure.

For other options or further details regarding options shown, see

> *Enclosed porch with screens*
>
> *Enclosed porch with sliding glass doors*
>
> *Farmer's porch*
>
> *Flooring**
>
> *Standard window installation*
>
> ** In Interior Home Improvement Costs.*

Enclosed Porch/Mudroom/Laundry

Description	Quantity/ Unit	Labor- Hours	Material
Door demolition, exterior, 3′ x 7′ high, single	1 Ea.	2.0	
Door frame demolition, including trim, wood	1 Ea.	1.0	
Siding demolition, shingles	96 S.F.	2.2	
Wall framing, studs and plates, 2 x 4 x 8′, 16″ O.C.	24 L.F.	3.8	101.38
Headers over openings, 2 x 6 stock	24 L.F.	1.1	16.42
Sheathing, plywood, 1/2″ thick, 4′ x 8′ sheets	192 S.F.	2.7	112.90
Housewrap, spun bonded polypropylene	240 S.F.	0.5	46.08
Ceiling joists, 2 x 6 x 12′, 16″ O.C.	60 L.F.	0.8	41.04
Insulation, ceiling & floor, 6″ thick, R-19, foil faced	144 S.F.	0.7	70.85
Walls, 3-1/2″ thick, R-11, foil faced	160 S.F.	0.8	65.28
Drywall, walls and ceiling, 1/2″ thick, 4′ x 8′, taped and finished	324 S.F.	5.4	97.20
Door, wood entrance, birch, solid-core, 2′-8″ x 6′-8″	1 Ea.	1.0	89.40
Lockset, residential, exterior	1 Ea.	1.0	141.60
Door, prehung exterior, comb. storm and screen, 6′-9″ x 2′-8″	1 Ea.	1.1	112.20
Windows, wood casement, insulating glass, 4′ x 4′	1 Ea.	0.7	648.00
Trim, colonial casing, windows, door and base	70 L.F.	2.8	118.44
Siding, vinyl embossed, white, 8″ wide, w/trim	40 S.F.	1.3	26.40
Flooring, vinyl sheet, 0.065″ thick	72 S.F.	2.9	216.86
Flooring, adhesive cement, 1 gallon covers 200 to 300 S.F.	1 Gal.		18.00
Recessed box with washer/dryer hook-up	1 Ea.	0.4	46.80
Rough-in, supply, waste and vent for washer box	1 Ea.	2.3	86.40
Dryer vent kit	1 Ea.	0.4	144.00
Electric baseboard heat	4 L.F.	4.8	235.20
Paint, trim, wood, incl. puttying, primer coat, oil base, brushwork	88 S.F.	1.1	2.11
Trim, wood, incl. puttying, 2 coats, oil base, brushwork	88 S.F.	1.8	5.28
Walls and ceilings, primer, oil base, brushwork, smooth finish	384 S.F.	2.7	23.04
Walls and ceilings, 2 coats, oil base, brushwork, smooth finish	384 S.F.	4.5	41.47
Totals		49.8	$2,506.35

Project Size	6′ x 12′	Contractor's Fee Including Materials	$6,037

Key to Abbreviations
C.Y.– cubic yard   Ea.– each   L.F.– linear foot   Pr.– pair   Sq.– square (100 square feet of area)
S.F.– square foot   S.Y.– square yard   V.L.F.– vertical linear foot   M.S.F.– thousand square feet

FARMER'S PORCH

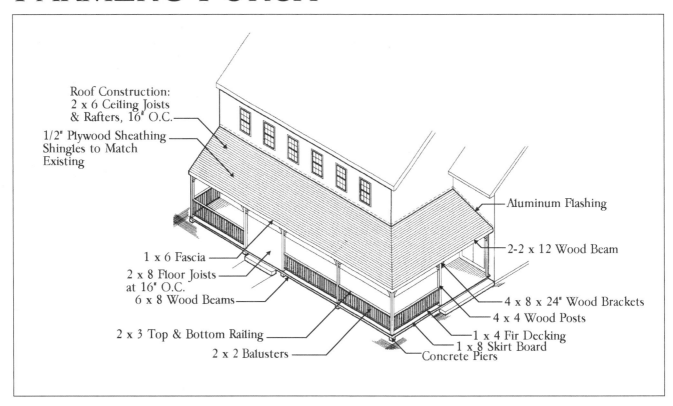

Roof Construction:
2 x 6 Ceiling Joists
& Rafters, 16" O.C.

1/2" Plywood Sheathing
Shingles to Match
Existing

Aluminum Flashing

2-2 x 12 Wood Beam

1 x 6 Fascia

2 x 8 Floor Joists
at 16" O.C.

6 x 8 Wood Beams

4 x 8 x 24" Wood Brackets

4 x 4 Wood Posts

1 x 4 Fir Decking

2 x 3 Top & Bottom Railing

1 x 8 Skirt Board

2 x 2 Balusters

Concrete Piers

New home buyers today are often looking for a home that represents values of an age gone by. Many new homes are designed to have an old-fashioned "main street" look. The addition of a farmer's porch to your existing house can greatly enhance the home's appearance, giving it charm, character, and additional outdoor living space.

MATERIALS

The materials used in this project are basic decking products available at most building supply outlets. These and any ornamental accessories can be made from a variety of wood species, and prices will depend on the type you choose. When you purchase the materials, select them yourself if possible, looking for straight deck boards and sound support members. You should also clarify the conditions of delivery at the time of purchase, since most suppliers will make curb deliveries free of charge for a complete deck order.

All of the wood products for this porch have been chosen for their durability in exterior applications where they are subjected to the elements. The support system for the porch consists of footings,

4 x 4 posts, and the frame, which is comprised of 2 x 6 joists and headers. Before you begin the excavation for the footings, lay out their locations accurately to ensure a square and level support system for the rest of the structure. Several methods can be employed to establish the location of the support posts and their footings, but one of the most accurate is to approximate the layout with batter boards at the four corners, then establish, by the use of stringlines, equal diagonals. Failure to place the footings and posts precisely at square with each other and with the side of your house will adversely affect the rest of the project.

Double-check the post locations before you set them in concrete and use cylindrical "tube" forms for the foundations. These are available at most building yards and are well worth the price. Review the layout of the 2 x 6 frame as well, to be sure the double joists are correctly aligned with the 4 x 4 supports.

The decking in this plan consists of 1 x 4 fir laid across the frame and fastened with galvanized nails. However, decking material of another thickness, width, or edge type can also be used. Remember that the decking itself is the most visible

component of the structure, so additional expense for aesthetic reasons may be a worthwhile investment. This porch plan includes a basic handrail and baluster-type railing system.

Fancy Victorian-style details such as gingerbread trim can be purchased at most home centers or renovation supply stores and easily added to the porch if the style of your house calls for this level of detail.

LEVEL OF DIFFICULTY

This farmer's porch should be undertaken only by an expert. Beginners and intermediates should hire a professional, or work alongside one for guidance and assistance. If you choose to install the porch yourselfer, add 100%–200% to the time estimates if you are an experienced do-it yourselfer, and 300% if you are a beginner working with someone more experienced.

WHAT TO WATCH OUT FOR

Because this porch is built at ground level with limited access to the ground underneath, precautions should be taken to restrict unwanted vegetation from growing in the area. Some weeds and hardy grasses can establish themselves under the porch, eventually restricting ventilation and creating an eyesore.

Porches that are built in sunny locations are particularly susceptible to this problem. To prevent unwanted growth, remove about 3" to 4" of topsoil from the area to be covered by the porch and then cover it with a layer of polyethylene. Place enough gravel over the plastic to bring the entire area to grade level. This procedure will increase the cost of the job, but will prevent a problem that is difficult to correct after the porch is in place.

Perhaps the most difficult part of the construction is tying the porch roof into the existing roof structure. This is where careful establishing of details and heights is critical. Depending on exactly where you choose to attach the new structure, the amount of work to be done to the existing structure will vary. Securing the new structure to a flat wall requires removal of siding, as well as cutting and patching to match existing surfaces. Be careful to plan the pitch of the roof so that there is room for proper flashing below second floor windows.

Farmer's Porch

Description	Quantity/Unit	Labor-Hours	Material
Layout, excavate post holes	1 C.Y.	2.5	
Concrete, field mix. 1 C.F. per bag, for posts	10 Bags		74.40
Forms, round fiber tube, 1 use, 8" diameter	30 L.F.	6.2	65.16
Porch material, fir posts, 4 x 4 x 8'	48 L.F.	1.7	123.26
Wood beam, 6 x 8	40 L.F.	1.1	124.80
Joists, 2 x 8	192 L.F.	8.5	156.67
Fir decking, 1 x 4	256 S.F.	7.5	583.68
Double 2 x 12 wood beam	32 L.F.	0.8	65.28
Rafters, 2 x 6, 16" O.C.	216 L.F.	3.5	147.74
Ceiling joists, 2 x 6, 16" O.C.	216 L.F.	2.4	147.74
Plywood sheathing, 1/2" CDX	360 S.F.	3.1	190.08
Fascia, 1 x 6, pine	32 L.F.	1.1	19.58
Skirt board, 1 x 8, pine	32 L.F.	1.1	52.22
Wood brackets	10 Ea.	1.9	366.00
Railing material, 2 x 3, fir, top and bottom rails	48 L.F.	1.3	112.32
Roofing material, asphalt felt paper, 15#	360 S.F.	0.8	8.64
Shingles, asphalt strip, organic, class C, 235 lb. per square	3.60 Sq.	5.8	162.00
Flashing, aluminum, 0.019" thick	40 S.F.	2.2	36.96
Brick steps	16 L.F.	5.3	37.80
Joist and beam hangers, 18 ga. galvanized	48 Ea.	2.3	24.77
Nails, galvanized	50 Lbs.		70.80
Bolts, 1/2" x 7-1/2" lag bolts, square head, w/nut and washer	24 Ea.	1.5	27.36
Totals		60.6	$2,597.26

Project Size 24' x 16'

Contractor's Fee Including Materials $6,635

Key to Abbreviations
C.Y.– cubic yard   Ea.– each   L.F.– linear foot   Pr.– pair   Sq.– square (100 square feet of area)
S.F.– square foot   S.Y.– square yard   V.L.F.– vertical linear foot   M.S.F.– thousand square feet

ROOFING OPTIONS
Cost per Square, Installed

Description	
Asphalt Roll Roofing	$144.00
Asphalt Shingles	$ 91.50
Cedar Shakes	$280.00
Clay Tile	$745.00

SUMMARY

Porches are functional entrances protected from rain and snow and are pleasant gathering places in warm weather for family and friends. Because they are architectural in nature, care must be taken to design a porch that complements your house yet satisfies your space requirements. Paying attention to height and width proportions is critical. A variety of materials can be used; low-maintenance products should be considered, since you want to spend time enjoying the porch, not maintaining it.

For other options or further details regarding options shown, see

Enclosed porch with windows

Enclosed porch with screens

Enclosed porch with sliding glass doors

Entryway steps

Main entry door

ENTRYWAY STEPS

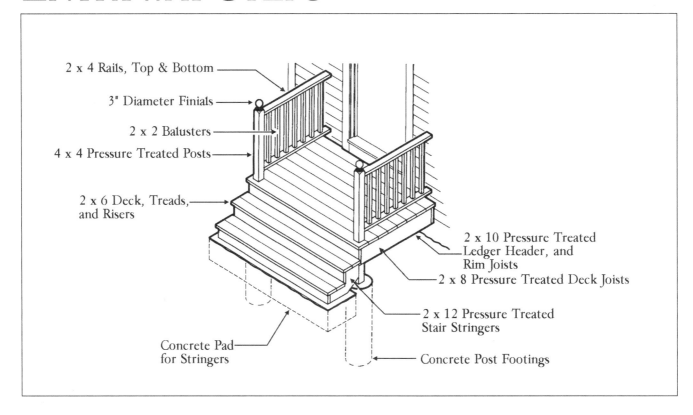

2 x 4 Rails, Top & Bottom

3" Diameter Finials

2 x 2 Balusters

4 x 4 Pressure Treated Posts

2 x 6 Deck, Treads, and Risers

Concrete Pad for Stringers

2 x 10 Pressure Treated Ledger Header, and Rim Joists

2 x 8 Pressure Treated Deck Joists

2 x 12 Pressure Treated Stair Stringers

Concrete Post Footings

If your house is over 25 years old, its wood porch, deck, and steps could very well be in need of repair or replacement. Prior to the development and marketing of pressure-treated wood, all such outdoor structures were built of standard construction-grade lumber, which is eventually destroyed by years of weather and wood-boring insects.

Moisture becomes trapped in joints, between deck boards and joists, and is absorbed by the end grain of cut boards. It may also collect around the bases of porch and deck posts, and can be absorbed directly from the earth wherever the wood contacts the ground. All this wet and dampness encourages the growth of wood-rotting fungi and bacteria, and attracts insects, such as termites and carpenter ants, which can make fast food out of your home's most massive structural members. Lumber treated with chromated copper arsenate (CCA) resists such moisture-related rot and bugs, and has for this reason become the standard material for use in the repair or replacement of outdoor wooden structures. Steps such as these are typical,

and can be customized in many ways to match the style of virtually any home.

MATERIALS

One of the negative points about pressure-treated lumber is its appearance. The most common CCA-treated wood is southern yellow pine, a grainy species the sapwood of which is easily penetrated by pressure treatment. The CCA preservative leaves the wood with a green tinge, which, combined with the broad, irregular grain, gives the pine a look that lacks elegance and one that many homeowners find unappealing and unattractive. Builders and remodelers often solve this problem by using pressure-treated lumber for the mostly hidden framing members and structural supports and using more attractive wood species, such as redwood or cedar, for the more visible decking, railings, and trim. Those pressure-treated components that do show – posts and end joists – can be covered with the other wood, left to weather (which will turn it a neutral tannish-gray), or simply stained or painted.

If you are replacing an existing set of steps, you can draw a plan of its construction details, taking the lines and measurements from the structure before you demolish it. Changes or improvements in the design can be made easily as you make the drawing. If you are building an entirely new structure, lay out the dimensions directly on the house wall below the door opening. Draw your design on paper, indicating all measurements and dimensions, visualizing as you do so the exact building sequence, and noting potential trouble spots.

Dig the post holes and set the points in place, plumbing and bracing them diagonally with lengths of 1 x 3 strapping and stakes. Mix and pour the concrete around the posts to anchor them securely. Taking careful measurements, lay out for the concrete pad that will serve as the footing for the stair stringers. After excavating, build a level form around the opening, and mix and pour the concrete.

Attach the ledgerboard to the house wall below the door opening, securing it with 6" lag screws. Construct the perimeter of the frame (end joists and header) and install the inner joists using steel hangers, 24" on center. Mark, cut, and attach the four stair stringers to the header, using the metal framing anchors. The cedar risers, treads, and deck boards can now be cut and installed.

The two balustrades each consist of a top and bottom rail, with a series of 2 x 2 balusters laid out with a spacing of about 1-1/2″ between each of them. Once assembled, the balustrades are attached to the house wall and to the posts, which can then be cut off at the appropriate height (about 38″ above the decking) and topped with the finials.

There are, of course, countless variations that can be worked out on this basic design. One of the more functional ones would be the addition of a small pitched roof over the deck, to provide some weather protection for the entry door and for the people using it.

A popular alternative to wood is some form of masonry step, the most durable being a poured concrete base overlaid with brick or stone. Metal or wrought iron railings are generally preferred with brick or masonry as they can be more securely anchored than a wood handrail system.

LEVEL OF DIFFICULTY

Deck building is a very popular do-it-yourself project, and this wooden step structure is really nothing more than a miniature deck. Careful planning and layout, as always, are essential to successful results. A beginner should have experienced help during this phase, and at least some general guidance during the construction, especially when marking and cutting the stair stringers.

Framing a deck is basic rough carpentry. Pressure-treated lumber can be difficult to work with; it tends to be splintery, is somewhat hard to cut (a carbide-tipped saw blade is essential), and nailing into it can result in a lot of bent nails and muttered imprecations (use a 20- or 25-ounce hammer). As is the case with many outdoor, heavy lumber structures, the finish work required here is not quite as fussily demanding as interior finish carpentry can be. This is not to excuse sloppy workmanship, but decks are somewhat forgiving, and you can get away with less than perfect cuts and joints that would be unacceptable inside the house. Adding a roof to this design would make it quite a bit more challenging, but still a potential do-it-yourself project.

Masonry steps call for skills the average homeowner is unlikely to possess, and so might be better left to a professional contractor.

A beginner should have some help with this project, and should add about 150% to the estimated times. An intermediate should add about 50%, and an expert, about 15%.

WHAT TO WATCH OUT FOR

For all exterior construction, you should use corrosion-resistant nails – stainless steel, aluminum alloy, or high-quality, hot-dipped galvanized. Avoid electroplated galvanized nails. Poor quality nails can react to the chemicals in pressure-treated lumber and to the natural resins of such woods as cedar or redwood. The resulting corrosion can produce unsightly stains and ultimately compromise the holding power of the nail itself.

Blunting the point of a nail before driving it helps prevent splitting when nailing the ends of the deck boards. Also, driving the nails at a slight angle will increase their holding power. To provide for surface drainage, leave a space of about 1/8″ between the deck boards; use 16d nails as spacers.

After sanding the wood to ease the sharp edges and to smooth the surface, a coat of clear wood preservative should be applied and allowed to soak in for about a week. The structure can then be finished with a top coat of opaque stain to match or blend with the color of the house or its trim.

SUMMARY

Replacing or adding a set of entryway steps to your home not only enhances its value visually, it shows your concern for the safety of your family members and guests. This stair structure is a sort of mini-deck, which can be adapted in numerous ways to blend with a variety of architectural layouts and styles, and which, without much trouble, can be built by the average homeowner.

For other options or further details regarding options shown, see

> *Elevated deck*
> *Farmer's porch*
> *Ground-level deck*
> *Main entry door*
> *Wheelchair ramp*

Entryway Steps

Description	Quantity/ Unit	Labor- Hours	Material
Pre-mixed concrete, for post anchors & footing pad, 70# bags	12 bags		89.28
Place concrete, spread footing, under 1 C.Y.	12 bags	0.9	
Ledger, end joists, and header, 2 x 10	22 L.F.	0.4	32.74
Inner joists, 2 x 8	8 L.F.	0.1	8.35
Posts, 4 x 4	16 L.F.	0.7	25.15
Stair stringer, 2 x 12	10 L.F.	1.2	20.40
Joist and beam hangers, galvanized, 2 x 6 to 2 x 10 joists	4 Ea.	0.2	2.06
Bolts, square head, nuts and washers incl., 1/2″ x 6″, galvanized	4 Ea.	0.3	4.56
Nails, 16d common, H.D. galvanized	10 lbs.		14.16
Decking, stair treads, risers, 2 x 6	90 L.F.	4.4	199.80
Rails and balusters	16 L.F.	5.8	172.80
Trim for top rail, cove molding, 3/8″ x 5/8″	16 L.F.	0.5	8.26
Finials, 3″ diameter	2 Ea.	0.5	13.44
Paint, railings, newels & spindles, stain, oil base, brushwork	32 L.F.	2.8	5.38
Stair stringers, rough sawn wood, stain, oil base, brushwork	116 L.F.	10.3	19.49
Totals		28.1	$615.87

Contractor's Fee Including Materials	$2,221

Key to Abbreviations
C.Y.– cubic yard   Ea.– each   L.F.– linear foot   Pr.– pair   Sq.– square (100 square feet of area)
S.F.– square foot   S.Y.– square yard   V.L.F.– vertical linear foot   M.S.F.– thousand square feet

WHEELCHAIR RAMP

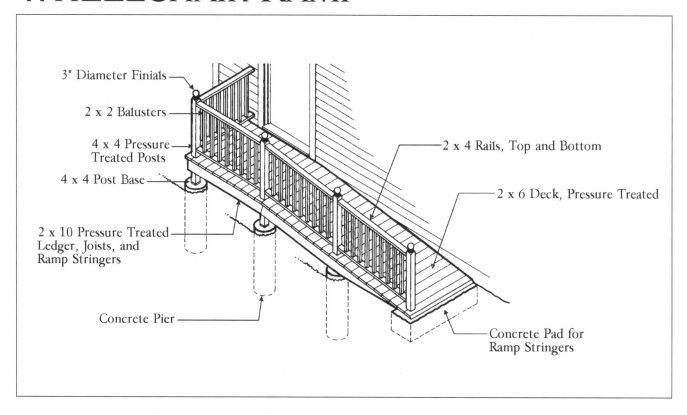

3" Diameter Finials

2 x 2 Balusters

4 x 4 Pressure Treated Posts

4 x 4 Post Base

2 x 10 Pressure Treated Ledger, Joists, and Ramp Stringers

Concrete Pier

2 x 4 Rails, Top and Bottom

2 x 6 Deck, Pressure Treated

Concrete Pad for Ramp Stringers

The most familiar mechanical aid for the disabled is, undoubtedly, the wheelchair. The standard arm-powered chair is still very much in use, but with advancing technology and changing social attitudes, wheelchairs have undergone numerous improvements, including the light and streamlined designs used in wheelchair racing. Perhaps most impressive are the many types of motorized chairs operable by hand, head, arm, chin, and even breath, which have dramatically redefined the limits implied by the very term "wheelchair." Although some models, known as "curb climbers," can surmount certain low obstacles and take fairly steep slopes, all chairs require ramps or lifts to enable their operators to go up or down from one level to another safely. For those who travel in wheelchairs, a ramp to an entry door is a prerequisite for an independent life.

MATERIALS

Most houses have, typically, two or three entry doors – front, back, and/or side. Where to locate the wheelchair ramp depends on a number of factors, mainly accessibility, appearance, and ease of installation. The latter is largely

determined by the height of the door threshold above grade and whether the wheelchair is manually propelled, motorized, or pushed by an assistant. The rule of thumb is that unassisted wheelchair use requires a ramp slope of 1" in 12" (1" of rise for every 12" of run), and assisted use, 2" in 12". Thus, a door 18" above grade calls for a ramp length (not including a landing) of 18' for unassisted use and 9' for assisted use. In either case, the principle is simple; the lower the door, the shorter the ramp.

In terms of accessibility, a paved and fairly level approach to the door is an obvious necessity. This could be a driveway or some sort of masonry walk. Building a wheelchair ramp is based on necessity, and pragmatic considerations should rule, but appearances should not be overlooked. Generally speaking, a ramp leading to the main entry door is not going to enhance the looks of the front of the house. Even if it means some extra work, the best choice for a ramp location, as far as appearance is concerned, is a side or back entry.

This design is similar to that of the entryway steps project that appears earlier in this book. Pressure-treated lumber is used for the posts and framing, while the

more attractive cedar is used for the decking and balustrade. The basic construction process is also similar to that used for the steps. First, lay out the landing. To provide space for the wheelchair and its turning radius, the landing should be at least 36" x 60" in front of an in-swing door, or 60" x 60" if the door swings out. Most exterior entry doors swing in, allowing for the installation of a storm door, which always swings out. To avoid building a more massive deck, it might be worthwhile to eliminate a storm door and, if necessary, replace an entry door with a good quality, weathertight, insulated model. Refer to the "Main Entry Door" project earlier in this book.

Next, lay out the posts, and build a form for the concrete pad that will support the bottoms of the ramp stringers. When you have finished excavating, align the posts square and parallel to the house. Plumb and brace them, and then pour the concrete. Attach the 2 x 10 ledgerboard to the house at the proper distance below the threshold, using nails and lag screws. Working out from the ledger, install the rim and deck joists. Cut the angles on the ramp stringers and attach them to the header joist with framing anchors. Use a straightedge to make sure the tops of the stringers line up. Nail solid blocking between the stringers, making sure to keep them parallel to the house.

After laying the deck boards, trim the tops of the posts about 38" above the deck. Assemble and install the balustrade sections, and cap the posts with the wooden finials. After the application of clear wood preservative, the balustrade can be stained and the decking coated with sand paint to provide good traction.

LEVEL OF DIFFICULTY

As mentioned above, the height of the door above grade determines the length of the ramp. A rise of 6" calls for a ramp of only 6'. Any number of variables could increase the difficulty of your specific project (for example, the need to remove shrubs or install a paved walkway), but building this mini-deck and ramp is not, in itself, that difficult. A beginner would, no doubt, need some help with the design and layout, and with figuring the angles

on the stringers, but once the relationship of the components is understood, the physical work of assembling them calls for no more than fairly basic carpentry, with the usual reminder to "measure twice, cut once."

Beginners should add about 150% to the time estimates. Intermediates should add 75%, and experts, 20%.

WHAT TO WATCH OUT FOR

Prior to starting this project it is recommended that you consult with the local building inspection department. Because of the permanent nature of most ramps and the large amount of area taken up by the actual structure, permits may be in order.

Most state and local building codes include sections on handicapped requirements. This is an excellent source of information for developing hand-rail heights and platform sizes.

Bolts and mechanical fasteners are recommended, as a nailed joint may loosen up and there is no way to tighten joints that shrink.

Determining the precise angles, top and bottom, for the ramp stringers can be a problem. Perhaps the best solution is to stretch a length of mason's cord along the house wall to indicate the top edge of the stringer. (Snapping a chalk line is likely to be difficult, if not impossible, because of the irregularity of the side wall and foundation surfaces). Make trial cuts on scraps of wood to get the proper angles. Once you have figured them out, measure and cut one stringer. Test it for fit at each of the layout locations; if it works, use it as a template for marking the others.

The deck lumber of the landing can be no lower than 1/2" below the threshold. If you lay the deck flush to the top of the threshold, you almost guarantee that rainwater from the deck will run in under the door. Be sure to check the finish level against the threshold before laying out the landing. Also, a small roof or overhang above the landing will help protect it, the door, and persons using the entrance from rain and snow.

SUMMARY

High-tech innovations have improved the design and capabilities of modern wheelchairs, and these in turn have given their owners a degree of mobility and independence that was unimaginable just a generation ago. Motorized models, compact and highly maneuverable, can be controlled in a number of ways, some of them quite astonishing. But the laws of physics still apply; ramps and lifts are necessary for wheelchair operators to get from one level to another. An entry door ramp is, for many physically disabled persons, a prerequisite for living independently. A typical home has two or three entries, and very likely one of these is, or can be, a suitable location for a ramp. Given other options, a front entry is not the best location for aesthetic reasons, but a thoughtfully designed and well constructed wheelchair ramp can blend very well with the house and it will, to say the least, greatly enhance the life of the person who uses it.

For other options or further details regarding options shown, see

Main entry door

Wheelchair Ramp

Description	Quantity/Unit	Labor-Hours	Material
Layout, excavate post holes	0.50 C.Y.	0.5	
Forms, round fiber tube, 8" diameter, for posts	12 L.F.	2.5	26.06
Concrete, field mix, 1 C.F. per bag, for posts	4 Ea.		29.76
Placing concrete, footings, under 1 C.Y.	0.50 C.Y.	0.4	
Post base, 4 x 4	4 Ea.	0.3	22.80
Framing, for ramp, 4 x 4 posts, treated	32 L.F.	1.3	50.30
Joists, 2 x 10, treated	80 L.F.	1.4	119.04
Joist hangers, galvanized, 2 x 10	4 Ea.	0.2	2.06
Post base 16 ga. galvanized, 2-piece, for 4 x 4	4 Ea.	0.4	34.20
Lag screws, 1/2" x 7-1/2", galvanized	5 Ea.	0.3	4.50
Decking, 2 x 6, treated	112 L.F.	5.4	248.64
Rails and balusters	36 L.F.	13.1	388.80
Finials	4 Ea.	1.0	26.88
Nails, 16d common, h.d. galvanized	10 lbs.		14.16
Nails, 10d box, h.d. galvanized	10 lbs.		14.16
Waterproofing, silicone, on decking, sprayed	60 S.F.	0.2	38.88
Paint, railings, cap, baluster, newels & spindles, stain, brushwork	40 L.F.	3.6	6.72
Deck, oil base, brushwork, 1 coat	112 L.F.	0.8	9.41
Sand paint, oil base, brushwork, 1 coat, for no-slip decking	112 L.F.	6.0	10.75
Totals		37.4	$1,047.12

Project Size	4' x 15'	Contractor's Fee Including Materials	**$3,301**

Key to Abbreviations
C.Y.– cubic yard   Ea.– each   L.F.– linear foot   Pr.– pair   Sq.– square (100 square feet of area)
S.F.– square foot   S.Y.– square yard   V.L.F.– vertical linear foot   M.S.F.– thousand square feet

Section Eight
ROOFING AND SIDING

If you find multiple roofing or siding problems such as leaks, worn sections, missing shingles, or cupped or bowed shingles, replacing your roof or siding may be a better idea than trying to repair it. Following are some tips to help you on your way to a successful roofing or siding project.

- Many roofing and siding materials are bought and sold by the *square*. One square is equal to an area that measures 100 square feet, or 10' x 10'.

- When replacing a section of siding, repeat the detailing and scale of the original siding as much as possible.

- Different siding textures and profiles can make a big difference in appearance, but not necessarily in cost. Check out all of your options. Most natural wood siding will discolor (age) differently depending on the exposure. Applying stain or preservatives can ensure consistent color. Rough-textured siding tends to hold stain better, and therefore may have a longer life than smooth siding.

- When re-siding, consider adding rigid insulation if your house is under-insulated and drafty.

- If you're replacing siding on an older house, you may find it difficult to find a suitable match. Your supplier may be able to special order replacement siding or suggest a source in your area.

- Surfaces exposed to direct sun for long periods of time may weather more quickly than faces exposed to moderate shade. Quality workmanship is even more crucial in these circumstances. Since there will be more chance of shrinkage, cupping and curling of the wood, care should be taken to properly select/install materials.

- Where different materials meet—at flashing penetrations on a roof, for example—sealants and caulkings provide extra protection against water penetration.

- Asphalt shingles are the most common roofing material. They are easy to install, economical, and come in a variety of colors and textures. Quality asphalt shingles are generally guaranteed for up to 20 years. Some brands and grades have a 30-year (or longer) guarantee.

- Asphalt shingles should be applied on a minimum slope of 2" in 12". Wood shingles should be applied on a minimum slope of 3" in 12", and wood shakes, 4" in 12".

- Before removing existing roofing material, be sure to check on the costs and appropriate methods of disposal, as well as any permits or other requirements of the town building department.

- If your existing roof has asbestos shingles, hire a professional to do the removal and repair.

- Check the condition of all your roof flashings at the time of roof replacement. Replace them if they are damaged.

- Re-nail roof sheathing before applying new roofing material. In some cases, sheathing that was not completely nailed down when originally installed may have warped over time.

- Avoid working on roofs in extreme heat or cold, as scuffing may occur.

- When working above the ground level where staging is required, be sure to use secure planks and brackets. When working from a ladder, place the ladder so the feet are away from the house at a distance equal to 1/4 the height of the ladder and make sure the legs are on a level surface. Check your local tool rental dealer.

- Protect your yard from falling debris by setting up plywood sheets against the house and over gardens and other yard elements.

ALUMINUM GUTTERS

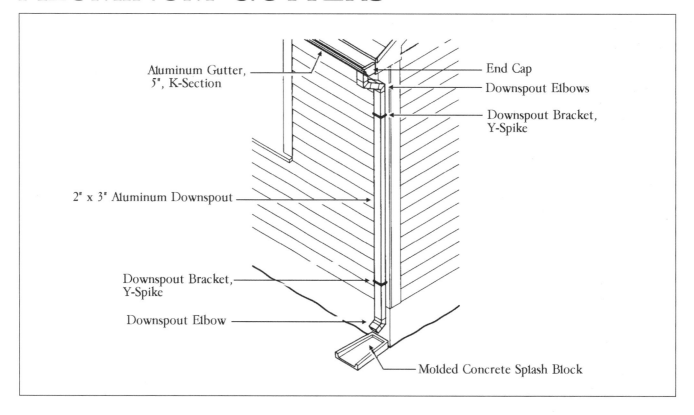

Aluminum Gutter, 5", K-Section

End Cap

Downspout Elbows

Downspout Bracket, Y-Spike

2" x 3" Aluminum Downspout

Downspout Bracket, Y-Spike

Downspout Elbow

Molded Concrete Splash Block

Among carpenters, few subjects are as controversial as gutters: the best material for them, their effectiveness, whether or not they are even needed. These topics are vigorously debated by builders and remodelers, and if you ask three of them what is best for your home, you are likely to get three different answers. Most experienced professionals do agree that gutters often cause problems – in many cases because the homeowner fails to clean and maintain them. Many people will, sometimes for years, allow water to cascade over a gutter without once climbing a ladder to simply remove the accumulated debris that is clogging the downspout. The resulting saturation of the fascia boards behind the gutters can cause rot that, if left alone, will migrate throughout the cornice and into the rafter ends, creating a damp welcome mat for ants and other unwanted crawling guests. A well-designed and properly functioning gutter system collects the water that drains off the roof, funnels it to the ground, and directs it away from the foundation to disperse harmlessly either as surface run-off or into a pipe, dry well, or storm drain. The theory is sound; its practical application can be quite problematic.

MATERIALS

There are principally four types of gutter used in residential construction: wood, plastic, copper, and aluminum. Gutters should blend into a house's facade and be almost invisible, and they will do that if the material and color are carefully chosen.

Wood gutters are very expensive, and are the most difficult to install. They must be securely fastened to a properly designed cornice, and carefully maintained with frequent cleanings and regular applications of waterproof sealant. A neglected wood gutter blocked with leaves will not only overflow, but also rot, causing the problems mentioned above. For some houses of traditional design and construction, especially authentic antiques, wood gutters are, in spite of their expense, the only architecturally correct choice.

There are a number of gutter styles available in plastic, including C-shaped PVC (polyvinyl chloride) with round downspouts. These are generally quite durable, but some inferior grades of plastic become brittle, both in cold weather and with age, which makes them

susceptible to cracking and chipping. Be sure to ask the salesperson about this if you are considering plastic gutters and, of course, consider how they will look on your house.

Copper gutters are custom made and are very good looking. When treated with an acid solution to darken them, they can virtually disappear against the trim of a dark stained house. They are expensive, and the cost of fabrication and installation is also high, but on the right house they can be very handsome indeed.

Aluminum gutters are by far the most popular type. They are relatively inexpensive, and not overly difficult to install. They come in several shapes: rectangular, beveled, K-section, and half round, and in a variety of sizes from 2-1/2" to 6" in height and from 3" to 8" in width. Stock lengths also vary, with 10' being the most common. Many building suppliers have machines that can roll-form gutter sections of virtually any length, and gutter contractors will usually bring one of these machines to the job site. Aluminum gutter components include inside and outside miter corners, joint connectors, outlets, end caps, and various elbows. The gutter can be attached with fascia brackets, strap hangers, or spikes and ferrules. Colors are white, bronze, and black.

Gutters, especially wood, are often run level for appearance, but a slight slope – about 1″ in 15′– toward the downspout is desirable for drainage. Downspouts should be spaced at no less than 20′ intervals, and no more than 50′ apart. This is a factor of roof area and pitch; a large roof with a steep pitch will require more downspouts to handle the volume and rate of flow than will a smaller, shallow-pitched roof. On a short run, a gutter can be installed to slope in one direction to a single downspout. On longer runs, the gutter should be crowned, or positioned slightly higher, in the center to drain the water to downspouts at both ends.

All connections should be made using aluminum sealant to prevent seepage through the joints. For spike-and-ferrule attachment, locate the rafter ends and determine their spacing. Mark these on the front lip of the gutter and drill holes for the spikes. The top outside edge of the gutter should be in line with the plane of the roof. Snap a line on the fascia board to mark the desired pitch for the bottom of the gutter. Hold the gutter to the line and drive the spikes through the ferrules into the rafter ends. Drive the Y-spikes into the corner board, using two for every ten feet of downspout. Connect the downspout to the gutter outlet and wire the spout to the brackets. Attach the elbow to the bottom of the downspout and place the splash block under it. You are now ready for rain.

LEVEL OF DIFFICULTY

Installing gutters of any kind is a two-person job, and very long lengths (over 20′) are more safely handled by three. Wood gutters are heavy and require a good deal of preparatory work. Half-inch wood spacers are nailed to the fascia board at the rafter ends. (These blocks provide for run-off of excess water caused by heavy rain or ice.) Lead end caps and gooseneck outlets are then tacked on and sealed with roofing tar, and the inside of the gutter is given a liberal application of waterproofing sealant. (Linseed oil is recommended.) Some builders recommend lining wood gutters with roofers' membrane.

Both plastic and aluminum gutters are much easier to install. They are lightweight and require little preparation before they are ready to go up. Spike-and-ferrule installation is the quickest; fascia brackets and strap hangers take a little more time to attach. Installing gutters means working from a ladder high above the ground. If you are inexperienced or uneasy about this aspect of the job, hire someone to do it. Even professionals are wary on extension ladders, but jittery nervousness can lead to an injurious fall. No job is worth such a risk.

Beginners and intermediates are advised to leave wood gutters to a professional carpenter, and for plastic or aluminum gutters, to work with someone experienced and comfortable on a ladder. A beginner should add at least 100% to the time estimates; an intermediate, at least 50%. An expert should figure on about 20% more time.

WHAT TO WATCH OUT FOR

An important part of a gutter replacement project is assessing the condition of the cornice – fascia, soffit, lookouts, and rafter ends – once the old gutter has been removed. Rotted wood must be replaced to prevent further damage and to provide solid backing to which to secure the new gutter. Make it clear to the contractor that inspecting the cornice and replacing rotted wood is part of the job, and that you are willing to pay for the necessary work and materials over and above the cost of the gutter. Money spent on a thorough job is never wasted. To make your new gutter last longer and provide trouble-free service, check with your local lumberyard for gutter accessories such as wire strainers and gutter screens that will eliminate clogging of downspouts and of the gutter itself. Keep in mind that you should still check your gutters from time to time and remove any debris.

Often, gutters are installed to prevent the buildup of water against a foundation. Sometimes the backfill or material used to fill the excavation has settled, creating a low spot. Introducing shrub beds next to a foundation can also create water collection problems against the foundation. Gutters prevent large amounts or run-off water from landing next to the foundation. The gutter collects the water, and the downspouts discharge the water away from the foundation. One gutter section may require two or more downspouts, depending on the area of the roof involved. Care should be taken not to discharge downspouts into shrub or flower beds located next to the foundation.

If you are adding gutters to a section of roof that currently does not have a gutter, keep in mind that you will be directing the total volume of water run-off to one location. Try to anticipate the amount of water you are redirecting and be sure you are not creating a safety hazard or a drainage problem at the end point of the gutter system.

Aluminum Gutter

Description	Quantity/ Unit	Labor- Hours	Material
Remove old gutter and downspout	20 L.F.	0.7	
Gutter, alum., stock, 5″ box, 0.026″ T, plain, incl. ends & outlets	20 L.F.	1.3	29.04
Downspout, aluminum, 2″ x 3″ including brackets	20 L.F.	0.8	17.76
Wire strainer, rectangular, 2″ x 3″	1 Ea.	0.1	1.94
Elbows, aluminum, 2″ x 3″	3 Ea.	0.2	3.28
Totals		3.1	$52.02

Project Size	20 L.F.		Contractor's Fee Including Materials	$235

Key to Abbreviations
C.Y.– cubic yard   Ea.– each   L.F.– linear foot   Pr.– pair   Sq.– square (100 square feet of area)
S.F.– square foot   S.Y.– square yard   V.L.F.– vertical linear foot   M.S.F.– thousand square feet

SUMMARY

It does not take long for water from overflowing gutters to cause rot and create the damp environment that is so attractive to house-hungry bugs. Choosing the appropriate type of gutter, whether wood, plastic, or metal, is very much a factor of the age and design of your house, as well as the size of your budget.

ATTIC VENTILATION

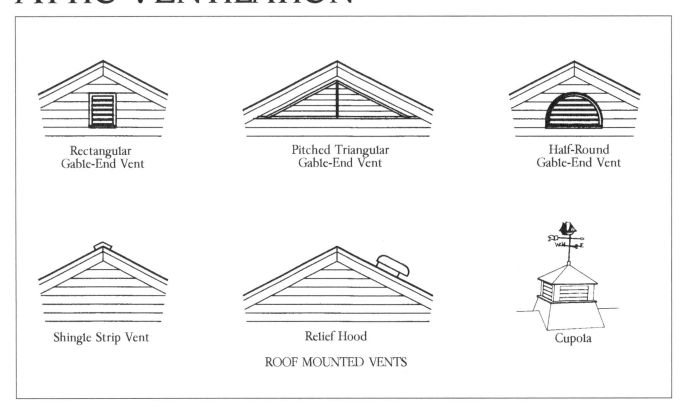

Rectangular
Gable-End Vent

Pitched Triangular
Gable-End Vent

Half-Round
Gable-End Vent

Shingle Strip Vent

Relief Hood

Cupola

ROOF MOUNTED VENTS

In an era of escalating energy costs, builders and homeowners are acutely aware of the need for sufficient insulation, properly installed. Often overlooked, however, is the equally important need for adequate ventilation to prevent the condensation of moisture that occurs in a tightly built, heavily insulated house. Moisture buildup can cause a number of problems, including wet (and therefore, useless) insulation, rotted sheathing and framing members, blistered and peeling exterior paint, ice dams, and leaks.

Very old homes were rarely provided with proper ventilation systems. On the other hand, they were usually built with little or no insulation and were anything but airtight. As a result, these houses "breathed" and condensation was hardly ever a problem. (Staying warm was!) Condensation becomes a problem when today's cost-conscious homeowner has insulation blown into the walls, adds a layer of insulation over the second-floor ceiling, caulks and weatherstrips doors and windows – in short, seals the house like King Tut's tomb, but neglects to add ventilation. Even newer houses have been built lacking adequate ventilation

by ignorant or dishonest builders working in areas where codes are not properly enforced.

It is important that areas on the cold side of the insulation be well ventilated to remove, by evaporation, condensed moisture. One of the places where this can be effectively accomplished is the attic.

MATERIALS

There are many types of ventilators. Wooden, louvered units are generally installed in the end walls of gable roofs, as close to the ridge as possible. The efficiency of these vents is enhanced by the addition of soffit ventilation. This can be either a series of 2-1/2" holes centered between each pair of rafters and covered with a louvered aluminum plug, or better, a 2"-wide slot the full length of each soffit, covered with a louvered aluminum strip.

Hip roofs should have air inlets at the soffit and outlets at, or near, the peak of the roof. Roof-mounted ventilators are designed to remove air from the attic without the use of motor-driven fans, either by convection or external wind action. Relief hoods may also be used for

air exhaust. Hoods and ventilators are usually made of galvanized steel or aluminum. They can be located below the peak on the rear slope of the roof so they are not visible from the front of the house. If the ridge is long enough, a continuous ridge vent may be used. This requires that a strip of sheathing be cut from both sides of the ridgeboard to produce an adequate opening. Over this is installed a vent suited to the design and style of the roofing material. A well-stocked building supply house can provide a variety of ridge vents with detailed directions for their installation.

The minimum free-air area (ventilator openings) for attic vents is based on the ceiling area of the rooms below. The free-air area should be 1/300 of the ceiling area. For example, if the ceiling area is 1,200 square feet, the minimum total free-air area of the ventilators should be four square feet. The actual area must be increased to allow for any restrictions such as louvers or wire screen mesh. Coarse screen should be used because dirt and airborne debris tend to clog fine mesh screens. When painting, be careful not to contact the screen and clog the mesh with paint. Ventilators should never be closed. In cold climates, it is especially important to leave them open in winter as well as summer.

LEVEL OF DIFFICULTY

The main difficulty involved in this project lies in the fact that most of the work has to be done from an extension ladder or on the roof. Safety considerations alone would put this work beyond recommendation for the beginner or even the moderately competent intermediate. Using hand and power tools at ground level and with solid footing can be hazardous enough, but working with them at heights and in awkward positions is best left to the expert or professional tradesperson.

This project also requires that holes be cut in the soffits, walls, and/or roof. Any improper installation of the vent unit or finished surface materials could result in leaks and potentially serious damage that would be costly to repair, further reasons why this work should not be undertaken by the novice.

The level of difficulty, least to most, for the various installation tasks is as follows: hole-type soffit vents; gable vents; slot-type soffit vents; roof-mounted vents; ridge vents.

Even if you are experienced enough to undertake all or part of this project, you should seek the advice of a professional to determine what type and size ventilating devices are suited to your situation. It would be most discouraging to complete the job only to discover that the venting system you labored mightily to install is wrong or undersized.

Experts should add about 30% to the professional time for all tasks.

WHAT TO WATCH OUT FOR

The soffit, sometimes called the planchia, is the finished underside of the box cornice – that part of the roof that overhangs the exterior wall. If hole-type soffit vents are, for whatever reason, inadequate, and the slot-type system is required, it will often be easier to remove and replace the soffit rather than to attempt cutting out the proper-sized strip.

Soffit material may be 3/4″ solid lumber, matched boards, plywood, corrugated and perforated aluminum, or special material with precut screened openings. Choose the material that best matches your house. Soffits are not highly visible from a distance but from close up they are, and there is no reason for them not to look good.

The simplest installation is to screen the soffit opening and nail up two narrow boards so that you leave a 2″-wide open strip the full length of each soffit. Make sure there is no insulation in the soffit; the insulation on the attic floor should end at the outer wall, not extend into the overhang.

Fiberglass screening may be easier to install than metal types. It also does not corrode or oxidize.

To guarantee the unobstructed flow of air from soffit vent to ridge vent, cardboard or styrofoam forms (pans) may be introduced to slightly depress the insulation at the wall line, providing a natural duct for air flow.

Framing bays (the space between rafters) are sometimes isolated by skylights or other roof penetrations. If possible, air flow should also be provided to these bays by boring a series of small holes through the rafter.

Any time a project involves disturbing existing roofing or siding material, the cutting and matching may involve a much greater area than you might have expected. Examine the condition of existing material before starting to ensure that you have enough to replace any area that might be disturbed during the renovation.

SUMMARY

Many homeowners are reluctant to spend money on projects such as this that provide no visible enhancement of the house's appearance or of the family's comfort. However, such shortsightedness can result in the long-term devaluation of a house because of the destruction wrought by condensed moisture. Installing a properly designed and adequately sized ventilation system is good insurance against this problem. This ounce of prevention is cheaper than a pound of cure.

Attic Ventilation System

Description	Quantity/ Unit	Labor-Hours	Material
Remove ridge shingles and cut vent opening at ridge and soffit	1 Job	4.0	
Ridge vent strip, polyethylene	40 L.F.	2.0	122.40
Under eaves vent, aluminum, mill finish, 16″ x 4″	10 Ea.	1.7	14.88
Totals		7.7	$137.28

Contractor's Fee Including Materials	$522

Key to Abbreviations
C.Y.– cubic yard   Ea.– each   L.F.– linear foot   Pr.– pair   Sq.– square (100 square feet of area)
S.F.– square foot   S.Y.– square yard   V.L.F.– vertical linear foot   M.S.F.– thousand square feet

PAINTING

There are few sorrier sights than a house badly in need of paint. A peeling, blistering, faded exterior can make the loveliest architectural gem look positively shabby. Both the expense of hiring a contractor and the enormous amount of labor in a paint-it-yourself project are dispiriting to contemplate, and can cause many a homeowner to postpone a much-needed paint job for years, during which time the house looks terrible and deterioration sets in. There is no denying the expense and labor involved in painting a home's exterior, but these must be viewed as investments necessary to protect, beautify, and maintain the value of your home.

MATERIALS

The disagreement among professional painters over the relative merits of stain vs. paint, and latex vs. oil, is enough to thoroughly confuse the inquiring homeowner. These ongoing debates will not be settled here. What follows is a brief summary of the opposing views. You must make your own choices aided by the advice and opinion of a local professional whose reputation and

workmanship is supported by the testimony of satisfied clients.

Stain is becoming a popular choice, especially for new, rough-sawn siding. Stain does not peel like paint and, therefore, never needs scraping. There are two types of stain: semi-transparent, usually oil-based, and opaque, usually latex. The former allows more of the grain and texture of the wood to show through; the latter is denser, having the consistency of very thin paint. Both types of stain are somewhat loose and watery and go on easier than paint, which is thicker and more viscous. Stain leaves little film on the wood's surface, and thus protects it less from sun and moisture than does paint; also, stain lasts only about half as long. Stained wood on the sunny side of a house should be given a new coating as regularly as every three years. Staining over paint is generally not recommended because the results can be unpredictable. Paint can, however, be applied over stain, with proper preparation.

Acrylic latex-based paint has better color retention, so it doesn't fade as quickly as oil-based paint. Latex remains more flexible after it dries, as opposed to oil paint, which dries harder, tends to be brittle, and thus is more likely to crack and

peel. Latex dries faster, cleans up with water, and is somewhat easier and more pleasant to work with.

Oil-based paint flows more easily off the brush, and tends to level better, leaving a smoother surface than latex. Over the years oil paint "chalks," or oxidizes, so that a new coat doesn't add much to the total thickness of the paint film. Latex, on the other hand, builds up with each repainting, resulting in a thicker film that becomes increasingly less flexible and resilient, and thus more apt to flake and chip. On bare wood, the linseed oil in oil paint permeates the fibers and helps prevent the siding's drying out; latex, however, tends to bond to just the surface of the wood, without adding moisture-restoring elements to it.

Although professionals will tell you latex can be put over oil, and vice versa, you are safer using the same type of paint as is already on the house. Be sure to use only premium-quality paint, and to prepare the surface well.

The key to a good paint job is good preparation. Even the best paint, if applied to an unscraped, unsanded, dirty surface will begin to flake off in a couple of years. Preparation involves scraping all loose, cracked, peeling, and blistering paint, and feathering the edges with sandpaper to blend the paint surface to the bare wood. If you use a power sander, be careful not to gouge or otherwise damage the siding or the trim. After scraping and sanding, the siding should be washed down with household detergent and a brush and thoroughly rinsed off. Areas of mold and mildew should be scrubbed off with a mixture of bleach and water. When the surface is dry, caulk any openings or cracks that might allow water, air, or insects to penetrate the walls. Also, thoroughly clean out any small areas of rot and cover them with exterior-grade spackling or epoxy filler. Badly rotted boards should be removed and replaced with new lumber. Remove any loose or cracked putty around the window panes, prime the bare wood, and reputty with glazing compound. Spot-prime all bare wood on the siding and trim, and wherever there is evidence of chalking. Use the type of primer recommended on the label of the topcoat paint. Place drop cloths over plants, shrubs, walkways, and other exterior features that you don't want splattered with paint, and mask or remove such items as door hardware.

Use a 2-1/2" angled sash brush for the windows and small trim. Paint windows, working from the glass outwards to the frame. Use a 3" or 4" brush for the wider trim and siding. Work from the top down, and try to paint entire areas that exist between natural breaks, such as between windows. In this way you will avoid uneven application and lap marks, which occur when wet paint is brushed over paint that has already started to dry.

Paint the bottom edge of the siding first, then the face. On wood shingles, get plenty of paint on their bottom edges to seal the porous end grain.

If there is little or no color change between the existing paint and the new coat, you may get by with one coat. Otherwise you will need a second coat for complete coverage and to give the paint job a professional finish.

LEVEL OF DIFFICULTY

Preparing and painting the exterior of a house, even one of modest dimensions, are physically demanding and time-consuming tasks. Before you decide to do it yourself, calculate the time required to prepare and paint your whole house, taking into account the number of doors, windows, and shutters; the number of different colors involved; the height and accessibility of the walls; the tools and equipment needed; the speed at which you can efficiently work; and the amount of time (assuming good weather) you have available to do it. The prospect of attempting a project of this size can be daunting even to an enthusiastic do-it-yourselfer. It can be done, and with good results, but it is wise to know what you are getting into: a difficult, dirty job that requires some time spent at heights and climbing up and down a ladder. An experienced, reputable painter knows how to set up a job and direct a crew to complete it quickly and in craftsmanlike manner.

Beginners should add at least 200% to the times estimated. Intermediates should add 100%, and experts, 50%.

WHAT TO WATCH OUT FOR

Scraping and sanding old paint is slow and tedious work, and people are often tempted to use alternative methods of removal to speed up the process. Be cautious. An electric heat gun works, but it uses a lot of electricity, and dragging around a long extension cord can be dangerous when working from a ladder. Propane blow torches are not recommended because of the obvious fire hazard.

For painting with acrylic latex paint, use a nylon bristle brush, and clean it with soap and water. For oil base paint, as well as enamel and varnish, use a natural bristle brush. Natural bristle tends to regulate the flow of oil paint, leaving a smooth finish and requiring less dipping. Clean this brush with mineral spirits only.

Spray painting machines have become very popular over the last few years, but effective spraying calls for a good deal of preparation and care in application to prevent the paint from hitting unintended targets. This involves a number of techniques that you don't want to learn by experimenting on your own house. You can save yourself time and effort by spraying the shutters, but generally, large area spray painting should be left to professionals.

When spot priming where a dark color will be the finish you may want to tint the primer; that is, darken it by adding a bit of the finish color to it. This will make the finish coats cover more evenly and eliminate the possibility of a streak of primer showing through.

SUMMARY

A wood-sided house needs either paint or stain, both for beauty and for protection from the effects of sunlight, rain, heat, and cold. There are cogent arguments on both sides of the debate over stain vs. paint, and oil base vs. latex. Ultimately, the homeowner must decide on the basis of an informed judgment, aided by the advice of an experienced, reputable painting professional. Careful consideration should be given the decision to either hire a contractor or to do the job yourself. The average homeowner will regret undertaking a project so physically demanding and time-consuming, unless budgetary constraints offer no alternative.

Painting

Description	Quantity/Unit	Labor-Hours	Material
Exterior siding, clapboard, oil base, primer, brushwork	2,000 S.F.	24.6	216.00
Exterior siding, clapboard, oil base, 2 coats, brushwork	2,000 S.F.	39.5	360.00
Panel door and frame, oil base, primer, brushwork	2 Ea.	2.7	4.22
Panel door and frame, oil base, 2 coats, brushwork	2 Ea.	5.3	13.68
Windows, oil base, primer, brushwork	16 Ea.	10.7	13.25
Windows, oil base, 2 coats, brushwork	16 Ea.	18.3	28.99
Gutters and downspouts, oil base, primer, brushwork	80 L.F.	1.0	9.60
Gutters and downspouts, oil base, 2 coats, brushwork	80 L.F.	1.6	21.12
Totals		103.7	$666.86

Project Size	2,000 S.F.		Contractor's Fee Including Materials	$5,667

Key to Abbreviations
C.Y.– cubic yard   Ea.– each   L.F.– linear foot   Pr.– pair   Sq.– square (100 square feet of area)
S.F.– square foot   S.Y.– square yard   V.L.F.– vertical linear foot   M.S.F.– thousand square feet

REROOFING — ASPHALT SHINGLES

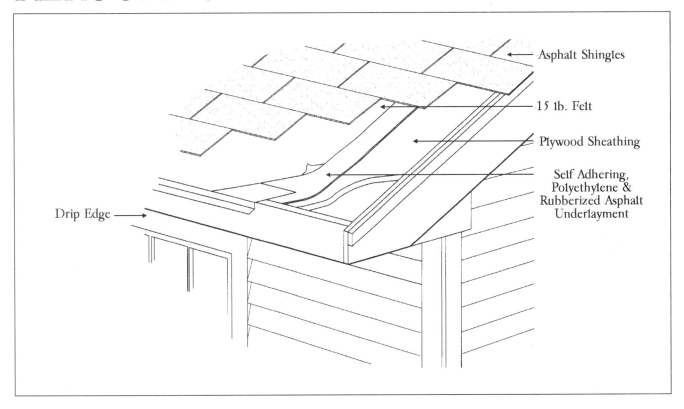

Asphalt Shingles

15 lb. Felt

Plywood Sheathing

Self Adhering, Polyethylene & Rubberized Asphalt Underlayment

Drip Edge

The primary function of a house is to shelter, and no single element of its construction is more necessary to that end than the roof. A well-designed roof clad, with high-quality, durable materials, can effectively shelter for many years with minimal maintenance. For most roofs, there comes a time when the wear and tear of the elements take their toll, and the roof covering must be replaced. The prudent homeowner will see to this renovation long before any leaks force the issue. A seemingly minor leak can cause major damage, so no evidence of rainwater penetrating the roof – such as stained ceilings and walls, peeling paint, or crumbling plaster – should be ignored. Maintaining a sound roof is not an option; it is a homeowner's obligation to the house and to those it shelters.

MATERIALS

The diverse aspects of roof design and roofing materials can, and do, fill volumes. The information presented here is merely an overview. Materials used for standard pitched-roof, residential construction include fiberglass, asphalt, wood, tile, and slate. For flat or low-pitched roofs, or certain custom designs, sheet materials

such as roll roofing, galvanized steel, aluminum, copper, and tin are sometimes used. The choice of materials and method of installation is determined by cost, roof slope, local climate, and the material's longevity, wind- and fire-resistance, and appearance.

If you are reroofing, you would probably do well to stick with the same material you are replacing, unless it has proved unsatisfactory. If you are roofing a newly built house, you have no existing roof to influence your choice, for better or for worse. In either case, get the advice of experienced local roofers, note the roofing materials prevalent in your area, and beware of investing in unproven products.

Slate roofing is very costly and the specialized skilled labor it requires is also expensive and not always easy to find. Slate is very heavy and demands a sturdy roof frame and sheathing system, but it looks great and, properly installed, can last beyond your lifetime. Although somewhat cheaper than slate, clay or concrete tiles also call for highly skilled (expensive) labor, and neither type of tile is recommended in areas that experience freeze/thaw cycles.

Wood roofing is available as shakes (split) or shingles (sawn). In either form, wood is expensive to purchase and, being labor intensive, expensive to install. Wood is susceptible to rot, moss, and fungus, and is highly flammable. In spite of all that, nothing beats it for enhancing the rugged, rustic looks of an outdoor-oriented contemporary design, or for blending with the folksy, nostalgic charm of a traditional antique.

Because they combine cost efficiency with acceptable looks, fiberglass shingles are the choice for most houses. These shingles are manufactured in several forms – giant individual rectangles, individual lock-down, and strip. The latter can be subdivided into the following styles: tabless; 2- and 3-tab hexagonal; 2- and 3-tab square butt; and "architectural" wood or tile appearance. All are produced from a fiberglass mat, coated with asphalt, and surfaced with variously colored mineral granules. Real asphalt shingles are made with an asphalt-saturated rag mat; but they are becoming hard to find, as they have been almost totally supplanted by the fiberglass varieties that, confusingly, are also called asphalt.

Fiberglass asphalt strip shingles will last from 15 to 30 years, depending on their design and weight, and local climatic conditions. Shingles come in several grades or weights, based on the thickness of the mineral surface and the weight of the base mat. Standard roof shingles weigh around 220 pounds per square (100 square feet of roof area – 3 bundles); heavyweight shingles, from 290 to 330 pounds per square. Shingles – either wood or asphalt – can be laid on any roof with a minimum slope of 4"-12"; that is, a roof with a vertical rise of 4" for every 12" of horizontal run. Roofs with less slope, or those that are flat, require some form of roll or sheet roofing.

In general terms, roofing begins at the eaves with a metal drip edge, followed by overlapping layers of 15-lb. roofing felt. In snowy climates, a wide strip of neoprene roofing membrane should be added along the eaves, extending far enough up the roof to seal it against leaks caused by ice dams. On a roof of average pitch, a 3'-wide strip usually suffices. Beginning with a double starter course, the shingles are laid, course by course, up the roof, staggering the vertical joints and maintaining straight lines in both directions, horizontally along the butts and vertically along the cutouts. At all intersections or obstructions – hips, valleys, walls, chimneys, skylights, soil stacks, etc. – an appropriate membrane or flashing should be installed, whether of roofing felt, aluminum, copper, or lead, and in some cases should also be sealed with caulk or roofing tar. The ridge is capped with either cut shingles or a ridge vent.

To estimate roofing materials, find the area of the roof in square feet and divide the total by 100 to determine the number of squares needed. For a simple, unobstructed roof, add another 10% for waste and cuts. A complicated roof – one with many dormers, valleys, and obstructions – requires that more be added.

LEVEL OF DIFFICULTY

Roofing a house should be undertaken only by an expert. Beginners and intermediates should hire a professional, or work with expert guidance and assistance. A good roofer is a skilled carpenter who possesses the combination of experience and intelligence that enables him to recognize both actual and potential problems and devise effective solutions to them. A roofer works in high places and often in awkward positions. Roofing is also a physically demanding task. You will find that stripping old shingles and cleaning up the resulting mess is hard, dirty work. You'll need to know how to set up safe, sturdy staging with pump and roof jacks. You must climb ladders with bundles of shingles weighing upwards of 70 pounds, and move them around a sloping roof. You'll have to work for long periods of time bent over in the heat and glare of the sun. You must be able to measure, cut, and nail shingles quickly and accurately, all the while keeping your balance and a healthy sense of self-preservation. It is not a job for the fainthearted.

Beginners and intermediates should hire a professional contractor, and for any

task they undertake, add 150% and 75%, respectively. An expert should have at least one helper, and add 25% to the time estimated to complete the project.

WHAT TO WATCH OUT FOR

It may be possible for you to lay the new roof over the existing shingles, thereby saving yourself a great deal of labor. Be sure that there is only one layer of shingles on the roof, and check with your building inspector to see if your local codes allow this kind of double coverage. Even if it is legal, it may not be advisable on a very old house or one whose framing and sheathing may not be strong enough to support the additional weight without sagging; but for most houses, it is a cost-saving option worth considering.

Some roofers and builders use a double run of red cedar shingles as a starter course at the eaves in place of a metal drip edge. The shingles are, in some ways, more attractive than the metal, but they are susceptible to rot and fungus. They can carry damaging moisture up into the sheathing, and they make wonderful nesting places for wasps and hornets. Metal drip edge is the better choice.

Narrow slots called cutouts divide asphalt strip shingles into two or three tabs that, when laid, give them the appearance of separate pieces. The vertical lines of the cutouts add visual interest to the roof, but they expose the shingles directly below them to more wear and weathering. For this reason, tab shingles have a shorter life than the tabless variety. If your roof is shallow pitched, screened by overhanging trees, or otherwise not highly visible, you might consider the durability of tabless shingles as an acceptable trade-off for their somewhat less attractive appearance.

Tab and architectural roof shingles are laid down in a pattern. Read and follow the instructions provided on the wrappers or on an instruction sheet to insure that the correct pattern is established.

SUMMARY

Unlike some other parts of a house, the roof cannot continue to function adequately if there is deterioration. Once it begins to leak, a roof has ceased to serve its primary purpose, which is to shelter and protect the house, its occupants, and their possessions.

Reroofing-Asphalt Shingles

Description	Quantity/Unit	Labor-Hours	Material
Roofing demolition, remove asphalt strip shingles	1,000 S.F.	11.4	
Self adhering polyethylene & rubberized asphalt underlayment	2.10 Sq.	0.8	93.24
Shingles, standard strip, class A, 210-235 lb. per square	10 Sq.	14.6	342.00
Asphalt felt sheathing paper, 15 lb.	1,100 S.F.	2.4	26.40
Drip edge, galvanized	70 L.F.	1.4	18.48
Nails, roofing, threaded, galvanized	50 lbs.		78.00
Totals		30.6	$558.12

Project Size	1,000 S.F.		Contractor's Fee Including Materials	$2,142

Key to Abbreviations
C.Y.– cubic yard   Ea.– each   L.F.– linear foot   Pr.– pair   Sq.– square (100 square feet of area)
S.F.– square foot   S.Y.– square yard   V.L.F.– vertical linear foot   M.S.F.– thousand square feet

RESIDING WITH WOOD CLAPBOARD

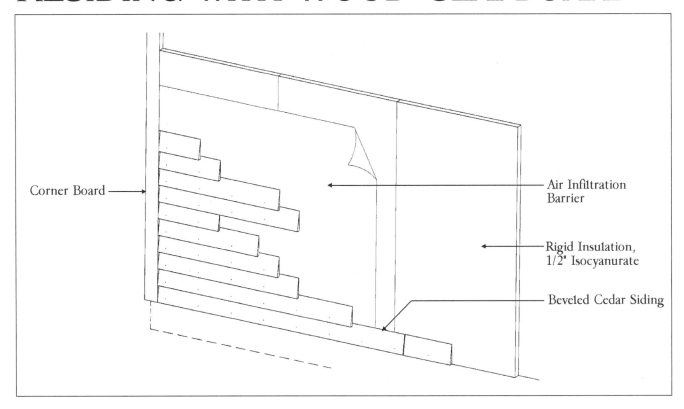

Corner Board

Air Infiltration Barrier

Rigid Insulation, 1/2" Isocyanurate

Beveled Cedar Siding

Siding is perhaps the single most visible component of a house, and should possess a number of qualities: It must be attractive, fit the architecture, hold its finish, withstand weathering and neglect, and protect everything under it from the elements. Over time, the ravages of sun, wind, rain, snow, and airborne pollution take their toll, and must be countered with continued maintenance. Repairing and repainting your home's siding can revive its looks. Residing — that is, removing the existing siding and replacing it with new — is like a complete facelift. It can dramatically transform your house, giving it a totally new and improved appearance.

A significant advantage to installing new siding on an older house is that you can improve the heat loss characteristics of your house. Drafts can be reduced by the installation of an air infiltration barrier (house-wrap). Heat loss can be reduced by the addition of 1/2" of rigid insulation.

MATERIALS

Exterior wall covering is called siding, except when the term refers to masonry

coverings such as stucco or brick veneer. The most common siding materials in residential construction are wood, metal, and vinyl.

The question of which material makes the best siding has fueled many a heated argument, and few members of the building trades are without strong opinions on this issue. This brief outline will highlight some of the pros and cons of each material. You should read up on the facts and consult with knowledgeable local building professionals, not just materials salespeople, to get their views on the best siding for your house.

Wood siding is available in a number of forms that can be applied either horizontally or vertically.

The most common traditional form of wood siding is clapboard. When properly installed, clapboards are inherently waterproof and, if properly maintained with paint or stain, can last for well over a hundred years. A high quality 1/2" x 6" vertical grain (VG) red cedar clapboard is about 1/2" thick at the bottom and 1/4" thick where the next board overlaps its upper edge. A thin board like this, cut to expose its vertical grain, is less likely to warp or split than a thicker board or

one cut to produce a flat grained surface. Cedar clapboards are very light and easy to work with. They can be installed smooth side out for painting, or rough side out for staining.

Although some building professionals claim that synthetic siding materials can trap moisture, thus causing rot and structural damage, most of the objections to vinyl and aluminum are based on aesthetics. Manufacturers have responded to complaints and criticism by producing such innovations as wood-grain textures and by making a wide variety of trim features and moldings that more or less replicate the look of wood.

Aluminum and steel siding are both available in a wide range of colors, can be repainted if the finish gets marred, and are fireproof. Steel is heavier and more rigid than aluminum and thus is more resistant to dents, which, in either material, are not easily removed or hidden. On both metals, the finish color is applied to the surface and can be easily scratched off. Because of its tendency to rust, steel is not a good choice near salt water or in areas of heavy air pollution, and scratches in steel siding should be touched up immediately before they begin to oxidize. Aluminum is very much lighter and, therefore, somewhat easier and quicker to install than steel.

Vinyl is becoming increasingly popular as a siding material, and technological advances have greatly improved both its quality and the variety of decorative options available. The color goes completely through a piece of vinyl so that scratches are not very visible. Vinyl also resists dents and is lightweight and less noisy in the wind than aluminum. If installed incorrectly, vinyl siding can buckle or crack, and with age it tends to fade.

The main advantage of synthetic siding is that maintenance is minimal; it requires only a wash down once or twice a year to keep it looking good. Most manufacturers guarantee their products for at least 20 years, but you should read your warranty carefully to find out exactly what is covered. Although it is possible to apply vinyl or aluminum over the existing siding, removal of the old siding is recommended.

LEVEL OF DIFFICULTY

Residing with wood clapboards is not a job for a beginner working alone, but it is certainly within the abilities of a competent intermediate. Stripping the old siding (installed horizontally) is best done from the top down. After making any necessary repairs to the exposed sheathing, cover it with an air infiltration barrier. The new clapboards can be installed either from the top down or

from the bottom up; whichever is more convenient. Putting up the air infiltration barrier, taking measurements, snapping chalk lines, and handling long pieces of clapboard are all much easier with a helper and with pump jack scaffolding instead of ladders, which have to be constantly moved and adjusted. Having a third helper to make the cuts on a fine-toothed chop saw will save you even more time and effort.

It can take six months or more to train a worker in the proper techniques of synthetic siding installation, so it is generally not recommended that the average homeowner undertake a vinyl or aluminum residing project.

Beginners should have help and guidance from preparation through completion, and should add 150% to the estimated times. Intermediates and experts will find the job much easier to accomplish with at least one helper. Intermediates should add about 75% to the times; experts, about 25%.

WHAT TO WATCH OUT FOR

It is good practice to back-prime (or paint the back side of) clapboards and other exterior wood trim pieces before putting them up. This prevents moisture from penetrating the boards from back to front, and helps to extend the life of both lumber and finish.

Despite the instructions found in many manuals, it is not good practice to place the nails high enough on the clapboard so they miss the top of the piece below. The idea that inspired this practice is sound enough: to let the boards expand and contract independently of one another. The problem is that that section of the clapboard is unsupported from behind, and you will split almost as many boards as you nail, no matter how careful you are. It is better to use 5d hot-dipped galvanized box nails and place them about 1/2" from the bottom edge of the clapboard, nailing through the board below, into the sheathing.

To estimate how much siding is needed, calculate the entire exterior surface area, deducting the area of all openings (window and doors) from the total. To calculate the area of a gable, multiply the length by the width, and divide the result by two.

SIDING OPTIONS
Cost per Square Foot, Installed

Description	
Wood Clapboard	$2.96
Cedar Shingles	$2.63
Vinyl	$1.88
T-111	$2.12
Stucco	$2.00

SUMMARY

A major exterior remodeling project like residing can completely transform a house. Of the many siding styles and materials, your choice should be based on: 1) practical considerations: the siding's proven record of durability, the sort of maintenance it requires, and its appropriateness to local climatic conditions; and 2) cost considerations. Wood, vinyl, and metal siding all have advantages and disadvantages, and the question of which is best is hotly debated by experienced and knowledgeable professionals, usually with strong feelings in favor of one type accompanied by equally strong feelings against the others. Remember that the results of your decision may outlast your ownership of the house, and will certainly affect its value for better or for worse.

For other options or further details regarding options shown see

Painting

Residing with Wood Clapboard

Description	Quantity/ Unit	Labor- Hours	Material
Siding demolition, clapboards, horizontal	2,200 S.F.	46.3	
Siding, 1/2" x 6" beveled cedar, A grade	2,200 S.F.	70.4	4,356.00
Housewrap, spun bonded polypropylene	2,200 S.F.	4.6	422.40
Insulation board, isocyanurate, 1/2", R-3.9	2,200 S.F.	22.0	765.60
Paint, back prime clapboards, oil base, primer, brushwork	2,200 S.F.	27.1	237.60
Clapboards, oil base, primer, brushwork	2,200 S.F.	27.1	237.60
Clapboards, oil base, 1 coat, brushwork	2,200 S.F.	43.5	396.00
Totals		241.0	$6,415.20

Project Size   16'-4" x 135'

Contractor's Fee Including Materials **$20,183**

Key to Abbreviations
C.Y.– cubic yard   Ea.– each   L.F.– linear foot   Pr.– pair   Sq.– square (100 square feet of area)
S.F.– square foot   S.Y.– square yard   V.L.F.– vertical linear foot   M.S.F.– thousand square feet

Section Nine
ROOM ADDITIONS

One of the most important considerations in planning an addition is to make sure it blends with the existing house's materials and geometry. Match exterior trim details; roof slopes; window type, style, and spacing; and dormers to the existing house. Consider also the following tips. In the early stages,

- Check building codes and any zoning restrictions for permit requirements and other issues that may limit or delay your building plans.

- It is always good to sketch out the additions before even consulting with a builder or designer.

- During the design stage go out in the yard and mark off the addition — get a feel for how it will affect your yard, your view, and the neighboring properties.

- In an addition — especially one that includes a public area like a family room — allow for lots of natural light.

- If you need to remove trees when clearing the site for an addition, keep in mind that certain trees may have a landscape resale value or firewood value. Also, stump disposal can be very expensive, particularly if the stumps cannot be buried on site.

- If excavating for the new addition involves large quantities of cut (removal of soil) or fill (additional soil), consider raising or lowering the site to accommodate your needs while reducing costs, but pay close attention to the associated effects on site drainage and utilities.

- Contact your local utility companies to verify the location of—and avoid damaging—underground electrical, gas, water, and septic lines.

- The price of lumber is quite sensitive to supply and demand in the marketplace. Therefore, it is advisable to call local suppliers for the latest market pricing.

- Think ahead about waste. Framing lumber, for example, is sold in even foot lengths (e.g., 10′, 12′, 14′, 16′). Depending on spans, wall heights, and grade of lumber, waste is inevitable. A rule of thumb for lumber waste is 5% to 10%, depending on material quality and the complexity of the project.

8' x 8' ENTRY ADDITION

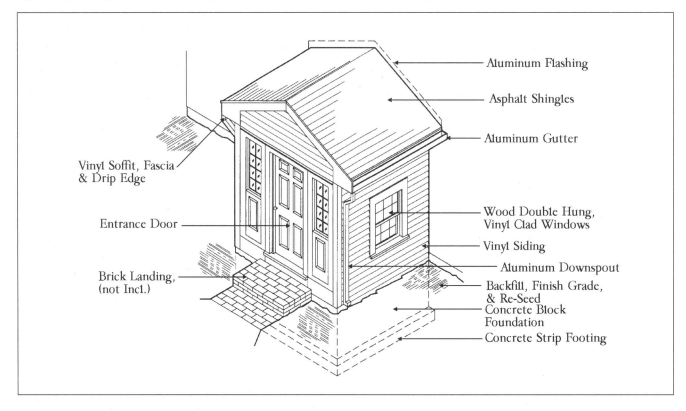

Aluminum Flashing

Asphalt Shingles

Aluminum Gutter

Vinyl Soffit, Fascia & Drip Edge

Wood Double Hung, Vinyl Clad Windows

Entrance Door

Vinyl Siding

Aluminum Downspout

Brick Landing, (not Incl.)

Backfill, Finish Grade, & Re-Seed

Concrete Block Foundation

Concrete Strip Footing

Room additions, like other projects involving attached structures, should complement the style and design of the house while meeting the need for new living area. Because of budget limitations, major remodeling projects like this one often require modifications in the structure's design and materials. This give-and-take process may lengthen the planning stages, but it helps ensure the livability, appearance, and value of the addition. Because this project is small and at first-floor level, do-it-yourselfers can take a hand in its construction and enjoy the benefits of lower costs.

MATERIALS

It is possible to dig the foundation for this addition by hand to its required 4' below grade depth; however, machine excavation is a practical alternative, and the cost is included in this estimate. The excavation should be wide enough to accommodate the footing and foundation walls and deep enough to rest on firm, frost-free subsoil. Once the footing has been carefully laid out and put into place, the concrete block wall can be built to the desired height. A formed and poured concrete wall is a slightly more expensive option. Generally, do-it-yourselfers should leave the excavation and masonry tasks to a professional contractor, as specialized skills, equipment, and considerable physical exertion are involved in foundation work.

The exterior of the structure is comprised of framing, sheathing, and finish materials common to wood-framed houses. The support for the floor of the room consists of 2 x 8 joists placed 16" on center and covered with 5/8" plywood sheathing. Before the floor frame is placed, a 2 x 8 pressure-treated sill should be lagged or bolted into the top of the foundation wall. A vent should be included in the foundation as protection against dampness and deterioration.

When planning the roof frame, be sure to match or complement the pitch of the house roof and the design of its soffit and fascia. This part of the project requires substantial carpentry skills and a basic knowledge of framing procedures. Seek advice and help if these tasks are new to you. The sheathing and exterior finish materials can be applied once the framing is completed. This plan includes economical asphalt strip shingles, vinyl siding, and aluminum soffits and fascias.

If other finish products are more appropriate to the style of your house, these substitutions will alter both the cost of materials and the estimated time for installation. Most do-it-yourselfers can accomplish this roof work if they are given instruction on laying it out and flashing it correctly at the house wall. But do allow plenty of time to work slowly if you are not an experienced roofer.

Inside, many different surface coverings and trim materials can be used, according to your home's decor and practical requirements. Be sure to take care of all the preliminary work, such as the electrical rough-in, insulating, and the placing of underlayment, before you apply the finish surfaces.

LEVEL OF DIFFICULTY

Much of this 8' x 8' entry can be completed by do-it-yourselfers. As noted earlier, the foundation work should be left to a contractor; but the framing and finish work can be performed by homeowners with a moderate level of remodeling skill. Beginners should get help before starting new procedures and seek guidance as they progress.

Homeowners should weigh the savings on labor costs against the investment of their time in jobs that require specialized skills and know-how. Siding, roofing, electrical work, and flooring can be done much faster and with better results by a specialty contractor than by the do-it-yourselfer who is unfamiliar with the tools, materials, and ins-and-outs of these jobs. Generally, beginners should add 100% to the labor-hours estimate for the basic tasks and hire a professional for the time-consuming, technically demanding specialty operations. Intermediates should add 50% to the time for basic tasks and carpentry procedures, and 70% for any specialty work that is new to them.

Experts should add about 20% to the professional time for all tasks in this project. Almost all do-it-yourselfers should hire a contractor to do the electrical rough-in, and even experts should leave the circuit tie-in to a qualified tradesperson.

8' x 8' Room Addition

Description	Quantity/Unit	Labor-Hours	Material
Excavate trench, with machine, 4' deep	1 C.Y.	0.1	
Footing, 8" thick x 16" wide x 20' long	1 C.Y.	2.9	120.00
Foundation, 8" concrete block, not reinf., 1/2" parged, 4' high	96 S.F.	9.0	193.54
Frame, floor, 2 x 8 joists, 16" O.C., 8' long	80 L.F.	1.2	83.52
Floor sheathing, 5/8" plywood, 4' x 8' sheets	64 S.F.	0.8	43.78
Frame, walls, 2 x 4 studs & plates, 8' long	20 L.F.	4.8	124.13
Header over window, 2 x 8 stock	8 L.F.	0.4	8.35
Wall sheathing, 1/2" x 4' x 8' plywood	192 L.F.	2.7	112.90
Frame, roof, 2 x 6 rafters, 16" O.C.	80 L.F.	1.3	54.72
Ridge board & sub-fascia, 2 x 8 x 8'	24 L.F.	0.9	25.06
Ceiling, 2 x 6 joists, 16" O.C.	48 L.F.	0.6	32.83
Roof sheathing, 1/2" x 4' x 8' plywood	96 L.F.	1.1	56.45
Siding, vinyl embossed, 8" wide, with trim	180 S.F.	5.8	142.56
Soffit, 12" wide, vented, fascia & drip edge	35 L.F.	4.7	62.16
Roofing, standard strip shingles, class C, 4 bundles per square	1 Sq.	2.0	58.80
Housewrap, spun bonded polypropylene	320 S.F.	0.7	61.44
Flashing, aluminum stop flashing, 0.019" thick	20 S.F.	1.1	18.48
Gutters, 5" aluminum, 0.027" thick	16 L.F.	1.1	23.23
Downspouts, 2" x 3", embossed aluminum, 0.020" thick	20 L.F.	0.8	17.76
Wall insulation, fiberglass, 3-1/2" thick, R-11	180 S.F.	0.9	73.44
Floor insulation, fiberglass, kraft faced, 6" thick, R-19	64 S.F.	0.4	25.34
Ceiling, insulation, fiberglass, 9" thick, R-30	50 S.F.	0.3	24.60
Window, double hung, vinyl clad, 2'-4" x 5'-4"	2 Ea.	1.6	715.20
Exterior door, steel, 3'-0" x 6'-8", prehung	1 Ea.	1.0	582.00
Sidelights	2 Ea.	1.1	744.00
Lockset, standard duty, cylindrical, keyed	1 Ea.	0.8	78.60
Drywall, 1/2" x 4' x 8', standard, taped and finished	320 S.F.	5.3	96.00
Underlayment, 5/8" plywood, underlayment grade	64 S.F.	0.7	53.76
Flooring, vinyl sheet, 0.125" thick	50 S.F.	2.0	214.20
Trim, ranch base & casing	50 L.F.	1.7	68.40
Paint, walls & ceilings, primer, oil base, roller	320 S.F.	1.3	15.36
Door and frame, primer, oil base, brushwork	1 Ea.	1.3	2.11
Window, incl. frame and trim, primer, oil base, brushwork	2 Ea.	1.1	0.67
Trim, primer, oil base, brushwork	50 L.F.	0.6	1.20
Walls & ceilings, 2 coats, oil base, roller	320 S.F.	1.3	15.36
Door and frame, 2 coats, oil base, brushwork	1 Ea.	1.3	2.11
Window, incl. frame and trim, 2 coats, oil base, brushwork	2 Ea.	1.1	0.67
Trim, 2 coats, oil base, brushwork	50 L.F.	0.6	1.20
Wiring, 4 outlets	4 Ea.	6.0	78.48
Wiring, 1 switch	1 Ea.	1.4	21.36
Backfill, by hand, no compaction, light soil	1 C.Y.	0.6	
Fine grading and seeding, incl. lime, fertilizer & seed	7 S.Y.	0.3	1.34
Totals		**74.7**	**$4,055.11**

Project Size	8' x 8'		Contractor's Fee Including Materials	**$9,510**

Key to Abbreviations
C.Y.– cubic yard   Ea.– each   L.F.– linear foot   Pr.– pair   Sq.– square (100 square feet of area)
S.F.– square foot   S.Y.– square yard   V.L.F.– vertical linear foot   M.S.F.– thousand square feet

WHAT TO WATCH OUT FOR

The type and placement of windows and doors are among the most important design features of any room addition. Although they are also some of the more expensive items in any home improvement project, it does not pay to compromise on the quality or design of these components. Whenever possible, purchase well-made thermopane window units. Not only do they provide ongoing energy savings, but they usually include screens, and do not require the extra expense of storm windows. The door used in this plan is a new fixture; however, the old exterior door may be relocated to the entryway of the new room. Existing window units can be relocated in the same way to reduce the materials cost. If you choose to reuse the old door and windows, be sure that they are in good enough condition to make their relocation worthwhile. While cosmetic deficiencies are usually fairly simple to correct, structural defects involve a considerable amount of repair time.

SUMMARY

This addition can add to the comfort and convenience of your home. It is an approachable project for the do-it-yourselfer because it is relatively small in size and is conveniently located.

For other options or further details regarding options shown, see

Main entry door

Standard window installation

8' x 12' ROOM ADDITION

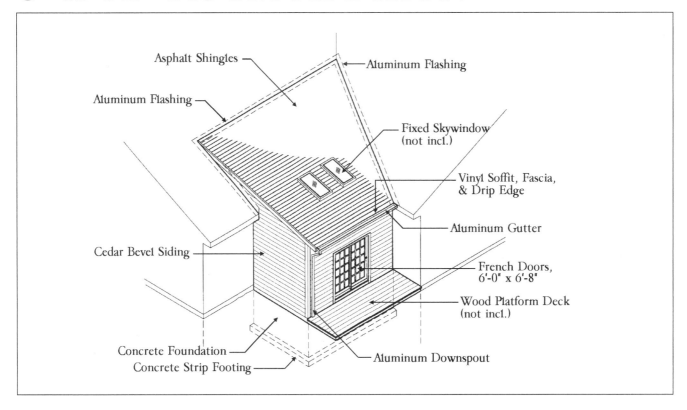

Asphalt Shingles

Aluminum Flashing

Aluminum Flashing

Fixed Skywindow
(not incl.)

Vinyl Soffit, Fascia,
& Drip Edge

Aluminum Gutter

French Doors,
6'-0" x 6'-8"

Cedar Bevel Siding

Wood Platform Deck
(not incl.)

Concrete Foundation

Concrete Strip Footing

Aluminum Downspout

The proximity of property boundaries, design of the existing house, and other factors often prevent the conventional construction of a room addition. In these circumstances, unusual designs may be called for, such as this 8' x 12' addition between the openings of the house. Although the structure is moderate in size, this is a challenging project for most do-it-yourselfers, as it calls for quite a few carpentry and building skills.

MATERIALS

One benefit of placing the addition in a corner location is that two of the walls are already in place. Thus, you can save on the materials and labor that are normally required to frame and cover these walls. For the new construction, a 4' foundation is required on the remaining two sides. The other materials are standard framing lumber, and interior and exterior coverings and finish products appropriate to the style of the house. Before placing the order, do some comparative pricing at several building supply outlets. Also, because of the size of the order and the number and weight of materials, trucking is required, so make arrangements for delivery before you close the deal.

The foundation suggested in this plan is constructed from poured concrete, but concrete blocks can also be used and will function as well. The extra time and expense required to install foundation vents are well invested, as they will help to increase air circulation and prevent moisture from accumulating underneath the structure. In most cases, excavations and foundation installations of this size are best left to the contractor. Accurate location and leveling of the forms, and skillful concrete placement, are vital to the success of the project.

The shell of this addition is comprised of standard materials commonly used in exterior framing and finishing. After the sill, floor joists, and floor sheathing have been placed, the two exterior walls are framed on the deck and raised into position. The roof framing of corner structures can be tricky, so plan ahead and get some help if you think you will need it. After the roof structure has been completed, the gable end must be framed with 2 x 4s placed individually. The sheathing for the walls in this plan is 3/4" rigid styrofoam, but standard 1/2" plywood can also be used. The roof also requires plywood sheathing before it is covered with a layer of felt paper and

asphalt shingles. Because of the amount of flashing required, most do-it-yourselfers should consider hiring a professional to do the roofing work.

Before the interior finish materials are applied, the exterior walls and the ceiling should be insulated at the recommended R-values for your area of the country. In colder climates, the floor should also be insulated before the sheathing is placed. Additional materials costs will have to be added to the project; but your floors will be warmer during the cold months, and your heating bill lower. After the electrical work has been roughed in, the interior finish materials can be installed. Cost will vary here according to your choice of products, the amount of trim, and decorative considerations. Standard 1/2" drywall requires know-how and skill to install and finish, but most intermediates can do the job if they are given some assistance at the outset. Underlayment should be installed beneath the carpet and laid at right angles to the floor sheathing, with no coincidental seams.

8' x 12' Room Addition

Description	Quantity/ Unit	Labor- Hours	Material
Excavate, with machine, 4' deep	14 C.Y.	0.2	
Footing, 8" thick x 16" wide x 20' long	1 C.Y.	2.9	120.00
Foundation walls, 6" thick, C.I.P. concrete	2 C.Y.	5.8	297.60
Frame, floor, 2 x 8 joists, 16" O.C., 8' long	120 L.F.	1.8	125.28
Floor, sheathing, 5/8" plywood, 4' x 8' sheets	96 S.F.	1.1	65.66
Frame, walls, 2 x 4 studs & plates, 8' long	20 L.F.	4.0	103.44
Header over door, 2 x 8	7 L.F.	0.3	7.31
Wall sheathing, 3/4" x 4'x8' plywood	160 S.F.	2.6	126.72
Frame, roof, 2 x 6 rafters, 16" O.C.	169 L.F.	2.7	115.60
Sub-fascia, 2 x 6	14 L.F.	0.2	9.58
Ceiling, 2 x 6 joists, 16" O.C.	90 L.F.	1.2	61.56
Support beam, double, 2 x 10	28 L.F.	0.8	83.33
Roof sheathing, 5/8" plywood, 4' x 8' sheets	230 L.F.	2.8	157.32
Siding, cedar bevel, 1/2" x 8", A-grade	160 S.F.	4.7	364.80
Soffit, 12" wide, vented, fascia & drip edge	22 L.F.	2.9	39.07
Roofing, standard strip shingles	2 Sq.	4.0	117.60
Building paper, asphalt felt sheathing paper, 15 lb.	50 S.F.	0.1	1.20
Flashing, at valley, aluminum, 0.019" thick	20 S.F.	1.1	18.48
Gutters, 5" aluminum, 0.027" thick	12 L.F.	0.8	17.42
Downspouts, 2" x 3", embossed aluminum, 0.020" thick	10 L.F.	0.4	8.88
Wall insulation, fiberglass, 3-1/2" thick, R-11	160 S.F.	0.8	65.28
Ceiling insulation, fiberglass, 9" thick, R-30	100 S.F.	0.6	85.20
Exterior door, french, 6'-0" x 6'-8"	1 Ea.	2.3	1,320.00
Lockset, std duty, cylindrical, keyed, single cylinder function	1 Ea.	0.8	78.60
Drywall, 1/2" x 4' x 8', standard, taped and finished	320 S.F.	5.3	96.00
Underlayment, 5/8" x 4' x 8' plywood, underlayment grade	96 S.F.	1.1	80.64
Carpet, 24 oz. nylon, light to medium traffic	108 S.F.	1.9	141.26
Prime quality urethane pad for carpet	108 S.F.	0.6	29.81
Trim, ranch base & casing	70 L.F.	2.3	95.76
Paint, walls & ceilings, primer, oil base, roller	320 S.F.	1.3	15.36
Door and frame, primer, oil base, brushwork	2 Ea.	2.7	4.22
Trim, primer, oil base, brushwork	70 L.F.	0.9	1.68
Walls & ceilings, 2 coats, oil base, roller	320 S.F.	1.3	15.36
Door and frame, 2 coats, oil base, brushwork	2 Ea.	2.7	4.22
Trim, 2 coats, oil base, brushwork	70 L.F.	0.9	1.68
Wiring, 4 outlets	4 Ea.	6.0	78.48
Wiring, 1 switch	1 Ea.	1.4	21.36
Backfill, by hand, no compaction, light soil	2 C.Y.	1.1	
Fine grading and seeding, incl. lime, fertilizer & seed	1 C.Y.	0.1	0.19
Totals		74.5	$3,975.95

Project Size	8' x 12'	Contractor's Fee Including Materials	**$9,343**

Key to Abbreviations
C.Y.– cubic yard   Ea.– each   L.F.– linear foot   Pr.– pair   Sq.– square (100 square feet of area)
S.F.– square foot   S.Y.– square yard   V.L.F.– vertical linear foot   M.S.F.– thousand square feet

LEVEL OF DIFFICULTY

This project requires a moderate to advanced level of skill in many construction procedures, including framing, roofing, siding, and exterior finish work. However, once the structure is closed in, the interior tasks can be completed, one at a time, over an extended period. Beginners can approach the interior work at a relaxed pace on weekends and evenings, but they should leave most of the exterior work to contractors.

For more basic tasks, like painting, sheathing, and landscaping, beginners should add 100% to the professional time. Intermediates should add 50% for all but the roof framing and finishing, electrical, and carpet installations. They should allow more time for these specialty tasks and, depending on their ability, should hire professionals to do the work. Experts should add 20% to allow for a slow and patient installation process. All do-it-yourselfers should hire a contractor for the excavation and foundation work.

WHAT TO WATCH OUT FOR

The unusual roof lines of structures like this corner room addition provide a good situation for a skylight. Adding this amenity to the project will increase the cost, but the installation will be much easier and faster if it is done while the roof is being built, rather than later. The good ones are expensive; but the extra cost for a quality product will be returned in energy efficiency and durability over the years.

SUMMARY

Corner room additions such as this design require imaginative planning and skillful installation, but offer a great potential for reward.

For other options or further details regarding options shown, see

> Bay window
> Main entry door
> Skylights
> Standard window installation

12′ x 16′ FAMILY ROOM ADDITION

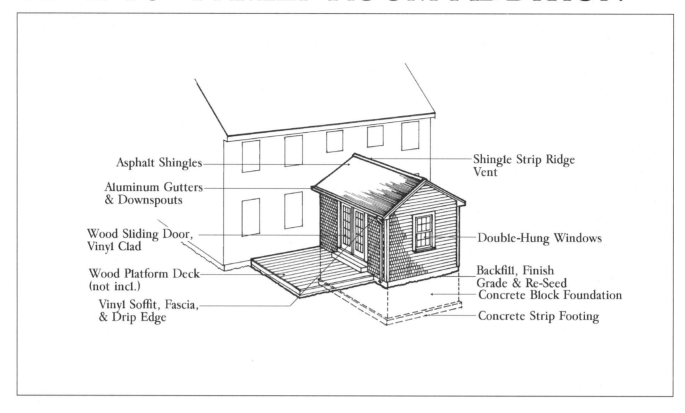

Asphalt Shingles

Aluminum Gutters & Downspouts

Wood Sliding Door, Vinyl Clad

Wood Platform Deck (not incl.)

Vinyl Soffit, Fascia, & Drip Edge

Shingle Strip Ridge Vent

Double-Hung Windows

Backfill, Finish Grade & Re-Seed

Concrete Block Foundation

Concrete Strip Footing

If you have sufficient yard area and adequate distance to property lines, an ell configuration is usually the most convenient and appropriate design to use for a room addition. The new structure, though attached, is essentially a miniature house with its own foundation, frame, and interior and exterior finish. Care should be taken in determining its location, roof lines, and exterior trim to ensure that it will become an integral part of the existing house. Just as important are local zoning regulations, which should be checked before you begin any work. Because this is a relatively large project, beginners and intermediates with limited remodeling experience should hire a contractor to do much of the work, especially on the exterior of the structure.

MATERIALS

The covering and finish materials for room additions and attached structures tend to vary with each project, as they should match or complement the style features of the existing house. However, the support and subsurface materials have little variation from structure to structure. The materials in this plan,

especially the basic subsurface components, are readily available at building supply outlets, and you can shop around for the best deal. Most do-it-yourselfers should have the excavation and foundation work for this addition done by a contractor. A substantial amount of earth has to be removed to allow for the 4′ foundation depth. Additional expense should be figured in if some of the excavated material has to be trucked from the site. This plan suggests a concrete block foundation, but a formed and poured foundation can also be used. If you plan to install the foundation yourself, but don't have any experience with masonry, get some professional advice before you start. A poorly constructed or misaligned foundation can cause major problems.

After the foundation has been placed, the framing, sheathing, and exterior finishing tasks can be accomplished. A 2 x 8 sill plate is secured to the top of the foundation wall, and the 2 x 10 floor joists installed. Be sure to use at least 5/8″ plywood for the decking and to stagger the termination seams for additional floor strengthening. Standard 2 x 4 framing with appropriate headers and jack studs is used for the door and window openings,

which are then covered with 1/2″ sheathing. If you are not experienced in framing, be sure to get some help for this part of the wall and roof work. The gable roof is basic in its framing design, but it requires skill and know-how to construct. The most critical aspect of the roof framing procedure is the layout of the rafter template, which must duplicate or complement the roof pitch and overhang features of the existing house. This plan also includes a ridge vent strip, which should be installed as part of the finish roofing procedure. The final steps in the exterior work include finishing the soffit and trim, placing the door and window units, and applying the siding.

Within the new room, preliminary work on the floor, ceiling, and walls has to be completed before the finish products are placed. The subfloor has to be covered with appropriate underlayment if carpeting or resilient flooring is to be installed. The ceiling and walls should be insulated and roughed in for the desired electrical fixtures. When installing the drywall, always hang the ceiling pieces first and then the wall sheets. This plan calls for wainscoting constructed from 1/4″ paneling on the lower part of the walls. Before the paneling is put into place, 3/8″

12' x 16' Room Addition

Description	Quantity/Unit	Labor-Hours	Material
Excavate, with machine, 4' deep	28 C.Y.	3.0	
Footing, 8" thick x 16" wide x 20' long	2 C.Y.	5.9	240.00
Foundation wall, 8" thick concrete block, 4' high	176 S.F.	16.6	354.82
Frame, floor, 2 x 10 joists, 16" O.C., 12' long	192 L.F.	3.4	285.70
Sill plates, 2 x 8, pressure treated	48 L.F.	1.7	84.48
Floor, sheathing, 5/8" plywood, 4' x 8' sheets	192 S.F.	2.3	131.33
Frame, walls, 2 x 4 plates, 12' long	144 L.F.	2.9	62.21
2 x 4 studs, 8' long	320 L.F.	4.7	138.24
Header over window & door, 2 x 8	36 L.F.	1.7	37.58
Wall sheathing, 1/2" plywood, 4' x 8' sheets	384 S.F.	5.5	225.79
Frame, roof, 2 x 6 rafters, 16" O.C., 8' long	208 L.F.	3.3	142.27
Ceiling, 2 x 6 joists, 16" O.C.	156 L.F.	2.0	106.70
Ridge board & sub-fascia, 2 x 8	50 L.F.	1.8	52.20
Roof sheathing, 5/8" plywood, 4' x 8' sheets	288 S.F.	3.6	196.99
Building paper, asphalt felt, 15 lb., on roof & walls	700 S.F.	1.5	16.80
Roofing, asphalt shingles, class A, 285 lb. premium	3 Sq.	6.9	178.20
Flashing at wall, 0.01 thick	16 S.F.	0.9	9.79
Ridge vent strip, polyethylene	16 L.F.	0.8	48.96
Siding, no. 1 red cedar shingles, 7-1/2" exposure, on walls	2.50 Sq.	9.8	303.00
Soffit, 12" wide vented, fascia & drip edge	48 L.F.	6.4	85.25
Gutters, 5" aluminum, 0.027" thick	34 L.F.	2.3	49.37
Downspouts, 2" x 3", embossed aluminum, 0.020" thick	20 L.F.	0.8	17.76
Wall insulation, fiberglass, 3-1/2" thick, R-11	400 S.F.	2.0	163.20
Ceiling insulation, fiberglass, 6" thick, R-19	200 S.F.	1.0	98.40
Sliding door, vinyl clad, 6'-0" x 6'-8"	1 Ea.	4.0	924.00
Dbl.-hung window, 36" x 48", insul., incl. frame, screen, & trim	3 Ea.	2.7	889.20
Drywall, 1/2" x 4' x 8', standard, taped and finished	384 S.F.	6.4	115.20
Paneling, 1/4" thick for wainscot, 4' x 8' sheets	160 S.F.	6.1	215.04
Plywood, 3/8" thick, backing for paneling, 4' x 8' sheets	160 S.F.	2.1	84.48
Carpet, 24 oz. nylon, light to medium traffic	207 S.F.	3.7	270.76
Prime quality, urethane pad for carpet	207 S.F.	1.2	57.13
Trim, ranch, base, casing & wainscoting	155 L.F.	5.2	212.04
Paint, walls & ceilings, primer, oil base, roller	420 S.F.	1.7	20.16
Door and frame, primer, oil base, brushwork	1 Ea.	1.3	2.11
Window, incl. frame and trim, primer, oil base, brushwork	3 Ea.	1.7	1.01
Trim, primer, oil base, brushwork	155 L.F.	1.9	3.72
Walls & ceilings, 2 coats, oil base, roller	420 S.F.	1.7	20.16
Door and frame, 2 coats, oil base, brushwork	1 Ea.	1.3	2.11
Window, incl. frame and trim, 2 coats, oil base, brushwork	3 Ea.	1.7	1.01
Trim, 2 coats, oil base, brushwork	155 L.F.	1.9	3.72
Wiring, 4 outlets	4 Ea.	6.0	78.48
Wiring, 1 switch	1 Ea.	1.4	21.36
Backfill, by hand, no compaction, light soil	2 C.Y	1.1	
Fine grading and seeding, incl. lime, fertilizer & seed	21 S.Y.	1.0	4.03
Totals		144.9	$5,954.76

Project Size	12' x 16'		Contractor's Fee Including Materials	**$15,669**

Key to Abbreviations
C.Y.– cubic yard   Ea.– each   L.F.– linear foot   Pr.– pair   Sq.– square (100 square feet of area)
S.F.– square foot   S.Y.– square yard   V.L.F.– vertical linear foot   M.S.F.– thousand square feet

plywood backing is applied to the studs as a support, furring, and fastening subsurface.

LEVEL OF DIFFICULTY

The exterior painting, landscaping, and other basic operations are tasks that beginners can tackle, but advanced procedures such as framing and roofing should be left to a contractor. Beginners should add at least 100% to the labor-hours estimate for all of the tasks they undertake. Intermediates should attempt only those exterior operations that they feel comfortable with. The roof framing and finish work are especially challenging undertakings that may require professional assistance. Intermediates should be able to handle all of the interior operations, except the carpeting and electrical installations, with 50% added to the professional time. Experts should add 20% for all interior and exterior jobs.

WHAT TO WATCH OUT FOR

The existing house wall that is included in the addition requires considerable preparation before it can be finished to blend in with the new room. The exterior siding has to be removed and any structural alterations made before the drywall or other finish material is applied. If an acceptable entryway from the house to the new room already exists, no structural work is required and the rough opening can simply be trimmed with the appropriate materials. If there is no access or an existing entryway has to be widened, the wall will have to be opened and new structural members added.

SUMMARY

This room addition will boost the value of your home by adding to its living space and enhancing its exterior appearance.

For other options or further details regarding options shown, see

Bay window

Main entry door

Skylights

Standard window installation

20' x 24' ROOM ADDITION

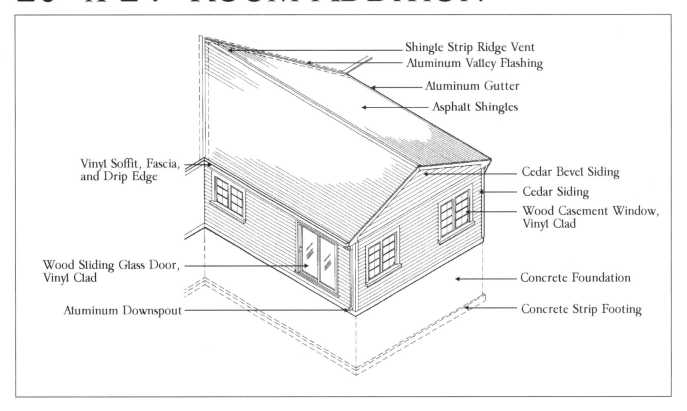

Shingle Strip Ridge Vent
Aluminum Valley Flashing
Aluminum Gutter
Asphalt Shingles

Vinyl Soffit, Fascia, and Drip Edge

Cedar Bevel Siding
Cedar Siding
Wood Casement Window, Vinyl Clad

Wood Sliding Glass Door, Vinyl Clad

Concrete Foundation

Aluminum Downspout

Concrete Strip Footing

This plan demonstrates how a large space can be attached to the eave end of a house at first-floor level. It has over 400 square feet of floor area, a full 20' x 24' basement, and a roof structure that is neatly tied in with the existing house. This area is large enough to accommodate a new master bedroom suite or family room/kitchen.

MATERIALS

One of the primary considerations in buying the components for this addition is the size of the order and the logistics involved in storing the materials. Unless you have a good-sized garage or sheltered enclosure, you should wait to have interior materials delivered until after the shell has been erected and made weathertight. Even the exterior items require a large outside area adjacent to the site for stockpiling. Your retailer may be willing to stagger the delivery of these goods to help you out if storage is a problem.

All of the foundation work should be completed by a professional contractor. The 9' excavation has to be performed by machine and the excess material hauled

away. The cost for the excavation will vary depending on the topographical features of the site. If the adjacent grade falls off to or below basement walk-out level, the amount of excavation may be reduced. The basement floor slab can be placed at any time after the foundation has been completed, providing there is proper access.

Because of the 20' width of this addition, a tripled 2 x 10 center support beam must be installed to bear the load of the floor and its frame. This primary structural component must be properly supported with lally columns or other suitable posts placed at regular intervals and set into a preformed notch in the foundation wall. After the sill has been placed and the 2 x 10 floor joists installed, 5/8" plywood sheathing will cover the frame. The 2 x 4 walls are then laid out, built on the deck, positioned, fastened, and covered with 1/2" sheathing. The ceiling joists and roof rafters are then placed and sheathed, and the gable end studded and sheathed to complete the framing operation. After the layout and placement of the new roof frame has been completed, the finished roofing, soffit and trim, window and door units, and siding are installed to close in the new structure.

All exterior and interior finish materials should be selected to complement the style of your house. The shell should first be thoroughly insulated on the ceiling, walls, and, in colder areas, the floor. The electrical rough-in should also be installed before the finish coverings are placed. The standard 1/2" sheetrock for the walls and ceiling can be taped and finished or, at extra cost, skim-coated with plaster. All of the interior finish products demand careful and patient installation; specialty contractors may have to be hired for this job.

LEVEL OF DIFFICULTY

Most do-it-yourselfers should plan to hire a general contractor or various subcontractors to do the heavy exterior work and the specialty installations. The insulating, drywalling, trim work, and painting are jobs that can be performed during weekends or evenings over an extended period. Generally, beginners should limit their work to the interior tasks mentioned above and to basic exterior work like painting and landscaping. They should add 100% to the professional time for these jobs and seek professional advice as they go. Most intermediates

20' x 24' Room Addition

Description	Quantity/Unit	Labor-Hours	Material
Excavate, trench, with machine, 9' deep	160 C.Y.	5.7	
Haul-off excess material	135 C.Y.	15.4	
Footing, 8" thick x 16" wide x 20' long	3 C.Y.	8.8	360.00
Foundation walls, 8" thick C.I.P. concrete	12 C.Y.	23.8	1,584.00
Basement floor slab, 4" thick conc., text. finish, no reinf., no form	480 S.F.	9.4	518.40
Frame floor, 2 x 10 joists, 16" O.C., 12' long	540 L.F.	9.6	803.52
Center support beam, triple 2 x 10	24 L.F.	0.7	71.42
Floor, sheathing, 5/8" plywood, 4' x 8' sheets	480 S.F.	5.7	328.32
Frame & trim, basement stairs, complete	1 Ea.	5.3	930.00
Frame, walls, 2 x 4 plates, 12' long, 2 x 4 studs, 8' long	96 L.F.	19.2	496.51
Header over window & door, 2 x 8	60 L.F.	2.8	62.64
Wall sheathing, 3/4" x 4' x 8' plywood	544 S.F.	8.9	430.85
Housewrap, spun bonded polypropylene	680 S.F.	1.4	130.56
Corner bracing, 16 ga. steel straps, 10' long	80 L.F.	1.1	57.60
Frame, roof, 2 x 8 rafters, 16" O.C., 14' long	1,008 L.F.	17.0	1,052.35
Ridge board & sub-fascia, 2 x 10	88 L.F.	3.5	130.94
Roof sheathing, 1/2" x 4' x 8' plywood	800 S.F.	9.1	470.40
Roofing, asphalt shingles, 285 lb. premium	8 Sq.	18.3	475.20
Building paper, asphalt felt sheathing paper, 15 lb., on roof	800 S.F.	1.7	19.20
Ridge vent strip, polyethylene	20 L.F.	1.0	61.20
Siding, cedar board & batten, pre-stained	390 S.F.	17.3	1,445.04
Cedar bevel, 1/2" x 8", grade A, pre-stained	60 S.F.	1.5	114.00
Soffit, 12" wide vented, with fascia	48 L.F.	6.4	85.25
Drip edge, aluminum, 0.016" thick, mill finish	48 L.F.	1.0	11.52
Flashing, at valley, 0.019" thick	80 S.F.	4.4	73.92
Gutters, 5" aluminum, 0.016" thick	68 L.F.	4.5	98.74
Downspouts, 5" aluminum, 0.025" thick	68 L.F.	3.9	123.22
Wall insulation, fiberglass, 3-1/2" thick, R-11	550 S.F.	2.8	224.40
Ceiling, insulation, fiberglass, 9" thick, R-30	480 S.F.	2.9	408.96
Door, sliding glass, 6' x 6'-8", vinyl clad, insul. glass	1 Ea.	4.0	924.00
Casement window, 24" x 52", vinyl clad, thermopane	4 Ea.	3.6	1,334.40
Casement window, 24" x 36", vinyl clad, thermopane	2 Ea.	1.6	400.80
Drywall, 1/2" x 4' x 8', standard, taped and finished	1,184 S.F.	19.6	355.20
Carpet, 22-oz. nylon, light to medium traffic	486 S.F.	8.7	635.69
Prime urethane pad for carpet	486 S.F.	2.9	134.14
Trim, ranch, base & casing	168 L.F.	5.6	229.82
Paint, walls & ceilings, primer, oil base, roller	1,200 S.F.	4.7	57.60
Window, incl. frame and trim, primer, oil base, brushwork	6 Ea.	1.5	2.02
Trim, primer, oil base, brushwork	168 L.F.	2.1	4.03
Walls & ceilings, 2 coats, oil base, roller	1,200 S.F.	4.7	57.60
Window, incl. frame and trim, 2 coats, oil base, brushwork	6 Ea.	2.4	4.32
Trim, 2 coats, oil base, brushwork	168 L.F.	2.1	4.03
Wiring, 8 outlets	8 Ea.	12.0	156.96
Wiring, 2 switches	2 Ea.	2.8	42.72
Backfill, by hand, no compaction, light soil	10 C.Y.	5.7	
Fine grading and seeding, incl. lime, fertilizer & seed	53 S.Y.	2.5	10.18
Totals		299.6	$14,921.67

Project Size	20' x 24'		Contractor's Fee Including Materials	$37,558

Key to Abbreviations
C.Y.– cubic yard   Ea.– each   L.F.– linear foot   Pr.– pair   Sq.– square (100 square feet of area)
S.F.– square foot   S.Y.– square yard   V.L.F.– vertical linear foot   M.S.F.– thousand square feet

should not tackle the framing on this project, and they should consider hiring a roofer unless they have substantial experience in this area. They should tack on 50% to the labor-hours for the other nonspecialty tasks. Experts should add 20% to all procedures except the roof framing and finish. Slightly more time should be allowed for these advanced exterior operations and for specialty interior work. All do-it-yourselfers should leave the foundation work and the basement floor installation to a professional contractor.

WHAT TO WATCH OUT FOR

The full basement included in this addition plan adds interior space that can be used for storage, a workshop, or living area. Access should be carefully planned to suit the function of the new room and the existing structure. The stairway from the new room does take away valuable floor space and may require a partition and a door to enclose the stairwell. Exterior cellar access is an option, but provisions for it will have to be included in the foundation design at extra cost. If the existing house already has a basement, it may be possible to gain access from it. Be aware that considerable expense may be incurred in hiring a concrete specialist to cut an opening in the foundation for the doorway.

DOOR OPTIONS

Cost Each, Installed

Description	
Steel Entrance, 3' wide	$ 235.00
Wood Entrance, 3' wide	$ 263.00
Sliding Door, Wood, 6' wide	$1,200.00
Sliding Door, Aluminum, 6' wide	$1,550.00
French Door, Wood, 5' wide	$1,275.00

SUMMARY

This room addition is a comprehensive project that requires a total commitment by the homeowner – both in the preliminary planning and in the carrying out of required construction standards.

For other options or further details regarding options shown, see

> *Bay windows*
> *Patio & sliding glass doors*
> *Skylights*
> *Standard window installation*

10′ x 10′ SECOND-STORY ADDITION

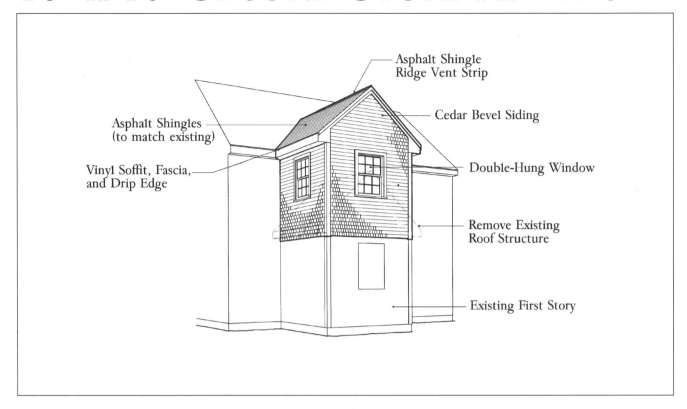

Asphalt Shingle Ridge Vent Strip

Asphalt Shingles (to match existing)

Cedar Bevel Siding

Vinyl Soffit, Fascia, and Drip Edge

Double-Hung Window

Remove Existing Roof Structure

Existing First Story

Second-floor living space can be increased by means of attic conversions or the placement of dormers, but still more area can be created with a complete room addition. If the conditions are right and there is an existing attached structure on which to build, adding a second story may be the answer. Because of the difficulties in removing the old roof, the potential for hidden problems, and the need to work rapidly, most do-it-yourselfers should hire a contractor to perform the exterior work.

MATERIALS

The materials used in this project are the same as those employed in first-floor room additions. Conventional framing, sheathing, and finish products are placed on top of the supporting first floor. What makes the installation of the new components difficult is the general inconvenience of the second-story location.

Before the new materials are placed, an extensive amount of preliminary planning has to be done, as well as preparation of the old structure to receive the new one. The first step is to evaluate the condition

of the existing wood structure and its foundation. Even expert do-it-yourselfers should hire a qualified inspector if there is any question about its strength or ability to support the load of the new materials. Foundation work and added lumber may have to be included in the estimate to reinforce the original structure before the new work begins. The next step is the removal of the old roof covering, sheathing, and rafters. As in dormer installations, this procedure can be time consuming, yet expediency is important because part of the interior of the house will be open to the elements. If you plan to do this part of the project without the services of a contractor, be sure to have plenty of help on hand to move the job quickly. Whenever possible, try to salvage the old lumber and sheathing. If they are in good condition, they may come in handy later in the project.

After the roof has been removed, the existing ceiling joists become the floor joists for the new room. If they are 2 x 8s or larger, they are probably strong enough as they are; but if they are 2 x 6s, they will need some reinforcing. Check with a professional to determine the size of required reinforcement materials and the methods of installing them. The

deck is then sheathed with 5/8″ plywood before the wall partitions are framed. Be sure to prepare the area of the existing house wall to receive the new framing. The roof is then framed with 2 x 6 rafters and a 2 x 8 ridge board tapered into the existing roof. The sheathing and finished roofing, selected to match the existing finish roof material, are then applied. After the window unit and siding are installed, the structure is ready for the interior work.

All of the operations involved in the exterior construction require a thorough knowledge of carpentry, skilled handling of tools, and safe equipment for work above ground level. Since complications tend to arise in this type of work, it is best to build in some extra time and, possibly, cost. Use this plan's price guide to get an accurate idea of what a similar structure would cost; then figure the expense of your own particular improvement by making any necessary modifications. This plan assumes that access to the new room will come through a finished second-story room or hallway. Whenever possible, existing window openings should be utilized as part of the future rough opening for a door. If you cannot use an existing window opening, then the wall will have to be opened at a convenient place and headers and jack studs placed to rough in the entryway. If you want to open the entire wall or a large portion of it, more support will be required. Additional cost and time may have to be included in the estimate for the entryway installation, according to its type, size, and conditions of placement. When the rough electrical work and insulating have been completed, the walls and ceiling can be drywalled and finished. Carpeting or other appropriate flooring material is put into place after the trim work and painting are done.

LEVEL OF DIFFICULTY

This project is one of the most challenging home improvements included in this book. The exterior operations require not only advanced carpentry and tool handling skills, but also fast and efficient work at roof level. The entire roof must be removed; the ceiling joists must be blocked off and, possibly, reinforced; and a new structure built and tied into the wall and roof of the existing house, all in a relatively short period of time. All levels of do-it-yourselfers, therefore, are encouraged to hire a specialty contractor for the exterior work. Even experts should consider professional help for the roof removal, and plan to have several experienced helpers on hand to assist with the framing and roofing tasks. Beginners and intermediates should restrict their efforts to the interior work only and tack on an additional 100%, and 50%, respectively, to the professional labor-hours. Experts should add 30% to the time estimates for all exterior tasks which they feel they can handle, and 20% to the interior operations.

10′ x 10′ Second-Story Addition

Description	Quantity/Unit	Labor-Hours	Material
Demolition, roofing built-up, shingles, asphalt strip	100 S.F.	1.1	
Demolition, gutters, aluminum or wood, edge hung	40 L.F.	1.3	
Demolition, roofing, sheathing, per 5 S.F.	20 Ea.	26.7	
Demolition, roofing, rafters, joists, & supports	100 S.F.	1.5	
Frame, boxing at existing joists, 2 x 6 stock	30 L.F.	0.4	20.52
Floor, sheathing, 5/8″ plywood, 4′ x 8′ sheets	100 S.F.	1.2	68.40
Frame, walls, 2 x 4 studs & plates, 8′ long	30 L.F.	6.0	155.16
Headers over windows, 2 x 8	24 L.F.	1.1	25.06
Gable end studs, 2 x 4	10 L.F.	0.2	4.32
Wall sheathing, 1/2″ x 4′ x 8′ plywood	270 S.F.	3.8	158.76
Housewrap, spun bonded polypropylene	300 S.F.	0.6	57.60
Frame, roof, 2 x 6 rafters, 16″ O.C.	192 L.F.	3.1	131.33
Sub-fascia, 2 x 8	20 L.F.	1.4	20.88
Ceiling, 2 x 6 joists, 16″ O.C.	90 L.F.	1.2	61.56
Roof sheathing, 1/2″ x 4′ x 8′ plywood	224 S.F.	2.6	131.71
Siding, cedar bevel, 1/2″ x 8″, A grade	270 S.F.	7.9	615.60
Soffit, 12″ wide vented, fascia & drip edge	20 L.F.	2.7	35.52
Roofing, standard asphalt strip shingles	3 Sq.	5.3	142.20
Building paper, asphalt felt sheathing paper, 15 lb.	270 S.F.	0.6	6.48
Ridge vent strip, polyethylene	10 L.F.	0.5	30.60
Flashing, at valley	40 S.F.	2.2	36.96
Gutters, 5″ aluminum, 0.027″ thick	20 L.F.	1.3	29.04
Downspouts, 2″ x 3″, embossed aluminum	36 L.F.	1.5	31.97
Wall insulation, fiberglass, 3-1/2″ thick, R-11	270 S.F.	1.4	110.16
Ceiling insulation, fiberglass, 9″ thick, R-30	100 S.F.	0.7	85.20
Window, double-hung, 3′ x 4′-6″, thermopane, clad	2 Ea.	1.8	732.00
Drywall, 1/2″ x 4′ x 8′, standard, taped and finished	416 S.F.	6.9	124.80
Underlayment, 3/8″ particleboard	100 S.F.	1.1	68.40
Carpet, Olefin, 22 oz., light traffic	11 S.Y.	0.2	8.71
Carpet padding, sponge rubber pad	11 S.Y.	0.1	4.09
Trim, ranch, base & casing	70 L.F.	2.3	95.76
Paint, walls and ceiling, primer, oil base, roller	512 S.F.	2.0	24.58
Trim, primer, oil base, brushwork	110 L.F.	1.4	2.64
Walls and ceiling, 2 coats, oil base, roller	512 S.F.	5.1	55.30
Trim, 2 coats, oil base, brushwork	110 L.F.	2.2	6.60
Wiring, 4 outlets	4 Ea.	6.0	78.48
Wiring, 1 switch	1 Ea.	1.4	21.36
Totals		106.8	$3,181.75

WHAT TO WATCH OUT FOR

The new attached structure should blend with the house and complement its style, and not look "tacked on." Roofing and siding lines are particularly important in this blending process. This plan, for example, requires that the roof line of the new room align with that of the existing house, and that the courses of asphalt roofing meet precisely at the valley. The new siding should follow the same spacing of the structure below, and its course lines should align at the seam between the house and the addition. The soffit and trim features, roof pitch, and door and window design should all be considered in the proper blending of old and new. The overall size and architectural proportion of the new room are equally important to the design.

SUMMARY

Although it requires a considerable effort to construct, this addition provides an alternative plan to attic conversion or dormer installation.

For other options or further details regarding options shown, see

Dormers

Skylights

Standard window installation

Project Size	10′ x 10′	Contractor's Fee Including Materials	$9,789

Key to Abbreviations
C.Y.– cubic yard   Ea.– each   L.F.– linear foot   Pr.– pair   Sq.– square (100 square feet of area)
S.F.– square foot   S.Y.– square yard   V.L.F.– vertical linear foot   M.S.F.– thousand square feet

20′ x 24′ Second-Story Addition

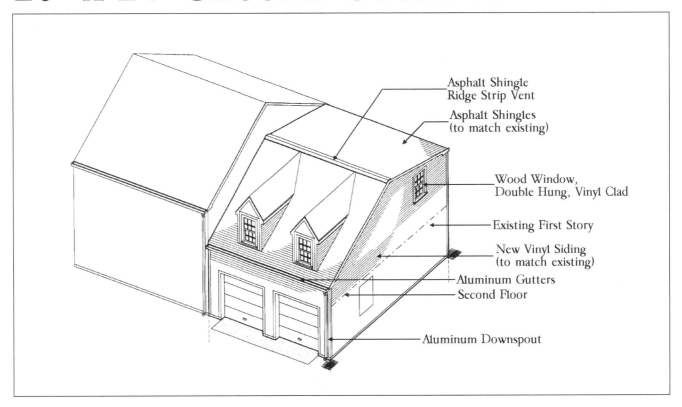

Asphalt Shingle
Ridge Strip Vent

Asphalt Shingles
(to match existing)

Wood Window,
Double Hung, Vinyl Clad

Existing First Story

New Vinyl Siding
(to match existing)

Aluminum Gutters

Second Floor

Aluminum Downspout

This room addition plan demonstrates how nearly 500 square feet of new interior space can be created above a two-car, attached garage. Such a project is an expensive, time-consuming undertaking that requires a major commitment on the part of the homeowner. The time schedule for the procedure has to be carefully planned to ensure expedient completion of the exterior enclosure. An architect can help you determine the style, design, and structural aspects of the project. Because this addition demands professional expertise and assistance, do-it-yourselfers should hire a contractor to complete the exterior work.

MATERIALS

This project involves costs for roof removal and, in most cases, for strengthening the support system for the new room. This expense is offset by the fact that a foundation already exists, and you can avoid the costs of its excavation and construction. Once the support system has been reconditioned to acceptable standards, the same materials used in other wood frame structures are employed in constructing this new room. All materials used in this project are readily

available at building supply outlets, so shop around for the best prices. In working out the delivery conditions, you might arrange to receive the exterior and interior products in two separate loads.

Before the new decking can be applied to the room, the existing garage roof has to be removed and the ceiling joists reinforced to safe standards. If the existing joists are 2 x 8s or larger and are supported by a center beam, the amount of reinforcement will be minimal. If they are 2 x 6s, a substantial amount of reinforcement will be required. The cost for this procedure will vary with each situation. After the 5/8″ plywood decking has been installed, the walls and roof are framed with standard 2 x 4 and 2 x 6 lumber and sheathed with 1/2″ plywood or comparable material. Care should be taken to blend the new structure with the old so that the roof line, the style of the overhang, and other design features complement one another. The new exterior finish products should also be selected to harmonize with the style of the house. For example, premium asphalt shingles are suggested in this plan, but standard asphalt or cedar shingles can also be used. Tile and slate are two more possibilities. Some of

the options for exterior walls are cedar shingles, clapboards, aluminum siding, stucco, or the vinyl siding suggested in this plan. All of these materials vary in price, and the selection of one product over another will have a considerable impact on the estimated cost.

After the exterior has been completed and the structure made weathertight, a doorway or opening into the new room must be created. If you are completing the interior work on your own and are unfamiliar with cutting and framing a wall opening, get some advice before you start. The placement of a standard-sized 30″ or 33″ door requires 2 x 8 headers and jack studs for the rough opening frame, but a 6′ or 8′ wall opening necessitates more support and, possibly, temporary ceiling bracing during the installation procedure. Remember to allow time for stripping the siding on the wall where the entry is placed.

One of the advantages of this and most other room addition projects is that the interior work can be completed by the do-it-yourselfer, one task at a time, when convenient. The insulating, rough electrical work, and any partitioning are completed first, and then the ceiling and walls are covered and finished.

Partitioning for closets or room dividers will add to the total cost and the installation time. The costs for interior coverings and their trim work will vary according to the style and quality of the products.

LEVEL OF DIFFICULTY

This large second-story addition is one of the most challenging home improvement projects presented in this book. The removal of the roof and subsequent framing and roofing are difficult operations that must be completed as expediently as possible to protect the interior from the elements. All do-it-yourselfers, therefore, are encouraged to leave the exterior work to a professional and concentrate their efforts on the inside tasks.

Do-it-yourselfers should be able to handle most of the interior operations, though they should leave the electrical and carpet installations to professionals. Beginners should add 100% to the labor-hours estimate for all of the tasks they undertake. Intermediates and experts should add 50% and 20%, respectively, to the estimate for the interior jobs, and more for the specialty work.

WHAT TO WATCH OUT FOR

In building above a garage, be sure that you check local building code requirements regarding fire separation. If it is not already in place, you probably will have to add 2 layers of 5/8" fire code gypsum board to the garage ceiling. You will probably also want to put insulation under the floor, especially if the new space will be used for bedrooms. By using vinyl siding and clad windows, you will avoid substantial costs for exterior painting. If you use other materials to match your existing siding, you may need to add costs for exterior painting.

SUMMARY

This addition requires the services of a professional contractor for much of the structural and exterior work. However, most do-it-yourselfers can substantially reduce the cost of the project by completing the interior work themselves. The combined efforts of contractor and homeowner can result in a lot of new living space at a reasonable cost.

For other options or further details regarding options shown, see

> *Bay window*
> *8' x 12' second-story addition*
> *Skylights*
> *Standard window installation*

20' x 24' Second-Story Addition

Description	Quantity/ Unit	Labor- Hours	Material
Demolition, roofing, built-up shingles, asphalt strip	480 S.F.	5.5	
Demolition, roofing, sheathing, per 5 S.F.	96 Ea.	128.0	
Demolition, roofing, rafters, joists, & supports	480 S.F.	7.1	
Frame, floor, 2 x 8 joists, 16" O.C., 12' long	540 L.F.	7.9	563.76
Floor sheathing, 3/4" plywood, 4' x 8' sheets	480 S.F.	6.1	380.16
Frame, walls, 2 x 4 plates, 12' long, 2 x 4 studs, 8' long	264 L.F.	52.8	1,365.41
Headers over windows, 2 x 8 stock	24 L.F.	1.1	25.06
Wall sheathing, 25/32" x 4' x 8' wood fiberboard	704 S.F.	9.4	295.68
Corner bracing, 16 ga. steel straps, 10' long	80 L.F.	1.1	57.60
Housewrap, spun bonded polypropylene	640 S.F.	1.4	122.88
Frame, roof, 2 x 6 rafters, 16" O.C., 14' long	600 L.F.	10.1	626.40
Ridge board, 2 x 8 stock	22 L.F.	0.9	32.74
Sub-fascia, 2 x 6 stock	40 L.F.	0.6	27.36
Ceiling, 2 x 6 joists, 16" O.C.	416 L.F.	5.3	284.54
Roof sheathing, 5/8" x 4' x 8' plywood	960 S.F.	11.8	656.64
Roofing, asphalt shingles, 285 lb. premium	10 Sq.	22.9	594.00
Felt paper, 15 lb., on roof	1,000 S.F.	2.2	24.00
Ridge vent strip, polyethylene	20 L.F.	1.0	61.20
Siding, vinyl panels, 8" - 10" wide	525 S.F.	16.5	434.70
Soffit, 12" wide vented, fascia & drip edge	90 L.F.	12.0	159.84
Gutters, 5" aluminum, 0.027" thick	68 L.F.	4.5	98.74
Downspouts, 2" x 3", embossed aluminum	36 L.F.	1.5	31.97
Wall insulation, fiberglass, 3-1/2" thick, R-11	525 S.F.	2.6	214.20
Ceiling insulation, fiberglass, 9" thick, R-30	480 S.F.	2.4	236.16
Window, double hung, vinyl clad, 3'-0" x 5'-0"	5 Ea.	5.0	1,890.00
Drywall, 1/2" x 4' x 8', standard, finished	1,184 S.F.	19.6	355.20
Carpet, 26-oz. nylon level loop, light to medium traffic	54 S.Y.	8.7	903.96
Carpet padding, sponge rubber pad	54 S.Y.	2.9	180.79
Trim, ranch, base & casing	186 L.F.	6.2	254.45
Jamb, 1 x 6 clear pine, at opening to house	19 L.F.	0.8	19.84
Paint, walls & ceiling, primer, oil base, roller	1,200 S.F.	4.7	57.60
Trim, primer, oil base, brushwork	186 L.F.	2.3	4.46
Walls & ceiling, 2 coats, oil base, roller	1,200 S.F.	12.0	129.60
Trim, 2 coats, oil base, brushwork	186 L.F.	3.7	11.16
Wiring, 8 outlets	8 Ea.	12.0	156.96
Wiring, 2 switches	2 Ea.	2.8	42.72
Totals		395.4	$10,299.78

Project Size	20' x 24'	Contractor's Fee Including Materials	$34,104

Key to Abbreviations
C.Y.– cubic yard   Ea.– each   L.F.– linear foot   Pr.– pair   Sq.– square (100 square feet of area)
S.F.– square foot   S.Y.– square yard   V.L.F.– vertical linear foot   M.S.F.– thousand square feet

LEAN-TO GREENHOUSE

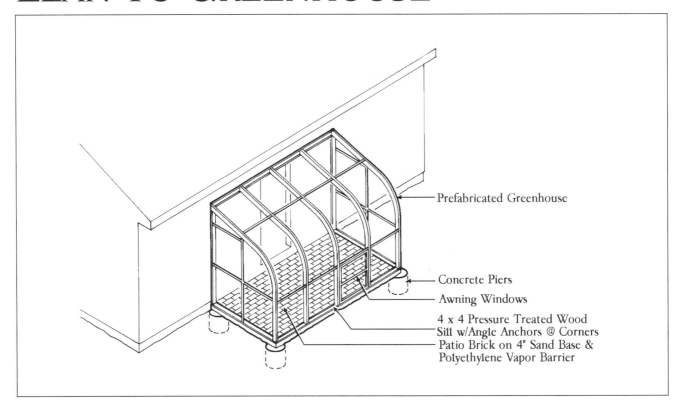

- Prefabricated Greenhouse
- Concrete Piers
- Awning Windows
- 4 x 4 Pressure Treated Wood Sill w/Angle Anchors @ Corners
- Patio Brick on 4" Sand Base & Polyethylene Vapor Barrier

Greenhouses can enhance the atmosphere of your home while adding to its comfort. Their primary function is as a practical and aesthetically pleasing enclosed garden area, but they can also increase interior living space, provide solar heat, and improve the lighting and ventilation of adjacent rooms. This plan presents both a small and a large greenhouse project.

MATERIALS

Like other attached exterior structures, greenhouses require a foundation or some other means of support, a basic framing system, flooring, and inside and outside finish work. If you plan to use a factory-prepared greenhouse kit, as this plan recommends, much of the framing and covering material for the structure has been taken care of by the manufacturer. The foundation, flooring, and finish materials are standard construction products that can be purchased at most building supply outlets. Additional materials and installation time may be needed if the new greenhouse requires an entryway into the house.

Before the greenhouse kit is assembled and placed, a considerable amount of preliminary work must be completed. Aluminum-framed greenhouses with fiberglass or tempered glass panels require a sturdy and deep foundation or ground support system.

If the foundation is not true in its placement, the greenhouse will not sit properly on the slab or against the house wall; consequently, the enclosure may develop leaks. In climates where deep frost occurs, the slab recommended in this plan may not be adequate, and a deeper foundation may be required. Any movement or heaving of the surface can cause the seams along the house and foundation to open, the mullions and weatherstripping in the greenhouse to separate, and, possibly, the glass panels to crack. Other floor and foundation systems could also be used; for example, redwood and pressure-treated pine are options for the base. These can be set in or on one of the following: concrete pedestals, concrete block foundations set on footings, or established supports like decks and patios.

The greenhouse itself is the most important and expensive component included in either project. Many designs

and options other than the ones presented here are available, so investigate all of the possibilities before you buy. Because of the craftsmanship involved and specialized materials required for their manufacture, greenhouse kits are expensive products. Further, their cost and quality vary widely, depending on the type of materials used in their frame and panels, and the engineering of their structural and tie-in systems. As with other specialized factory-made home building products, the selection of a quality item with a proven reputation for durability and low maintenance is usually worth the extra cost. Some additional expense will be incurred if you choose to include thermal shades, screens, vents, fans, and other options. Further costs will be incurred if any electrical, plumbing, or heating components are to be included in the project.

When erecting the new facility, be sure to fasten it correctly to the base frame and tie it into the house with acceptable flashing methods. If there is not already a doorway, you might consider installing one. In colder climates, especially, arranging direct access from the house makes sense because of convenience and the potential benefits of passive solar heating. Like other attached structures, the new greenhouse should blend with the rest of the house, so some compromises may have to be made in determining its location. Factors to consider include the design of the house, the extent to which the greenhouse is to be used, and its orientation to the sun.

LEVEL OF DIFFICULTY

This project presents an opportunity for do-it-yourselfers of all ability levels to take part in its construction. While beginners may find the pedestal and frame work out of reach, they should add 100% to the labor-hours for the brick floor installation and for any other finish work. Intermediates should figure on an additional 40% time allowance; and experts, about 10%. Because the greenhouse kit is expensive and intricate in its materials and design, all do-it-yourselfers should allow extra time for its assembly and installation.

WHAT TO WATCH OUT FOR

A greenhouse can collect a considerable amount of passive solar heat in the course of a winter day. The floor, house wall, and the plants themselves absorb heat from the sunlight, store it, and then provide a source of radiant warmth after the sun has gone down. With a little ingenuity and a modest investment, this energy can be put to use to heat the greenhouse at night, and can even help to heat the house. One way of collecting more solar heat is to paint the house wall and other surfaces, like plant racks and floor borders, a dark color. Another easy way to collect more heat is to place portable dark-colored containers filled with water against the house wall. To transfer the heat from the greenhouse to the main house during the day and, sometimes, even at night, you might install a small exhaust fan in the house wall near the ceiling of the greenhouse. For a modest extra investment, you can purchase a thermostatic fan control that will turn it on and off to ensure a stable temperature. All of these means of collecting and controlling passive solar heat energy are economical to install and their benefits can be surprisingly great.

SUMMARY

This greenhouse project offers rewards for gardeners and homeowners who want to enhance the atmosphere of their homes. For a small additional investment, you can also reap the benefits of passive solar heat.

For other options or further details regarding options shown, see

> *Brick or flagstone patio in sand*
>
> *Interior doors**
>
> *Patio & sliding glass doors*
>
> ** In Interior Home Improvement Costs*

Small Lean-To Greenhouse*

Description	Quantity/ Unit	Labor- Hours	Material
Excavate, post holes, incl. layout	0.50 C.Y.	0.5	
Forms, round fiber tube, 8" diameter, for posts	12 L.F.	2.5	26.06
Concrete, field mix, 1 C.F. per bag, for posts	4 Bags		29.76
Placing concrete, footings, under 1 C.Y.	0.50 C.Y.	0.4	
Angle anchors for 4 x 4 wood frame	4 Ea.	0.2	1.97
Wood frame for base, 4 x 4 treated lumber	36 L.F.	1.3	62.64
Polyethylene vapor barrier, 0.006" thick	72 S.F.	15.6	253.15
Washed sand for 4" base	1 C.Y.	0.1	14.34
Grade and level sand base	60 S.F.	2.1	
Patio brick, 4" x 8" x 1-1/2" thick, laid flat	60 S.F.	9.6	154.80
Greenhouse, prefab, 5'-6" x 10'-6" x 7' high, w/2 awning windows	58 SF Flr	16.9	5,533.20
Flashing, aluminum, 0.016" thick	24 S.F.	1.3	22.18
Totals		50.5	$6,098.10

Project Size	5'-6" x 10'-6"	Contractor's Fee Including Materials	**$11,131**

Key to Abbreviations
C.Y.– cubic yard   Ea.– each   L.F.– linear foot   Pr.– pair   Sq.– square (100 square feet of area)
S.F.– square foot   S.Y.– square yard   V.L.F.– vertical linear foot   M.S.F.– thousand square feet
*For 8' x 16' greenhouse, add $6,959 to the contractor's fee.

SUNROOM-GREENHOUSE

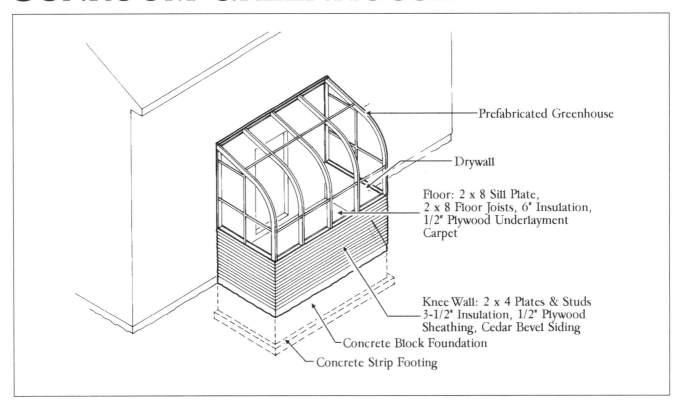

Prefabricated Greenhouse

Drywall

Floor: 2 x 8 Sill Plate,
2 x 8 Floor Joists, 6" Insulation,
1/2" Plywood Underlayment
Carpet

Knee Wall: 2 x 4 Plates & Studs
3-1/2" Insulation, 1/2" Plywood
Sheathing, Cedar Bevel Siding

Concrete Block Foundation

Concrete Strip Footing

Combination sunroom-greenhouses combine the beauty of plants and open natural lighting with the practical benefits of passive solar heating and increased year-round interior living area. If the location and access are appropriate, the greenhouse can become an integral part of your home's floor plan and provide an attractive improvement. A sunroom can be costly to install, but the benefits to your lifestyle, energy savings, and increased property value can make it worthwhile. This project provides estimates for both a small and a large sunroom-greenhouse.

MATERIALS

The materials used in this project include standard construction components for the foundation, floor, and knee wall in the new structure, and for the entryway into the house. The biggest item is the factory-prepared sunroom-greenhouse kit, which consists of an insulated aluminum frame and double-layered tempered glass. The basic structural and finish materials are readily available from building supply retailers, but the greenhouse kit is usually a special-order product. Before you purchase the materials for this project, do some

research and shop around, especially for the greenhouse unit. Because there are many brand names and variations in quality and cost, you can take advantage of competitive pricing while selecting the greenhouse unit that meets your practical needs and aesthetic standards.

Other than the fabrication and placement of the greenhouse itself, the most important operation in the project is the construction of the foundation and floor systems. Before the foundation work begins, a substantial amount of excavation must be completed, especially for the large greenhouse. A perimeter trench must be dug to receive the form for the concrete. This task may have to be done by hand if the location of the site prohibits machine excavation. After the concrete has been allowed to set up for several days, a gravel subsurface, reinforcing, and expansion joints are installed, and the slab poured. The surface of the slab is troweled and then broom-finished to provide an acceptable base surface for the floor covering. In most cases, a contractor should be hired to complete these operations. Mistakes in concrete work can be difficult and expensive to correct, and a precisely set

foundation is a must for the fixed dimensions of the greenhouse.

Insulating the foundation and floor is strongly advised, as temperature variations and drafts are detrimental to the operation and comfort of the greenhouse.

Before selecting the type and design of the greenhouse kit, consider its thermal and aesthetic impact on the house. Because the greenhouse kit is being used to create a year-round sunroom, the extra investment for insulating glass and frames is recommended. Some kits come complete with screens, shades, and vent windows, while others offer these items at additional cost. When assembling the greenhouse, work carefully to maintain precise alignment. The base fasteners for the greenhouse are usually included in the kit, but flashing for the base, roof, and walls is extra.

Before the finish work can begin on the interior of the small sunroom-greenhouse, the doorway from the house must be installed or reconditioned. In most cases, the exterior siding must be removed and, in the large greenhouse plan, a window opening enlarged and framed to receive a 6' sliding glass door unit. If the new sunroom is to be incorporated into the house's interior for year-round use, the opening can be enlarged even more and then trimmed with a jamb and facings. In this way you can deduct the expense of the slider unit, but you may have to add slightly to the cost of preparing the rough opening. Standard interior or indoor/outdoor carpeting provides comfort and additional warmth to the floor. The interior design and the selection of finish materials offer a wide range of choice and the opportunity to use your imagination in blending the new facility with your home's decor.

LEVEL OF DIFFICULTY

Compared to a basic attached greenhouse, these combined greenhouse-and-sunroom facilities require more time, skill, and expense to install. Mistakes can be very costly. Beginners and inexperienced intermediates should consider hiring a contractor to perform most operations, but might finish the project on their own. Because of the out-of-the-way location of most greenhouse sites, the light finish work can be completed at a comfortable pace and under convenient conditions. Beginners should add 100% to the labor-hour estimates for basic finish work,

and should seek professional advice. Intermediates and experts should add 40% and 10%, respectively, to the professional time for all tasks in the smaller greenhouse. Intermediates should probably stick to the interior finishing for the larger model at a rate of 50% over the estimated time. If they attempt to lay the tile or brick floor, they should add 100% to the professional time to allow for slow and careful work. Experts should leave the foundation and slab installations to a specialist and consider hiring a contractor to erect the larger greenhouse. For the basic tasks for the large greenhouse, they should add 20% to the estimated labor-hours, and for the finished floor, about 40%. All do-it-yourselfers should plan to add extra time to assemble and install the greenhouse kit, as the cost of the product and the intricacy of its design demand careful workmanship.

WHAT TO WATCH OUT FOR

Before finalizing the master plan for your greenhouse project, be sure to make provisions for electrical and plumbing fixtures. Although these items are not listed in this plan, you might consider installing at least one electrical outlet and, perhaps, a light fixture for the interior wall. For an extra cost, a small sink or other fixture can also be included in the plan. If these electrical and plumbing conveniences are roughed in during the construction process, the procedure will cost less and look better.

SUMMARY

With imagination, creative design, and some extra expense, a basic attached greenhouse can become a delightful sunroom as well as an interior garden.

For other options or further details regarding options shown, see

Electrical system: Outlets*

Electrical sytem: Receptacles*

Interior doors*

Lean-to greenhouse

* In Interior Home Improvement Costs

Small Sunroom-Greenhouse*

Description	Quantity/ Unit	Labor-Hours	Material
Excavation, trench, by hand, pick and shovel, to 6' deep	2 C.Y.	2.0	
Footing, 6" thick x 16" wide x 32' long	2 C.Y.	5.5	218.40
Foundation, 8" concrete block, 2'-8" high, parged 1/2" thick	105 S.F.	9.9	211.68
Perimeter insulation, molded bead board, 1" thick, R-4	100 S.F.	1.2	22.80
Sill plate, 2 x 8 x 12'	22 L.F.	0.8	38.28
Floor joists, 2 x 8, 16" O.C.	132 L.F.	1.9	137.81
Floor insulation, fiberglass blankets, foil-backed, 6" thick, R-19	60 S.F.	0.4	29.52
Underlayment, plywood, 1/2" thick, 4' x 8' sheets	60 S.F.	0.6	35.28
Framing for knee wall, 2 x 4 plates and studs, 3' high, 16" O.C.	108 L.F.	2.7	46.66
Knee wall insulation, fiberglass, foil-backed, 3-1/2" thick, R-11	70 S.F.	0.4	28.56
Sheathing, plywood on knee walls, 1/2" thick	96 S.F.	1.4	56.45
Greenhouse, prefab., 5'-6" x 10'-6" x 6'-6", insul. dbl. glass	58 SF Flr	9.6	5,150.40
Flashing, aluminum, 0.019" thick	25 S.F.	1.4	23.10
Remove window for opening, to 25 S.F., in existing building	1 Ea.	0.4	
Remove siding for opening in existing building	60 S.F.	1.4	
Frame opening, 2 x 8 stock	24 L.F.	1.1	25.06
Jamb and casing for opening, 9/16" x 3-1/2", interior and exterior	48 L.F.	1.6	65.66
Drywall for interior knee wall and sides of opening, 1/2" thick	128 S.F.	2.1	38.40
Paneling, plywood, prefinished, 1/4" thick	128 S.F.	4.1	113.66
Flooring, carpet, 26 oz., nylon level loop, light to medium traffic	60 S.F.	1.1	111.60
Flooring, sponge rubber pad, maximum	60 S.F.	0.4	61.92
Siding for exterior knee wall, cedar, beveled, rough-sawn, stained	80 S.F.	2.7	288.96
Totals		52.7	$6,704.20

Project Size	5'-6" x 10'-6"	Contractor's Fee Including Materials	$12,088

Key to Abbreviations
C.Y.– cubic yard   Ea.– each   L.F.– linear foot   Pr.– pair   Sq.– square (100 square feet of area)
S.F.– square foot   S.Y.– square yard   V.L.F.– vertical linear foot   M.S.F.– thousand square feet
*For 10-6"x18'-0" Sunroom-greenhouse, add $20,529 to contractor's fee.

Section Ten
SECURITY MEASURES

Simply keeping your doors locked when you are away, and even when you are home, is one of the most effective measures you can take against burglary. This project provides some additional measures as well as details for installing new locks. Following are some additional tips.

- The first security device that police and other experts recommend is a deadbolt lock.

- Before boring a hole in a door for a deadbolt lock or door viewer, check your templates carefully and review the size of the holes required.

- If possible, get keyed window locks from a single manufacturer so that one key can fit them all. Keep a key near each window, hidden from outside view, so the window can be opened quickly in an emergency. Check local code.

- Eliminate trees and shrubbery near doors and windows that obstruct visibility from the street and provide hiding places for burglars.

- More dramatic security measures than those described here include installing security shutters, window grilles, or a security gate.

- Prior to undertaking any security project, make sure you have the correct tools and equipment. Often, minor fitting is required — sharp tools make a difference.

- Check all products for any special tool requirements. Some hardware uses special screws and fasteners requiring special drivers, etc.

SECURITY MEASURES

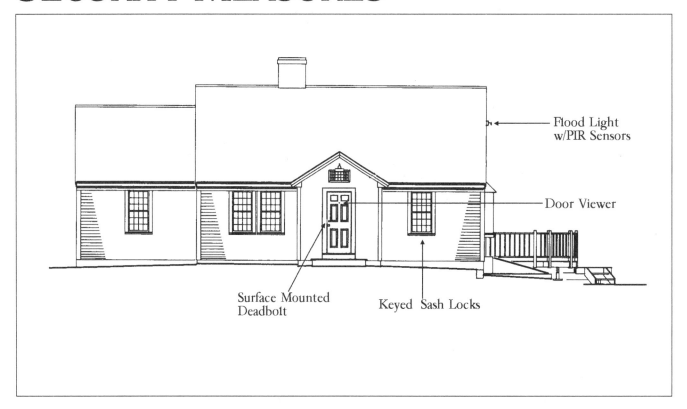

Flood Light w/PIR Sensors

Door Viewer

Surface Mounted Deadbolt

Keyed Sash Locks

Whether you live in the city, the suburbs, or a rural area, you are not immune to the growing epidemic of crime, and providing some form of minimal security system for your home and family is prudent.

MATERIALS

This project provides some basic security measures that are relatively easy to incorporate in any home. It does not provide every possible alternative. A totally integrated security system includes such features as lighting, landscaping, fencing, windows, doors, locks, and alarms. However, setting up a reasonably effective security system for your home does not necessarily mean spending large amounts of money on expensive, high tech components. Your local police department can provide you with valuable advice on home security, and many economical suggestions and tips. Keep in mind that a smart, determined, and well-equipped "professional" burglar can outwit virtually any system you devise. In fact, any thief will seek out the house that offers the least amount of resistance to his efforts and causes him the least amount of trouble.

Doors are the most obvious points of entry, especially those in back of the house, screened by shrubs, tucked in a corner of the building, or otherwise hidden from view. Every exterior entrance door should be equipped with a deadbolt lock. The surface-mounted type is the easiest to install and generally provides the best protection. Also very common and effective is the cylinder deadbolt. Either style is available as a single cylinder (thumb turn inside and key outside) or double cylinder (key inside and out). A deadbolt or a door latch can be strengthened with a reinforcing strike plate attached to the door jamb with 3" screws. Such a plate will reduce the chances of the door's being kicked in or pried open. The length of the screws allows them to go through the jamb and penetrate the studs that form the rough opening. A brass wrap-around knob reinforcer helps protect the somewhat weak knob area of the door from similar criminal abuse. A peephole-type door viewer can also be installed to allow you to see who is at the door without having to open it.

Windows, especially the double-hung variety, are also obvious and vulnerable openings for burglars. Keyed locks or pins

can be installed in the sashes, as well as steel bars across the opening, to delay or forestall entry. Bars are very unattractive and should only be used in extreme circumstances, when no other alternative exists.

A drop-down steel bar and a keyed lock should be installed on every sliding patio door. The design of a slider makes it relatively easy to pry open, so special care should be taken to secure it.

There are various types of electronic security systems available. One of the most effective and least expensive involves the installation of outdoor passive infrared (PIR) sensors, which detect motion and can switch on lights and activate alarms. The most common PIR floodlight fixtures resemble standard double floodlights, but have sensors mounted between the bulbs. On some models, the sensor's viewing angle can be adjusted to control the size of the area to be monitored. When the unit detects motion, the sensor will turn on the lights and, after an adjustable, preset delay period, shut them off. This type of fixture can usually be installed in place of a standard floodlight, and can be controlled by the same interior switch.

PIR sensors can be mounted separately from the floodlight, located almost anywhere, and arranged to detect motion in a wide area. These sensors operate the light switch with a radio signal and are powered by batteries, so no exterior wiring is needed. Some PIR sensors can send signals to the home's electrical system to activate interior lights or alarms.

Improvements and innovations are constantly being made in PIR technology. Portable units that are currently available incorporate a photovoltaic solar panel, a battery, a floodlight, an alarm, and a PIR sensor. During daylight hours the battery is charged with solar energy, which powers the unit at night.

Strategically located PIR sensors, combined with floodlights and alarms, make an excellent first line of defense for your home, detecting intruders as they approach the house and, it is hoped, scaring them off before they attempt to break in.

The same technology used for outdoor PIR sensor systems is available for indoor applications to enhance a home's overall security. Prices vary with the system's level of sophistication. For example, a single battery-operated motion detector can be purchased and installed by a homeowner for around $60. A multi-zone, programmable system connected to the house wiring, with a backup battery, can cost several hundred dollars, depending on factors such as the site of the house and number of components required. Complex security systems are usually purchased for an installed price, as they require specific expertise on the part of the installer. They are not an appropriate do-it yourself project for the homeowner.

LEVEL OF DIFFICULTY

All the security products described here come with detailed instructions for installation and use, and none require other than common household tools — drill, wood chisel, screwdriver, etc. When you shop for deadbolt locks, strike plates, and knob reinforcers, be sure you know the material construction and dimensions of your door and of their existing hardware, to ensure that you come home with the proper items and sizes. Get help if you are unsure of exactly what you need. Hooking up PIR fixtures in place of existing floodlights is not too difficult. If you need to run new wiring, it's best to leave the work to an expert or a professional electrician. You would also be well advised to get an expert's opinion on how best to arrange an outdoor light/alarm system to maximize its effectiveness. Except for electrical wiring, these tasks can be handled successfully by the average homeowner. Security is important enough to justify hiring professional installers. They can design a system that works best for your home and will work within your budget.

Beginners should add 100% to the time estimates provided; intermediates, about 50%; experts, about 15%.

WHAT TO WATCH OUT FOR

If you install double cylinder deadbolts on exterior doors, be sure that there is always a key near every lock, and that everyone in the house knows exactly where all the keys are and can locate them, even in the dark. Deadbolt locks are effective deterrents against forced entry, but should not prevent your rapid exit in the case of a fire or other emergency.

Valuables should be marked, indelibly and in an unobtrusive place, for purposes of identification. Most police departments have marking and etching tools available on loan for citizens to use. If your security system fails to deter a thief and these items are stolen, your identifying marks are valuable aids to the police in their efforts to trace them.

The installation of some devices may actually weaken the construction of a door or window. Before purchasing security devices, check on the construction of the door, jamb, sash, and frame. Be sure that boring a hole through the door doesn't pose a weakness problem. Also check the backset (distance from the edge of the door to the center of the hole through the door for the device) to ensure you have room for all the hardware trim. Finally, you can purchase a wireless security system that you can install yourself, including a control panel and sensors for two doors, at your local home center for about $200.

SUMMARY

No amount of high tech hardware can guarantee total protection from a professional burglar. However, since most robberies are committed by amateurs, you greatly increase protection for your family, home, and possessions by such common sense measures as simply keeping your doors and windows locked, and by installing any device that makes forced entry difficult. All burglars shy away from light and noise. An outdoor system linking lights and alarms to strategically located passive infrared (PIR) motion detectors is adaptable to almost any layout, fairly easy to install, relatively inexpensive, and is generally a reliable and effective deterrent.

Alarm systems are a good deterrent themselves, but work even better when linked to a monitoring service.

For other options or further details regarding options shown, see

Post light & GFI receptacle

Security System

Description	Quantity/ Unit	Labor-Hours	Material
Locks, surface-mounted, deadbolt, reinf. strike plate, 3" screws	2 Ea.	2.0	304.80
Knob reinforcers, brass, wrap-around	2 Ea.	2.0	49.20
Door viewers	2 Ea.		35.28
Window locks, keyed sash locks, brass	12 Ea.	3.0	138.24
Remove existing floodlights	3 Ea.	1.0	
Floodlights, 2-bulb w/PIR sensors, to replace exist. floodlights	3 Ea.	3.0	100.80
Totals		11.0	$628.32

Project Size	Per Job	Contractor's Fee Including Materials	$1,468

Key to Abbreviations
C.Y.– cubic yard   Ea.– each   L.F.– linear foot   Pr.– pair   Sq.– square (100 square feet of area)
S.F.– square foot   S.Y.– square yard   V.L.F.– vertical linear foot   M.S.F.– thousand square feet

Part Three
DETAILS

This part supplements the preceding project plans and descriptions. It deals with the cost of unit items rather than that of whole, predetermined projects. You can use this section to add or substitute components in one of the example projects, or to put together your own model that will suit your specific space requirements, the style of your house, your budget, and your personal taste. This information will also help you to arrive at the total cost of a smaller project, such as adding a shutters or a combination door.

"Details" includes a variety of available materials and their comparative costs, as well as the labor-hours involved in their installation and the corresponding contractors' fees. Everything from kitchen cabinets to insulation is listed in this part of the book, with a range of quality and price in each category. By knowing the prices of the available options, as well as their installation times and contractors' charges, you can make informed choices about the fixtures and materials you need and want for a given home improvement project. You can also determine how much of the work you are willing to take on yourself and how much should be done by a contractor. This section also has illustrations identifying

the various components. This feature makes planning and purchasing that much easier and helps you to visualize the renovation in progress.

Using the information in this part can be a bit more challenging than reading through the projects described in Part Two. For example, different standards of measure (such as square foot, linear foot, and set) are used according to the type of material; and the quantity of building components must be calculated with the measurements you make yourself. A "How To Use" page follows this introduction and explains the format in more detail. Some organization is required when you use the "Details" to draw up plans and make estimates, and a bit of imagination is helpful, too, when it comes to putting the separate elements together. The benefits of this independent approach are the ability to create your own remodeling system, to expand or reduce the size of a given system, and to change certain aspects within that system for reasons of economy, practicality, or personal preference. As a result, you can be the designer of your own room or facility and can make plans in advance according to the commitment of time, effort, and money that is right for you.

HOW TO USE PART THREE

Part Three, "Details," is provided as a quick, efficient reference for estimating the cost of home improvements. It enables you to closely match dollars and cents to your ideas before you commit yourself to a project.

Basic building elements are grouped together to create a complete "system." These "systems" can then be used to make up a proposed project. For example, the "Gable End Roofing Systems," lists all the basic elements required to shingle a roof. The result is the total cost per square foot of area to be shingled.

The systems can also be combined to create an estimate for a more complex

project. For example, if you need to replace the roof structure before shingling, you can consult the "Gable End Roof Framing Systems," to calculate the cost per square foot for framing a new roof. Then, add in the cost for shingling from the "Gable End Roofing Systems." In this manner, an entire project, matched to your specifications for size and proposed design, can be cost estimated.

Two types of pages are found in "Details." The most common format is composed of a graphic illustration of the construction process and a list of the building elements that make it up. In many cases, several different options are presented. For example, the "Gable End Roofing

Systems" pages show the cost per square foot for asphalt as well as for wood shingles.

The second type of page contains the same kind of information in a slightly different format. These pages list the cost for items that do not require a grouping of building elements. For example, the "Insulation Systems" page lists the cost per square foot for various types of building insulation.

The illustration below defines terms that appear in this section and will help you use the information presented. The illustration on the facing page shows how this part of the book is organized.

Key to Abbreviations

C.Y. – cubic yard
Ea. – each
L.F. – linear foot
Pr. – pair
Sq. – square (100 square feet of area)
S.F. – square foot
S.Y. – square yard
V.L.F. – vertical linear foot

System Descriptions uniquely define the system being priced. Descriptions range from item-by-item component lists to single entries differentiating systems by size or other relevant characteristics.

Quantity of each component per system unit

Unit of measure for each item. A Key to Abbreviations appears to the left on this page.

Material Cost for each component of the system. In systems comprised of a group of components, the total material cost for one unit of the system appears in a box at the bottom of the column.

System Description	QUAN.	UNIT	LABOR-HOURS	MAT. COST	COST PER S.F.
ASPHALT, ROOF SHINGLES, CLASS A					
Shingles, inorganic class A, 210-235 lb./sq., 4/12 pitch	1.160	S.F.	.017	.37	
Drip edge, metal, 5" wide	.150	L.F.	.003	.04	
Building paper, #15 felt	1.300	S.F.	.002	.03	
Ridge shingles, asphalt	.042	L.F.	.001	.03	
Soffit & fascia, white painted aluminum, 1' overhang	.083	L.F.	.012	.20	
Rake trim, 1" x 6"	.040	L.F.	.002	.03	
Rake trim, prime and paint	.040	L.F.	.002	.01	
Gutter, seamless, aluminum painted	.083	L.F.	.006	.10	Contractor's Fee
Downspouts, aluminum painted	.035	L.F.	.002	.04	**3.62**
TOTAL			.047	.85	

Total Cost per unit for each system. This figure includes the price of materials (without sales tax), labor costs, and the contractor's overhead expenses and profit.

Labor-Hours required to install the system. In systems made up of components, the total labor-hours to install one unit of the system appears in a box at the bottom of the column.

PART THREE
TABLE OF CONTENTS

Information in Part Three is organized into six divisions listed here. The schematic diagram pictured below shows how the components of a typical house fit into these Part Three divisions.

Note: For other divisions, see *Interior Home Improvement Costs.*

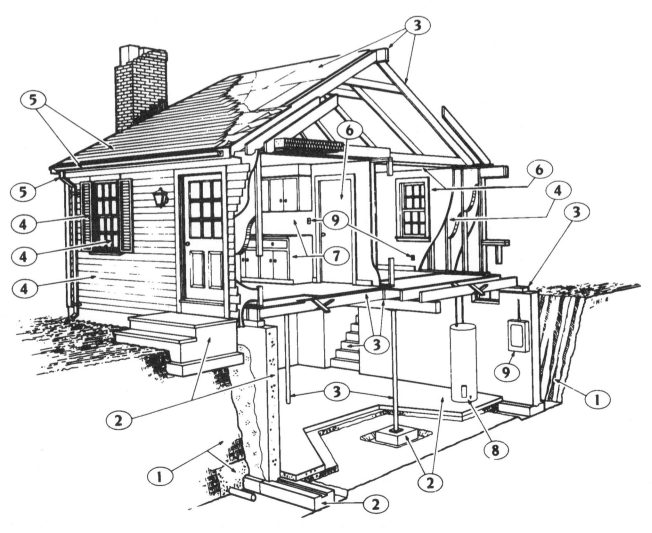

Division 1
SITE WORK

Footing Excavation Systems

Backfill

Excavate

System Description	QUAN.	UNIT	LABOR-HOURS	MAT. COST	COST EACH
BUILDING, 24' X 38', 4' DEEP					
Clear and strip, dozer, light trees, 30' from building	.190	Acre	9.120		
Excavate, backhoe	174.000	C.Y.	2.319		
Backfill, dozer, 4" lifts, no compaction	87.000	C.Y.	.580		
Rough grade, dozer, 30' from building	87.000	C.Y.	.580		**Contractor's Fee**
TOTAL			12.599		**1271.67**

Foundation Excavation Systems

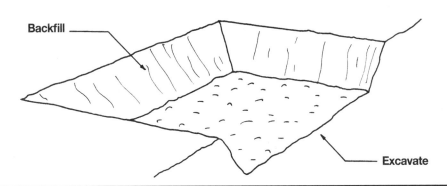

Backfill

Excavate

System Description	QUAN.	UNIT	LABOR-HOURS	MAT. COST	COST EACH
BUILDING, 24' X 38', 8' DEEP					
Clear & grub, dozer, medium brush, 30' from building	.190	Acre	2.027		
Excavate, track loader, 1-1/2 C.Y. bucket	550.000	C.Y.	7.860		
Backfill, dozer, 8" lifts, no compaction	180.000	C.Y.	1.201		
Rough grade, dozer, 30' from building	280.000	C.Y.	1.868		**Contractor's Fee**
TOTAL			12.956		**1711.93**

Be sure to read pages 170-171 for proper use of this section.

Utility Trenching Systems

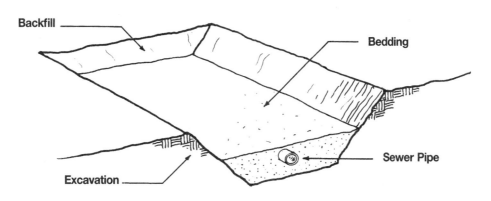

System Description	QUAN.	UNIT	LABOR-HOURS	MAT. COST	COST PER L.F.
2′ DEEP					
Excavation, backhoe	.296	C.Y.	.032		
Bedding, sand	.111	C.Y.	.044	1.45	
Utility, sewer, 6″ cast iron	1.000	L.F.	.283	13.12	**Contractor's Fee**
Backfill, incl. compaction	.185	C.Y.	.044		
TOTAL			.403	14.57	**41.76**

Sidewalk Systems

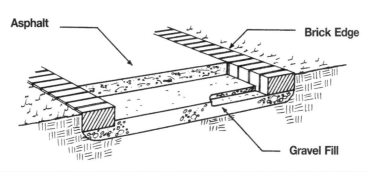

System Description	QUAN.	UNIT	LABOR-HOURS	MAT. COST	COST PER S.F.
ASPHALT SIDEWALK SYSTEM, 3′ WIDE WALK					
Gravel fill, 4″ deep	1.000	S.F.	.001	.59	
Compact fill	.012	C.Y.	.001		
Handgrade	1.000	S.F.	.004		
Walking surface, bituminous paving, 2″ thick	1.000	S.F.	.007	.45	
Edging, brick, laid on edge	.670	L.F.	.079	1.33	**Contractor's Fee**
TOTAL			.092	2.37	**7.81**

Be sure to read pages 170-171 for proper use of this section.

Driveway Systems

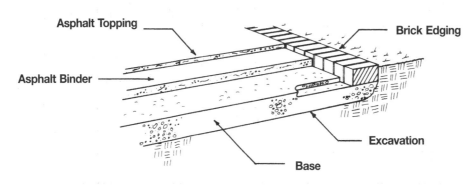

System Description	QUAN.	UNIT	LABOR-HOURS	MAT. COST	COST PER S.F.
ASPHALT DRIVEWAY TO 10' WIDE					
Excavation, driveway to 10' wide, 6" deep	.019	C.Y.	.001		
Base, 6" crushed stone	1.000	S.F.	.001	.87	
Handgrade base	1.000	S.F.	.004		
2" thick base	1.000	S.F.	.002	.45	
1" topping	1.000	S.F.	.001	.24	
Edging, brick pavers	.200	L.F.	.024	.40	
					Contractor's Fee
					4.69
TOTAL			.033	1.96	

Septic Systems

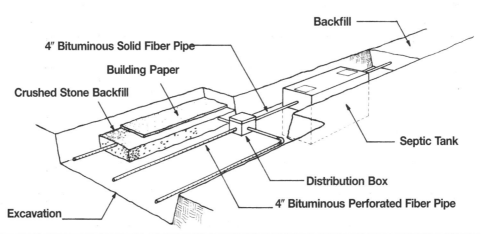

System Description	QUAN.	UNIT	LABOR-HOURS	MAT. COST	COST EACH
SEPTIC SYSTEM WITH 1000 S.F. LEACHING FIELD, 1000 GALLON TANK					
Tank, 1000 gallon, concrete	1.000	Ea.	3.500	570.00	
Distribution box, concrete	1.000	Ea.	1.000	104.00	
4" PVC pipe	25.000	L.F.	1.600	39.25	
Tank and field excavation	119.000	C.Y.	6.565		
Crushed stone backfill	76.000	C.Y.	12.160	1710.00	
Backfill with excavated material	36.000	C.Y.	.240		
Building paper	125.000	S.Y.	2.430	22.50	
4" PVC perforated pipe	145.000	L.F.	9.280	227.65	
4" pipe fittings	2.000	Ea.	1.939	14.00	
					Contractor's Fee
					6648.56
TOTAL			38.714	2687.40	

Be sure to read pages 170-171 for proper use of this section.

Chain Link Fence Price Sheet

SYSTEM DESCRIPTION	QUAN.	UNIT	LABOR-HOURS	COST PER UNIT	
				MAT.	TOTAL
Chain link fence					
Galv.9ga. wire, 1-5/8"post 10'O.C., 1-3/8"top rail, 2"corner post, 3'hi	1.000	L.F.	.130	4.59	12.11
4' high	1.000	L.F.	.141	7.00	16.26
6' high	1.000	L.F.	.209	7.90	20.29
Add for gate 3' wide 1-3/8" frame 3' high	1.000	Ea.	2.000	41.00	140.39
4' high	1.000	Ea.	2.400	51.00	171.53
6' high	1.000	Ea.	2.400	92.00	235.08
Add for gate 4' wide 1-3/8" frame 3' high	1.000	Ea.	2.667	48.00	176.86
4' high	1.000	Ea.	2.667	63.00	200.10
6' high	1.000	Ea.	3.000	116.00	295.14
Alum.9ga. wire, 1-5/8"post, 10'O.C., 1-3/8"top rail, 2"corner post,3'hi	1.000	L.F.	.130	5.50	13.51
4' high	1.000	L.F.	.141	6.30	15.18
6' high	1.000	L.F.	.209	8.10	20.59
Add for gate 3' wide 1-3/8" frame 3' high	1.000	Ea.	2.000	54.00	160.51
4' high	1.000	Ea.	2.400	73.50	207.14
6' high	1.000	Ea.	2.400	110.00	263.01
Add for gate 4' wide 1-3/8" frame 3' high	1.000	Ea.	2.400	73.50	207.14
4' high	1.000	Ea.	2.667	98.00	254.34
6' high	1.000	Ea.	3.000	153.00	352.44
Vinyl 9ga. wire, 1-5/8"post 10'O.C., 1-3/8"top rail, 2"corner post,3'hi	1.000	L.F.	.130	4.90	12.57
4' high	1.000	L.F.	.141	8.20	18.12
6' high	1.000	L.F.	.209	9.40	22.61
Add for gate 3' wide 1-3/8" frame 3' high	1.000	Ea.	2.000	61.50	172.81
4' high	1.000	Ea.	2.400	79.50	216.45
6' high	1.000	Ea.	2.400	123.00	283.06
Add for gate 4' 1'wide 1-3/8" frame 3' high	1.000	Ea.	2.400	83.50	222.64
4' high	1.000	Ea.	2.667	110.00	272.96
6' high	1.000	Ea.	3.000	159.00	361.82
Tennis court, chain link fence, 10' high					
Galv.11ga.wire, 2"post 10'O.C., 1-3/8"top rail, 2-1/2"corner post	1.000	L.F.	.253	12.25	28.67
Add for gate 3' wide 1-3/8" frame	1.000	Ea.	2.400	153.00	329.69
Alum.11ga.wire, 2"post 10'O.C., 1-3/8"top rail, 2-1/2"corner post	1.000	L.F.	.253	17.15	36.34
Add for gate 3' wide 1-3/8" frame	1.000	Ea.	2.400	196.00	396.37
Vinyl 11ga.wire,2"post 10' O.C.,1-3/8"top rail,2-1/2"corner post	1.000	L.F.	.253	14.70	32.46
Add for gate 3' wide 1-3/8" frame	1.000	Ea.	2.400	221.00	435.18

Be sure to read pages 170-171 for proper use of this section.

Wood Fence Price Sheet

SYSTEM DESCRIPTION	QUAN.	UNIT	LABOR-HOURS	COST PER UNIT	
				MAT.	TOTAL
Basketweave, 3/8"x4" boards, 2"x4" stringers on spreaders, 4"x4" posts					
No. 1 cedar, 6' high	1.000	L.F.	.150	8.10	18.32
Treated pine, 6' high	1.000	L.F.	.160	9.85	21.39
Board fence, 1"x4" boards, 2"x4" rails, 4"x4" posts					
Preservative treated, 2 rail, 3' high	1.000	L.F.	.166	6.00	15.63
4' high	1.000	L.F.	.178	6.60	17.06
3 rail, 5' high	1.000	L.F.	.185	7.45	18.66
6' high	1.000	L.F.	.192	8.55	20.65
Western cedar, No. 1, 2 rail, 3' high	1.000	L.F.	.166	6.55	16.48
3 rail, 4' high	1.000	L.F.	.178	7.75	18.84
5' high	1.000	L.F.	.185	8.95	20.99
6' high	1.000	L.F.	.192	9.80	22.60
No. 1 cedar, 2 rail, 3' high	1.000	L.F.	.166	9.85	21.60
4' high	1.000	L.F.	.178	11.20	24.19
3 rail, 5' high	1.000	L.F.	.185	12.95	27.20
6' high	1.000	L.F.	.192	14.45	29.81
Shadow box, 1"x6" boards, 2"x4" rails, 4"x4" posts					
Fir, pine or spruce, treated, 3 rail, 6' high	1.000	L.F.	.160	11.05	23.25
No. 1 cedar, 3 rail, 4' high	1.000	L.F.	.185	13.55	28.13
6' high	1.000	L.F.	.192	16.75	33.44
Open rail, split rails, No. 1 cedar, 2 rail, 3' high	1.000	L.F.	.150	5.45	14.21
3 rail, 4' high	1.000	L.F.	.160	7.35	17.51
No. 2 cedar, 2 rail, 3' high	1.000	L.F.	.150	4.24	12.35
3 rail, 4' high	1.000	L.F.	.160	4.83	13.63
Open rail, rustic rails, No. 1 cedar, 2 rail, 3' high	1.000	L.F.	.150	3.40	11.03
3 rail, 4' high	1.000	L.F.	.160	4.55	13.17
No. 2 cedar, 2 rail, 3' high	1.000	L.F.	.150	3.26	10.80
3 rail, 4' high	1.000	L.F.	.160	3.44	11.46
Rustic picket, molded pine pickets, 2 rail, 3' high	1.000	L.F.	.171	4.81	14.06
3 rail, 4' high	1.000	L.F.	.197	5.55	16.16
No. 1 cedar, 2 rail, 3' high	1.000	L.F.	.171	6.55	16.77
3 rail, 4' high	1.000	L.F.	.197	7.55	19.28
Picket fence, fir, pine or spruce, preserved, treated					
2 rail, 3' high	1.000	L.F.	.171	4.18	13.05
3 rail, 4' high	1.000	L.F.	.185	4.94	14.71
Western cedar, 2 rail, 3' high	1.000	L.F.	.171	5.25	14.75
3 rail, 4' high	1.000	L.F.	.185	5.35	15.41
No. 1 cedar, 2 rail, 3' high	1.000	L.F.	.171	10.45	22.82
3 rail, 4' high	1.000	L.F.	.185	12.20	26.03
Stockade, No. 1 cedar, 3-1/4" rails, 6' high	1.000	L.F.	.150	9.85	21.03
8' high	1.000	L.F.	.155	12.80	25.82
No. 2 cedar, treated rails, 6' high	1.000	L.F.	.150	9.85	21.03
Treated pine, treated rails, 6' high	1.000	L.F.	.150	9.65	20.72
Gates, No. 2 cedar, picket, 3'-6" wide 4' high	1.000	Ea.	2.667	52.50	184.46
No. 2 cedar, rustic round, 3' wide, 3' high	1.000	Ea.	2.667	66.50	206.22
No. 2 cedar, stockade screen, 3'-6" wide, 6' high	1.000	Ea.	3.000	58.00	205.15
General, wood, 3'-6" wide, 4' high	1.000	Ea.	2.400	51.00	171.46
6' high	1.000	Ea.	3.000	63.50	214.32

Be sure to read pages 170-171 for proper use of this section.

Division 2
FOUNDATIONS

Footing Systems

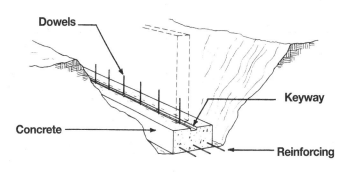

System Description	QUAN.	UNIT	LABOR-HOURS	MAT. COST	COST PER L.F.
8″ THICK BY 18″ WIDE FOOTING					
Concrete, 3000 psi	.040	C.Y.		3.04	
Place concrete, direct chute	.040	C.Y.	.016		
Forms, footing, 4 uses	1.330	SFCA	.103	.81	
Reinforcing, 1/2″ diameter bars, 2 each	1.380	Lb.	.011	.44	
Keyway, 2″ x 4″, beveled, 4 uses	1.000	L.F.	.015	.21	
Dowels, 1/2″ diameter bars, 2′ long, 6′ O.C.	.166	Ea.	.006	.07	**Contractor's Fee**
TOTAL			.151	4.57	**14.01**

Block Wall Systems

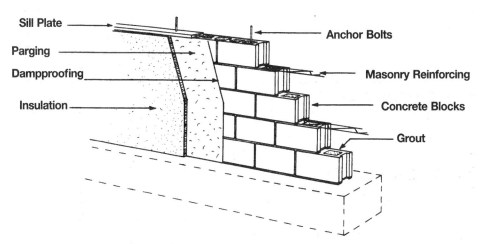

System Description	QUAN.	UNIT	LABOR-HOURS	MAT. COST	COST PER S.F.
8″ WALL, GROUTED, FULL HEIGHT					
Concrete block, 8″ x 16″ x 8″	1.000	S.F.	.094	1.85	
Masonry reinforcing, every second course	.750	L.F.	.002	.09	
Parging, plastering with portland cement plaster, 1 coat	1.000	S.F.	.014	.23	
Dampproofing, bituminous coating, 1 coat	1.000	S.F.	.012	.07	
Insulation, 1″ rigid polystyrene	1.000	S.F.	.010	.37	
Grout, solid, pumped	1.000	S.F.	.059	.94	
Anchor bolts, 1/2″ diameter, 8″ long, 4′ O.C.	.060	Ea.	.002	.04	
Sill plate, 2″ x 4″, treated	.250	L.F.	.007	.14	**Contractor's Fee**
TOTAL			.200	3.73	**15.24**

Be sure to read pages 170-171 for proper use of this section.

Concrete Wall Systems

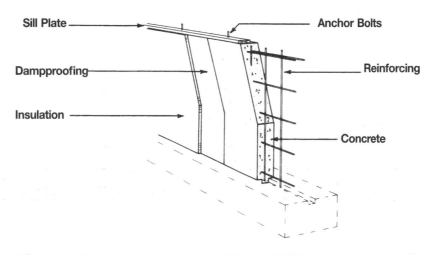

System Description	QUAN.	UNIT	LABOR-HOURS	MAT. COST	COST PER S.F.
8″ THICK, POURED CONCRETE WALL					
Concrete, 8″ thick , 3000 psi	.025	C.Y.		1.90	
Forms, prefabricated plywood, 4 uses per month	2.000	SFCA	.076	1.22	
Reinforcing, light	.670	Lb.	.004	.21	
Placing concrete, direct chute	.025	C.Y.	.013		
Dampproofing, brushed on, 2 coats	1.000	S.F.	.016	.11	
Rigid insulation, 1″ polystyrene	1.000	S.F.	.010	.37	
Anchor bolts, 1/2″ diameter, 12″ long, 4′ O.C.	.060	Ea.	.003	.08	
Sill plates, 2″ x 4″, treated	.250	L.F.	.007	.14	Contractor's Fee
TOTAL			.129	4.03	**12.20**

Wood Wall Foundation Systems

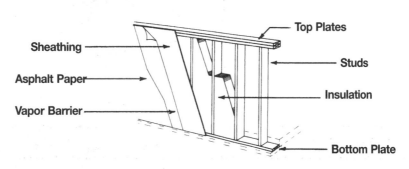

System Description	QUAN.	UNIT	LABOR-HOURS	MAT. COST	COST PER S.F.
2″ X 4″ STUDS, 16″ O.C., WALL					
Studs, 2″ x 4″, 16″ O.C., treated	1.000	L.F.	.015	.55	
Plates, double top plate, single bottom plate, treated, 2″ x 4″	.750	L.F.	.011	.41	
Sheathing, 1/2″, exterior grade, CDX, treated	1.000	S.F.	.014	.73	
Asphalt paper, 15# roll	1.100	S.F.	.002	.02	
Vapor barrier, 4 mil polyethylene	1.000	S.F.	.002	.03	
Insulation, batts, fiberglass, 3-1/2″ thick, R 11	1.000	S.F.	.005	.25	Contractor's Fee
TOTAL			.049	1.99	**5.61**

Be sure to read pages 170-171 for proper use of this section.

Floor Slab Systems

Concrete Slab — Expansion Material — Bank Run Gravel — Welded Wire Fabric — Vapor Barrier

System Description	QUAN.	UNIT	LABOR-HOURS	MAT. COST	COST PER S.F.
4″ THICK SLAB					
Concrete, 4″ thick, 3000 psi concrete	.012	C.Y.		.91	
Place concrete, direct chute	.012	C.Y.	.005		
Bank run gravel, 4″ deep	1.000	S.F.	.001	.66	
Polyethylene vapor barrier, .006″ thick	1.000	S.F.	.002	.03	
Edge forms, expansion material	.100	L.F.	.005	.05	
Welded wire fabric, 6 x 6, 10/10 (W1.4/W1.4)	1.100	S.F.	.005	.09	
Steel trowel finish	1.000	S.F.	.015		
					Contractor's Fee
TOTAL			.033	1.74	**4.27**

Be sure to read pages 170-171 for proper use of this section.

Division 3
FRAMING

Floor Framing Systems

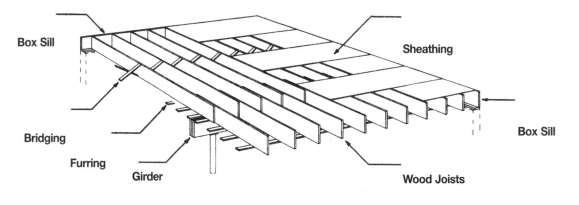

System Description	QUAN.	UNIT	LABOR-HOURS	MAT. COST	COST PER S.F.
2" X 8", 16" O.C.					
Wood joists, 2" x 8", 16" O.C.	1.000	L.F.	.015	.95	
Bridging, 1" x 3", 6' O.C.	.080	Pr.	.005	.02	
Box sills, 2" x 8"	.150	L.F.	.002	.14	
Concrete filled steel column, 4" diameter	.125	L.F.	.002	.09	
Girder, built up from three 2" x 8"	.125	L.F.	.013	.36	
Sheathing, plywood, subfloor, 5/8" CDX	1.000	S.F.	.012	.62	**Contractor's Fee**
Furring, 1" x 3", 16" O.C.	1.000	L.F.	.023	.21	**7.36**
TOTAL			.072	2.39	

Exterior Wall Framing Systems

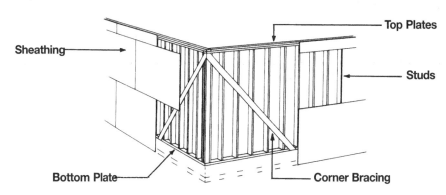

System Description	QUAN.	UNIT	LABOR-HOURS	MAT. COST	COST PER S.F.
2" X 4", 16" O.C.					
2" x 4" studs, 16" O.C.	1.000	L.F.	.015	.39	
Plates, 2" x 4", double top, single bottom	.375	L.F.	.005	.15	
Corner bracing, let-in, 1" x 6"	.063	L.F.	.003	.04	
Sheathing, 1/2" plywood, CDX	1.000	S.F.	.011	.54	**Contractor's Fee**
TOTAL			.034	1.12	**3.49**

Be sure to read pages 170-171 for proper use of this section.

Gable End Roof Framing Systems

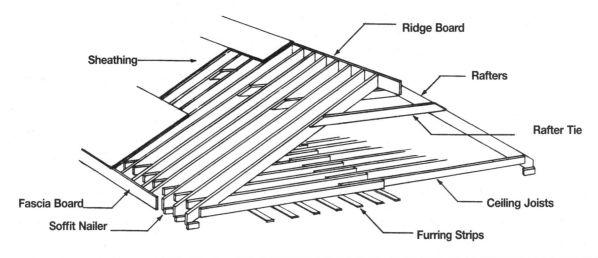

System Description	QUAN.	UNIT	LABOR-HOURS	MAT. COST	COST PER S.F.
2″ X 6″ RAFTERS, 16″ O.C., 4/12 PITCH					
Rafters, 2″ x 6″, 16″ O.C., 4/12 pitch	1.170	L.F.	.019	.73	
Ceiling joists, 2″ x 4″, 16″ O.C.	1.000	L.F.	.013	.39	
Ridge board, 2″ x 6″	.050	L.F.	.002	.03	
Fascia board, 2″ x 6″	.100	L.F.	.005	.07	
Rafter tie, 1″ x 4″, 4′ O.C.	.060	L.F.	.001	.02	
Soffit nailer (outrigger), 2″ x 4″, 24″ O.C.	.170	L.F.	.004	.07	
Sheathing, exterior, plywood, CDX, 1/2″ thick	1.170	S.F.	.013	.63	
Furring strips, 1″ x 3″, 16″ O.C.	1.000	L.F.	.023	.21	Contractor's Fee
TOTAL			.080	2.15	**7.41**

Truss Roof Framing Systems

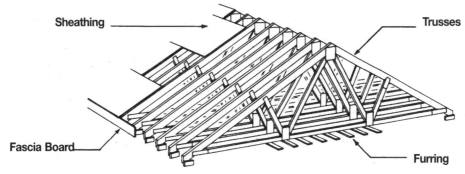

System Description	QUAN.	UNIT	LABOR-HOURS	MAT. COST	COST PER S.F.
TRUSS, 16″ O.C., 4/12 PITCH, 1′ OVERHANG, 26′ SPAN					
Truss, 40# loading, 16″ O.C., 4/12 pitch, 26′ span	.030	Ea.	.021	2.04	
Fascia board, 2″ x 6″	.100	L.F.	.005	.07	
Sheathing, exterior, plywood, CDX, 1/2″ thick	1.170	S.F.	.013	.63	
Furring, 1″ x 3″, 16″ O.C.	1.000	L.F.	.023	.21	Contractor's Fee
TOTAL			.062	2.95	**8.13**

Be sure to read pages 170-171 for proper use of this section.

Hip Roof Framing Systems

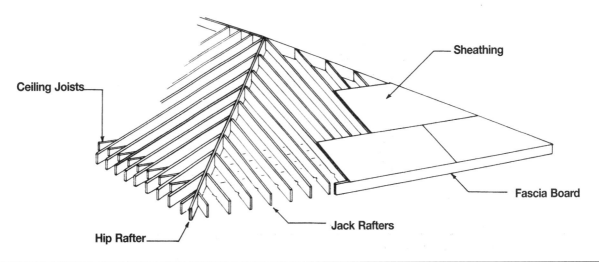

System Description	QUAN.	UNIT	LABOR-HOURS	MAT. COST	COST PER S.F.
2" X 6", 16" O.C., 4/12 PITCH					
Hip and valley rafters, 2" x 6", ordinary	.160	L.F.	.003	.10	
Jack rafters, 2" x 8", 16" O.C., 4/12 pitch	1.430	L.F.	.038	.89	
Ceiling joists, 2" x 6", 16" O.C.	1.000	L.F.	.013	.39	
Fascia board, 2" x 8"	.220	L.F.	.016	.21	
Soffit nailer (outrigger), 2" x 4", 24" O.C.	.220	L.F.	.006	.09	
Sheathing, 1/2" exterior plywood, CDX	1.570	S.F.	.018	.85	
Furring strips, 1" x 3", 16" O.C.	1.000	L.F.	.023	.21	Contractor's Fee
TOTAL			.117	2.74	**10.14**

Gambrel Roof Framing Systems

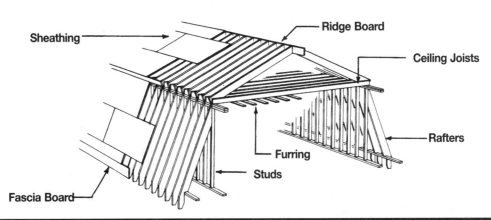

System Description	QUAN.	UNIT	LABOR-HOURS	MAT. COST	COST PER S.F.
2" X 6" RAFTERS, 16" O.C.					
Roof rafters, 2" x 6", 16" O.C.	1.430	L.F.	.029	.89	
Ceiling joists, 2" x 6", 16" O.C.	.710	L.F.	.009	.44	
Stud wall, 2" x 4", 16" O.C., including plates	.790	L.F.	.012	.31	
Furring strips, 1" x 3", 16" O.C.	.710	L.F.	.016	.15	
Ridge board, 2" x 8"	.050	L.F.	.002	.05	
Fascia board, 2" x 6"	.100	L.F.	.006	.08	
Sheathing, exterior grade plywood, 1/2" thick	1.450	S.F.	.017	.78	Contractor's Fee
TOTAL			.091	2.70	**8.75**

Be sure to read pages 170-171 for proper use of this section.

Mansard Roof Framing Systems

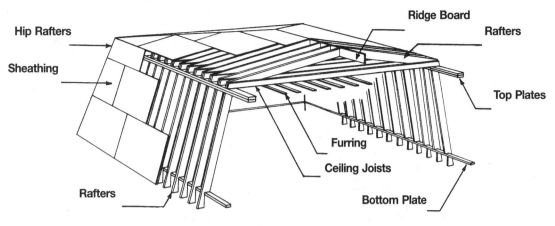

System Description	QUAN.	UNIT	LABOR-HOURS	MAT. COST	COST PER S.F.
2″ X 6″ RAFTERS, 16″ O.C.					
Roof rafters, 2″ x 6″, 16″ O.C.	1.210	L.F.	.033	.75	
Rafter plates, 2″ x 6″, double top, single bottom	.364	L.F.	.010	.23	
Ceiling joists, 2″ x 4″, 16″ O.C.	.920	L.F.	.012	.36	
Hip rafter, 2″ x 6″	.070	L.F.	.002	.04	
Jack rafter, 2″ x 6″, 16″ O.C.	1.000	L.F.	.039	.62	
Ridge board, 2″ x 6″	.018	L.F.	.001	.01	
Sheathing, exterior grade plywood, 1/2″ thick	2.210	S.F.	.025	1.19	
Furring strips, 1″ x 3″, 16″ O.C.	.920	L.F.	.021	.19	**Contractor's Fee**
TOTAL			.143	3.39	**12.51**

Shed/Flat Roof Framing Systems

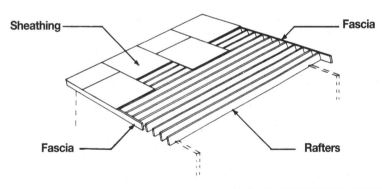

System Description	QUAN.	UNIT	LABOR-HOURS	MAT. COST	COST PER S.F.
2″ X 6″,16″ O.C., 4/12 PITCH					
Rafters, 2″ x 6″, 16″ O.C., 4/12 pitch	1.170	L.F.	.019	.73	
Fascia, 2″ x 6″	.100	L.F.	.006	.08	
Bridging, 1″ x 3″, 6′ O.C.	.080	Pr.	.005	.02	
Sheathing, exterior grade plywood, 1/2″ thick	1.230	S.F.	.014	.66	**Contractor's Fee**
TOTAL			.044	1.49	**4.51**

Be sure to read pages 170-171 for proper use of this section.

Gable Dormer Framing Systems

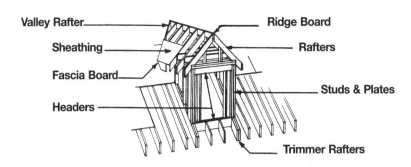

System Description	QUAN.	UNIT	LABOR-HOURS	MAT. COST	COST PER S.F.
2" X 6", 16" O.C.					
Dormer rafter, 2" x 6", 16" O.C.	1.330	L.F.	.036	.82	
Ridge board, 2" x 6"	.280	L.F.	.009	.17	
Trimmer rafters, 2" x 6"	.880	L.F.	.014	.55	
Wall studs & plates, 2" x 4", 16" O.C.	3.160	L.F.	.056	1.23	
Fascia, 2" x 6"	.220	L.F.	.012	.16	
Valley rafter, 2" x 6", 16" O.C.	.280	L.F.	.009	.17	
Cripple rafter, 2" x 6", 16" O.C.	.560	L.F.	.022	.35	
Headers, 2" x 6", doubled	.670	L.F.	.030	.42	
Ceiling joist, 2" x 4", 16" O.C.	1.000	L.F.	.013	.39	
Sheathing, exterior grade plywood, 1/2" thick	3.610	S.F.	.041	1.95	Contractor's Fee
TOTAL			.242	6.21	**21.90**

Shed Dormer Framing Systems

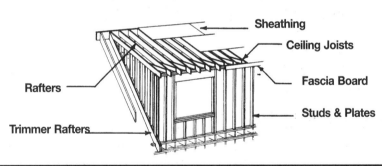

System Description	QUAN.	UNIT	LABOR-HOURS	MAT. COST	COST PER S.F.
2" X 6" RAFTERS, 16" O.C.					
Dormer rafter, 2" x 6", 16" O.C.	1.080	L.F.	.029	.67	
Trimmer rafter, 2" x 6"	.400	L.F.	.006	.25	
Studs & plates, 2" x 4", 16" O.C.	2.750	L.F.	.049	1.07	
Fascia, 2" x 6"	.250	L.F.	.014	.18	
Ceiling joist, 2" x 4", 16" O.C.	1.000	L.F.	.013	.39	
Sheathing, exterior grade plywood, CDX, 1/2" thick	2.940	S.F.	.034	1.59	Contractor's Fee
TOTAL			.145	4.15	**13.77**

Be sure to read pages 170-171 for proper use of this section.

Partition Framing Systems

System Description	QUAN.	UNIT	LABOR-HOURS	MAT. COST	COST PER S.F.
2″ X 4″, 16″ O.C.					
2″ x 4″ studs, #2 or better, 16″ O.C.	1.000	L.F.	.015	.39	
Plates, double top, single bottom	.375	L.F.	.005	.15	
Cross bracing, let-in, 1″ x 6″	.080	L.F.	.004	.04	**Contractor's Fee**
TOTAL			.024	.58	**2.13**
2″ X 4″, 24″ O.C.					
2″ x 4″ studs, #2 or better, 24″ O.C.	.800	L.F.	.012	.31	
Plates, double top, single bottom	.375	L.F.	.005	.15	
Cross bracing, let-in, 1″ x 6″	.080	L.F.	.003	.04	**Contractor's Fee**
TOTAL			.020	.50	**1.79**
2″ X 6″, 16″ O.C.					
2″ x 6″ studs, #2 or better, 16″ O.C.	1.000	L.F.	.016	.62	
Plates, double top, single bottom	.375	L.F.	.006	.23	
Cross bracing, let-in, 1″ x 6″	.080	L.F.	.004	.04	**Contractor's Fee**
TOTAL			.026	.89	**2.72**
2″ X 6″, 24″ O.C.					
2″ x 6″ studs, #2 or better, 24″ O.C.	.800	L.F.	.013	.50	
Plates, double top, single bottom	.375	L.F.	.006	.23	
Cross bracing, let-in, 1″ x 6″	.080	L.F.	.003	.04	**Contractor's Fee**
TOTAL			.022	.77	**2.29**

Be sure to read pages 170-171 for proper use of this section.

Division 4
EXTERIOR WALLS

Block Masonry Systems

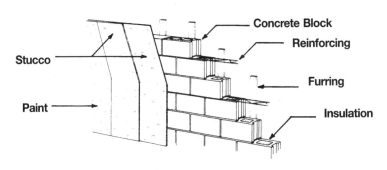

System Description	QUAN.	UNIT	LABOR-HOURS	MAT. COST	COST PER S.F.
8″ THICK CONCRETE BLOCK WALL					
8″ thick concrete block, 8″ x 8″ x 16″	1.000	S.F.	.107	1.59	
Masonry reinforcing, truss strips every other course	.625	L.F.	.002	.08	
Furring, 1″ x 3″, 16″ O.C.	1.000	L.F.	.016	.21	
Masonry insulation, poured vermiculite	1.000	S.F.	.018	.79	
Stucco, 2 coats	1.000	S.F.	.069	.19	
Masonry paint, 2 coats	1.000	S.F.	.016	.18	Contractor's Fee
TOTAL			.228	3.04	**15.35**

Brick/Stone Veneer Systems

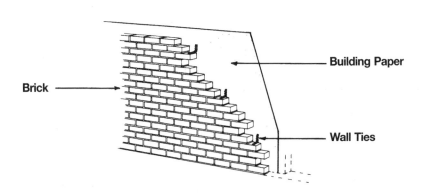

System Description	QUAN.	UNIT	LABOR-HOURS	MAT. COST	COST PER S.F.
SELECT COMMON BRICK					
Brick, select common, running bond	1.000	S.F.	.174	2.96	
Wall ties, 7/8″ x 7″, 22 gauge	1.000	Ea.	.008	.05	
Building paper, #15 asphalt	1.100	S.F.	.002	.02	
Trim, pine, painted	.125	L.F.	.004	.08	Contractor's Fee
TOTAL			.188	3.11	**13.56**

Be sure to read pages 170-171 for proper use of this section.

Wood Siding Systems

System Description	QUAN.	UNIT	LABOR-HOURS	MAT. COST	COST PER S.F.
1/2" X 6" BEVELED CEDAR SIDING, "A" GRADE					
1/2" x 6" beveled cedar siding	1.000	S.F.	.032	1.82	
#15 asphalt felt paper	1.100	S.F.	.002	.02	
Trim, cedar	.125	L.F.	.005	.18	
Paint, primer & 2 coats	1.000	S.F.	.017	.17	Contractor's Fee **6.19**
TOTAL			.056	2.19	
1/2" X 8" BEVELED CEDAR SIDING, "A" GRADE					
1/2" x 8" beveled cedar siding	1.000	S.F.	.029	2.09	
#15 asphalt felt paper	1.100	S.F.	.002	.02	
Trim, cedar	.125	L.F.	.005	.18	
Paint, primer & 2 coats	1.000	S.F.	.017	.17	Contractor's Fee **6.46**
TOTAL			.053	2.46	
1" X 4" TONGUE & GROOVE, REDWOOD, VERTICAL GRAIN					
Redwood, clear, vertical grain, 1" x 10"	1.000	S.F.	.018	3.62	
#15 asphalt felt paper	1.100	S.F.	.002	.02	
Trim, redwood	.125	L.F.	.005	.18	
Sealer, 1 coat, stain, 1 coat	1.000	S.F.	.013	.11	Contractor's Fee **8.03**
TOTAL			.038	3.93	
1" X 6" TONGUE & GROOVE, REDWOOD, VERTICAL GRAIN					
Redwood, clear, vertical grain, 1" x 10"	1.000	S.F.	.019	3.73	
#15 asphalt felt paper	1.100	S.F.	.002	.02	
Trim, redwood	.125	L.F.	.005	.18	
Sealer, 1 coat, stain, 1 coat	1.000	S.F.	.013	.11	Contractor's Fee **8.21**
TOTAL			.039	4.04	

Be sure to read pages 170-171 for proper use of this section.

Shingle Siding Systems

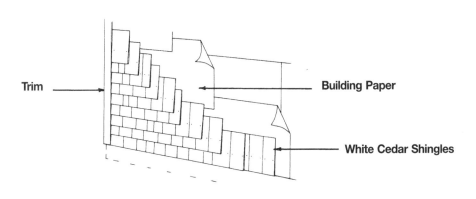

Trim → Building Paper → White Cedar Shingles

System Description	QUAN.	UNIT	LABOR-HOURS	MAT. COST	COST PER S.F.
WHITE CEDAR SHINGLES, 5" EXPOSURE					
White cedar shingles, 16" long, grade "A", 5" exposure	1.000	S.F.	.033	1.18	
#15 asphalt felt paper	1.100	S.F.	.002	.02	
Trim, cedar	.125	S.F.	.005	.18	Contractor's Fee
Paint, primer & 1 coat	1.000	S.F.	.017	.09	
TOTAL			.057	1.47	**5.14**
NO. 1 PERFECTIONS, 5-1/2" EXPOSURE					
No. 1 perfections, red cedar, 5-1/2" exposure	1.000	S.F.	.029	1.76	
#15 asphalt felt paper	1.100	S.F.	.002	.02	
Trim, cedar	.125	S.F.	.005	.18	Contractor's Fee
Stain, sealer & 1 coat	1.000	S.F.	.017	.09	
TOTAL			.053	2.05	**5.83**
RESQUARED & REBUTTED PERFECTIONS, 5-1/2" EXPOSURE					
Resquared & rebutted perfections, 5-1/2" exposure	1.000	S.F.	.027	2.19	
#15 asphalt felt paper	1.100	S.F.	.002	.02	
Trim, cedar	.125	S.F.	.005	.18	Contractor's Fee
Stain, sealer & 1 coat	1.000	S.F.	.017	.09	
TOTAL			.051	2.48	**6.38**
HAND-SPLIT SHAKES, 8-1/2" EXPOSURE					
Hand-split red cedar shakes, 18" long, 8-1/2" exposure	1.000	S.F.	.040	1.07	
#15 asphalt felt paper	1.100	S.F.	.002	.02	
Trim, cedar	.125	S.F.	.005	.18	Contractor's Fee
Stain, sealer & 1 coat	1.000	S.F.	.017	.09	
TOTAL			.064	1.36	**5.30**

Be sure to read pages 170-171 for proper use of this section.

Metal & Plastic Siding Systems

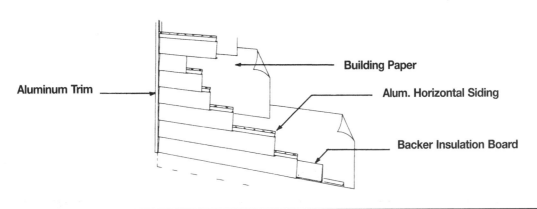

Aluminum Trim

Building Paper

Alum. Horizontal Siding

Backer Insulation Board

System Description	QUAN.	UNIT	LABOR-HOURS	MAT. COST	COST PER S.F.
ALUMINUM CLAPBOARD SIDING, 8″ WIDE, WHITE					
Aluminum horizontal siding, 8″ clapboard	1.000	S.F.	.031	1.38	
Backer, insulation board	1.000	S.F.	.008	.45	
Trim, aluminum	.600	L.F.	.016	.62	Contractor's Fee
Paper, #15 asphalt felt	1.100	S.F.	.002	.02	
					6.73
TOTAL			.057	2.47	
ALUMINUM VERTICAL BOARD & BATTEN, WHITE					
Aluminum vertical board & batten	1.000	S.F.	.027	1.54	
Backer insulation board	1.000	S.F.	.008	.45	
Trim, aluminum	.600	L.F.	.016	.62	Contractor's Fee
Paper, #15 asphalt felt	1.100	S.F.	.002	.02	
					6.78
TOTAL			.053	2.63	
VINYL CLAPBOARD SIDING, 8″ WIDE, WHITE					
PVC vinyl horizontal siding, 8″ clapboard	1.000	S.F.	.032	.65	
Backer, insulation board	1.000	S.F.	.008	.45	
Trim, vinyl	.600	L.F.	.014	.39	Contractor's Fee
Paper, #15 asphalt felt	1.100	S.F.	.002	.02	
					5.21
TOTAL			.056	1.51	
VINYL VERTICAL BOARD & BATTEN, WHITE					
PVC vinyl vertical board & batten	1.000	S.F.	.029	1.51	
Backer, insulation board	1.000	S.F.	.008	.45	
Trim, vinyl	.600	L.F.	.014	.39	Contractor's Fee
Paper, #15 asphalt felt	1.100	S.F.	.002	.02	
					6.39
TOTAL			.053	2.37	

Be sure to read pages 170-171 for proper use of this section.

Double Hung Window Systems

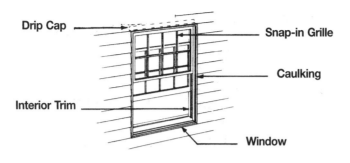

Drip Cap — Snap-in Grille

Caulking

Interior Trim

Window

System Description	QUAN.	UNIT	LABOR-HOURS	MAT. COST	COST EACH
BUILDER'S QUALITY WOOD WINDOW 2' X 3', DOUBLE HUNG					
Window, primed, builder's quality, 2' x 3', insulating glass	1.000	Ea.	.800	243.00	
Trim, interior casing	11.000	L.F.	.367	8.25	
Paint, interior & exterior, primer & 2 coats	2.000	Face	1.778	2.02	
Caulking	10.000	L.F.	.323	1.60	
Snap-in grille	1.000	Set	.333	43.00	
Drip cap, metal	2.000	L.F.	.040	.44	Contractor's Fee
TOTAL			3.641	298.31	**638.94**
PLASTIC CLAD WOOD WINDOW 3' X 4', DOUBLE HUNG					
Window, plastic clad, premium, 3' x 4', insulating glass	1.000	Ea.	.889	271.00	
Trim, interior casing	15.000	L.F.	.500	11.25	
Paint, interior, primer & 2 coats	1.000	Face	.889	1.01	
Caulking	14.000	L.F.	.452	2.24	
Snap-in grille	1.000	Set	.333	43.00	Contractor's Fee
TOTAL			3.063	328.50	**663.73**
METAL CLAD WOOD WINDOW, 3' X 5', DOUBLE HUNG					
Window, metal clad, deluxe, 3' x 5', insulating glass	1.000	Ea.	1.000	289.00	
Trim, interior casing	17.000	L.F.	.567	12.75	
Paint, interior, primer & 2 coats	1.000	Face	.889	1.01	
Caulking	16.000	L.F.	.516	2.56	
Snap-in grille	1.000	Set	.235	119.00	
Drip cap, metal	3.000	L.F.	.060	.66	Contractor's Fee
TOTAL			3.267	424.98	**820.14**

Be sure to read pages 170-171 for proper use of this section.

Casement Window Systems

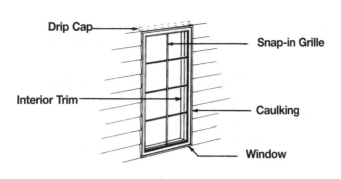

Drip Cap · Snap-in Grille · Interior Trim · Caulking · Window

System Description	QUAN.	UNIT	LABOR-HOURS	MAT. COST	COST EACH
BUILDER'S QUALITY WINDOW, WOOD, 2' BY 3', CASEMENT					
Window, primed, builder's quality, 2' x 3', insulating glass	1.000	Ea.	.800	224.00	
Trim, interior casing	11.000	L.F.	.367	8.25	
Paint, interior & exterior, primer & 2 coats	2.000	Face	1.778	2.02	
Caulking	10.000	L.F.	.323	1.60	
Snap-in grille	1.000	Ea.	.267	24.00	
Drip cap, metal	2.000	L.F.	.040	.44	Contractor's Fee
TOTAL			3.575	260.31	**576.53**
PLASTIC CLAD WOOD WINDOW, 2' X 4', CASEMENT					
Window, plastic clad, premium, 2' x 4', insulating glass	1.000	Ea.	.889	287.00	
Trim, interior casing	13.000	L.F.	.433	9.75	
Paint, interior, primer & 2 coats	1.000	Ea.	.889	1.01	
Caulking	12.000	L.F.	.387	1.92	
Snap-in grille	1.000	Ea.	.267	24.00	Contractor's Fee
TOTAL			2.865	323.68	**644.59**
METAL CLAD WOOD WINDOW, 2' X 5', CASEMENT					
Window, metal clad, deluxe, 2' x 5', insulating glass	1.000	Ea.	1.000	256.00	
Trim, interior casing	15.000	L.F.	.500	11.25	
Paint, interior, primer & 2 coats	1.000	Ea.	.889	1.01	
Caulking	14.000	L.F.	.452	2.24	
Snap-in grille	1.000	Ea.	.250	35.00	
Drip cap, metal	12.000	L.F.	.040	.44	Contractor's Fee
TOTAL			3.131	305.94	**629.26**

Be sure to read pages 170-171 for proper use of this section.

Awning Window Systems

Drip Cap

Interior Trim

Snap-in Grille

Caulking

Window

System Description	QUAN.	UNIT	LABOR-HOURS	MAT. COST	COST EACH
BUILDER'S QUALITY WINDOW, WOOD, 34" X 22", AWNING					
Window, builder quality, 34" x 22", insulating glass	1.000	Ea.	.800	238.00	
Trim, interior casing	10.500	L.F.	.350	7.88	
Paint, interior & exterior, primer & 2 coats	2.000	Face	1.778	2.02	
Caulking	9.500	L.F.	.306	1.52	
Snap-in grille	1.000	Ea.	.267	19.60	
Drip cap, metal	3.000	L.F.	.060	.66	Contractor's Fee
TOTAL			3.561	269.68	**590.12**
PLASTIC CLAD WOOD WINDOW, 40" X 28", AWNING					
Window, plastic clad, premium, 40" x 28", insulating glass	1.000	Ea.	.889	299.00	
Trim interior casing	13.500	L.F.	.450	10.13	
Paint, interior, primer & 2 coats	1.000	Face	.889	1.01	
Caulking	12.500	L.F.	.403	2.00	
Snap-in grille	1.000	Ea.	.267	19.60	Contractor's Fee
TOTAL			2.898	331.74	**655.74**
METAL CLAD WOOD WINDOW, 48" X 36", AWNING					
Window, metal clad, deluxe, 48" x 36", insulating glass	1.000	Ea.	1.000	335.00	
Trim, interior casing	15.000	L.F.	.500	11.25	
Paint, interior, primer & 2 coats	1.000	Face	.889	1.01	
Caulking	14.000	L.F.	.452	2.24	
Snap-in grille	1.000	Ea.	.250	28.50	
Drip cap, metal	4.000	L.F.	.080	.88	Contractor's Fee
TOTAL			3.171	378.88	**743.06**

Be sure to read pages 170-171 for proper use of this section.

Sliding Window Systems

System Description	QUAN.	UNIT	LABOR-HOURS	MAT. COST	COST EACH
BUILDER'S QUALITY WOOD WINDOW, 3' X 2', SLIDING					
Window, primed, builder's quality, 3' x 2', insul. glass	1.000	Ea.	.800	187.00	
Trim, interior casing	11.000	L.F.	.367	8.25	
Paint, interior & exterior, primer & 2 coats	2.000	Face	1.778	2.02	
Caulking	10.000	L.F.	.323	1.60	
Snap-in grille	1.000	Set	.333	23.00	
Drip cap, metal	3.000	L.F.	.060	.66	Contractor's Fee
					522.43
TOTAL			3.661	222.53	
PLASTIC CLAD WOOD WINDOW, 4' X 3'-6", SLIDING					
Window, plastic clad, premium, 4' x 3'-6", insulating glass	1.000	Ea.	.889	650.00	
Trim, interior casing	16.000	L.F.	.533	12.00	
Paint, interior, primer & 2 coats	1.000	Face	.889	1.01	
Caulking	17.000	L.F.	.548	2.72	
Snap-in grille	1.000	Set	.333	23.00	Contractor's Fee
					1223.14
TOTAL			3.192	688.73	
METAL CLAD WOOD WINDOW, 6' X 5', SLIDING					
Window, metal clad, deluxe, 6' x 5', insulating glass	1.000	Ea.	1.000	640.00	
Trim, interior casing	23.000	L.F.	.767	17.25	
Paint, interior, primer & 2 coats	1.000	Face	.889	1.01	
Caulking	22.000	L.F.	.710	3.52	
Snap-in grille	1.000	Set	.364	48.00	
Drip cap, metal	6.000	L.F.	.120	1.32	Contractor's Fee
					1292.99
TOTAL			3.850	711.10	

Be sure to read pages 170-171 for proper use of this section.

Bow/Bay Window Systems

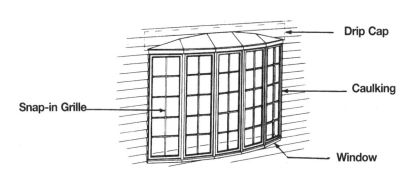

Drip Cap

Caulking

Snap-in Grille

Window

System Description	QUAN.	UNIT	LABOR-HOURS	MAT. COST	COST EACH
AWNING TYPE BOW WINDOW, BUILDER'S QUALITY, 8' X 5'					
Window, primed, builder's quality, 8' x 5', insulating glass	1.000	Ea.	1.600	1250.00	
Trim, interior casing	27.000	L.F.	.900	20.25	
Paint, interior & exterior, primer & 1 coat	2.000	Face	3.200	10.80	
Drip cap, vinyl	1.000	Ea.	.533	77.50	
Caulking	26.000	L.F.	.839	4.16	
Snap-in grilles	1.000	Set	1.067	96.00	**Contractor's Fee**
TOTAL			8.139	1458.71	**2646.47**
CASEMENT TYPE BOW WINDOW, PLASTIC CLAD, 10' X 6'					
Window, plastic clad, premium, 10' x 6', insulating glass	1.000	Ea.	2.286	1925.00	
Trim, interior casing	33.000	L.F.	1.100	24.75	
Paint, interior, primer & 1 coat	1.000	Face	1.778	2.02	
Drip cap, vinyl	1.000	Ea.	.615	84.50	
Caulking	32.000	L.F.	1.032	5.12	
Snap-in grilles	1.000	Set	1.333	120.00	**Contractor's Fee**
TOTAL			8.144	2161.39	**3747.64**
DOUBLE HUNG TYPE, METAL CLAD, 9' X 5'					
Window, metal clad, deluxe, 9' x 5', insulating glass	1.000	Ea.	2.667	1150.00	
Trim, interior casing	29.000	L.F.	.967	21.75	
Paint, interior, primer & 1 coat	1.000	Face	1.778	2.02	
Drip cap, vinyl	1.000	Set	.615	84.50	
Caulking	28.000	L.F.	.903	4.48	
Snap-in grilles	1.000	Set	1.067	96.00	**Contractor's Fee**
TOTAL			7.997	1358.75	**2512.14**

Be sure to read pages 170-171 for proper use of this section.

Fixed Window Systems

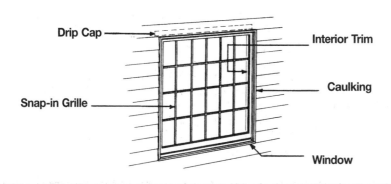

Drip Cap — Interior Trim

Snap-in Grille — Caulking

Window

System Description	QUAN.	UNIT	LABOR-HOURS	MAT. COST	COST EACH
BUILDER'S QUALITY PICTURE WINDOW, 4' X 4'					
Window, primed, builder's quality, 4' x 4', insulating glass	1.000	Ea.	1.333	300.00	
Trim, interior casing	17.000	L.F.	.567	12.75	
Paint, interior & exterior, primer & 2 coats	2.000	Face	1.778	2.02	
Caulking	16.000	L.F.	.516	2.56	
Snap-in grille	1.000	Ea.	.267	137.00	Contractor's Fee
Drip cap, metal	4.000	L.F.	.080	.88	
TOTAL			4.541	455.21	**931.23**
PLASTIC CLAD WOOD WINDOW, 4'-6" X 6'-6"					
Window, plastic clad, prem., 4'-6" x 6'-6", insul. glass	1.000	Ea.	1.455	670.00	
Trim, interior casing	23.000	L.F.	.767	17.25	
Paint, interior, primer & 2 coats	1.000	Face	.889	1.01	
Caulking	22.000	L.F.	.710	3.52	
Snap-in grille	1.000	Ea.	.267	137.00	Contractor's Fee
TOTAL			4.088	828.78	**1486.13**
METAL CLAD WOOD WINDOW, 6'-6" X 6'-6"					
Window, metal clad, deluxe, 6'-6" x 6'-6", insulating glass	1.000	Ea.	1.600	550.00	
Trim interior casing	27.000	L.F.	.900	20.25	
Paint, interior, primer & 2 coats	1.000	Face	1.600	5.40	
Caulking	26.000	L.F.	.839	4.16	
Snap-in grille	1.000	Ea.	.267	137.00	Contractor's Fee
Drip cap, metal	6.500	L.F.	.130	1.43	
TOTAL			5.336	718.24	**1373.68**

Be sure to read pages 170-171 for proper use of this section.

Entrance Door Systems

Drip Cap — Door — Frame & Exterior Casing — Interior Casing — Sill

System Description	QUAN.	UNIT	LABOR-HOURS	MAT. COST	COST EACH
COLONIAL, 6 PANEL, 3′ X 6′-8″, WOOD					
Door, 3′ x 6′-8″ x 1-3/4″ thick, pine, 6 panel colonial	1.000	Ea.	1.067	385.00	
Frame, 5-13/16″ deep, incl. exterior casing & drip cap	17.000	L.F.	.725	126.65	
Interior casing, 2-1/2″ wide	18.000	L.F.	.600	13.50	
Sill, 8/4 x 8″ deep	3.000	L.F.	.480	43.95	
Butt hinges, brass, 4-1/2″ x 4-1/2″	1.500	Pr.		13.73	
Lockset	1.000	Ea.	.571	31.00	
Weatherstripping, metal, spring type, bronze	1.000	Set	1.053	17.20	Contractor's Fee
Paint, interior & exterior, primer & 2 coats	2.000	Face	1.778	10.70	
TOTAL			6.274	641.73	**1307.33**
SOLID CORE BIRCH, FLUSH, 3′ X 6′-8″					
Door, 3′ x 6′-8″, 1-3/4″ thick, birch, flush solid core	1.000	Ea.	1.067	90.50	
Frame, 5-13/16″ deep, incl. exterior casing & drip cap	17.000	L.F.	.725	126.65	
Interior casing, 2-1/2″ wide	18.000	L.F.	.600	13.50	
Sill, 8/4 x 8″ deep	3.000	L.F.	.480	43.95	
Butt hinges, brass, 4-1/2″ x 4-1/2″	1.500	Pr.		13.73	
Lockset	1.000	Ea.	.571	31.00	
Weatherstripping, metal, spring type, bronze	1.000	Set	1.053	17.20	Contractor's Fee
Paint, Interior & exterior, primer & 2 coats	2.000	Face	1.778	9.98	
TOTAL			6.274	346.51	**846.90**

Be sure to read pages 170-171 for proper use of this section.

Sliding Door Systems

System Description	QUAN.	UNIT	LABOR-HOURS	MAT. COST	COST EACH
WOOD SLIDING DOOR, 8' WIDE, PREMIUM					
Wood, 5/8" thick tempered insul. glass, 8' wide, premium	1.000	Ea.	5.333	1225.00	
Interior casing	22.000	L.F.	.733	16.50	
Exterior casing	22.000	L.F.	.733	16.50	
Sill, oak, 8/4 x 8" deep	8.000	L.F.	1.280	117.20	
Drip cap	8.000	L.F.	.160	1.76	
Paint, interior & exterior, primer & 2 coats	2.000	Face	2.816	14.96	Contractor's Fee
TOTAL			11.055	1391.92	**2717.87**
ALUMINUM SLIDING DOOR, 8' WIDE, PREMIUM					
Aluminum, 5/8" tempered insul. glass, 8' wide, premium	1.000	Ea.	5.333	1375.00	
Interior casing	22.000	L.F.	.733	16.50	
Exterior casing	22.000	L.F.	.733	16.50	
Sill, oak, 8/4 x 8" deep	8.000	L.F.	1.280	117.20	
Drip cap	8.000	L.F.	.160	1.76	
Paint, interior & exterior, primer & 2 coats	2.000	Face	2.816	14.96	Contractor's Fee
TOTAL			11.055	1541.92	**2952.45**

Be sure to read pages 170-171 for proper use of this section.

Residential Garage Door Systems

Exterior Trim — Drip Cap — Door — Weatherstripping — Jamb

210

System Description	QUAN.	UNIT	LABOR-HOURS	MAT. COST	COST EACH
OVERHEAD, SECTIONAL GARAGE DOOR, 9′ X 7′					
Wood, overhead sectional door, std., incl. hardware, 9′ x 7′	1.000	Ea.	2.000	415.00	
Jamb & header blocking, 2″ x 6″	25.000	L.F.	.901	15.50	
Exterior trim	25.000	L.F.	.833	18.75	
Paint, interior & exterior, primer & 2 coats	2.000	Face	3.556	21.40	
Weatherstripping, molding type	1.000	Set	.767	17.25	Contractor's Fee
Drip cap	9.000	L.F.	.180	1.98	
TOTAL			8.237	489.88	**1157.68**
OVERHEAD, SECTIONAL GARAGE DOOR, 16′ X 7′					
Wood, overhead sectional, std., incl. hardware, 16′ x 7′	1.000	Ea.	2.667	830.00	
Jamb & header blocking, 2″ x 6″	30.000	L.F.	1.081	18.60	
Exterior trim	30.000	L.F.	1.000	22.50	
Paint, interior & exterior, primer & 2 coats	2.000	Face	5.333	32.10	
Weatherstripping, molding type	1.000	Set	1.000	22.50	Contractor's Fee
Drip cap	16.000	L.F.	.320	3.52	
TOTAL			11.401	929.22	**1992.92**
OVERHEAD, SWING-UP TYPE, GARAGE DOOR, 16′ X 7′					
Wood, overhead, swing-up, std., incl. hardware, 16′ x 7′	1.000	Ea.	2.667	615.00	
Jamb & header blocking, 2″ x 6″	30.000	L.F.	1.081	18.60	
Exterior trim	30.000	L.F.	1.000	22.50	
Paint, interior & exterior, primer & 2 coats	2.000	Face	5.333	32.10	
Weatherstripping, molding type	1.000	Set	1.000	22.50	Contractor's Fee
Drip cap	16.000	L.F.	.320	3.52	
TOTAL			11.401	714.22	**1659.54**

Be sure to read pages 170-171 for proper use of this section.

Aluminum Window Systems

Drywall → ← Finish Drywall

Window →

Corner Bead →

Sill →

System Description	QUAN.	UNIT	LABOR-HOURS	MAT. COST	COST EACH
SINGLE HUNG, 2′ X 3′ OPENING					
Window, 2′ x 3′ opening, enameled, insulating glass	1.000	Ea.	1.600	173.00	
Blocking, 1″ x 3″ furring strip nailers	10.000	L.F.	.146	2.10	
Drywall, 1/2″ thick, standard	5.000	S.F.	.040	1.15	
Corner bead, 1″ x 1″, galvanized steel	8.000	L.F.	.160	.88	
Finish drywall, tape and finish corners inside and outside	16.000	L.F.	.233	1.12	Contractor's Fee
Sill, slate	2.000	L.F.	.400	16.30	
TOTAL			2.579	194.55	**451.10**
SLIDING, 3′ X 2′ OPENING					
Window, 3′ x 2′ opening, enameled, insulating glass	1.000	Ea.	1.600	178.00	
Blocking, 1″ x 3″ furring strip nailers	10.000	L.F.	.146	2.10	
Drywall, 1/2″ thick, standard	5.000	S.F.	.040	1.15	
Corner bead, 1″ x 1″, galvanized steel	7.000	L.F.	.140	.77	
Finish drywall, tape and finish corners inside and outside	14.000	L.F.	.204	.98	Contractor's Fee
Sill, slate	3.000	L.F.	.600	24.45	
TOTAL			2.730	207.45	**477.57**
AWNING, 3′-1″ X 3′-2″					
Window, 3′-1″ x 3′-2″ opening, enameled, insul. glass	1.000	Ea.	1.600	199.00	
Blocking, 1″ x 3″ furring strip, nailers	12.500	L.F.	.182	2.63	
Drywall, 1/2″ thick, standard	4.500	S.F.	.036	1.04	
Corner bead, 1″ x 1″, galvanized steel	9.250	L.F.	.185	1.02	
Finish drywall, tape and finish corners, inside and outside	18.500	L.F.	.269	1.30	Contractor's Fee
Sill, slate	3.250	L.F.	.650	26.49	
TOTAL			2.922	231.48	**524.26**

Be sure to read pages 170-171 for proper use of this section.

Storm Door & Window Systems

Aluminum Window

Aluminum Door

SYSTEM DESCRIPTION	QUAN.	UNIT	LABOR-HOURS	COST EACH	
				MAT.	TOTAL
Storm door, aluminum, combination, storm & screen, anodized, 2'-6" x 6'-8"	1.000	Ea.	1.067	154.00	292.87
2'-8" x 6'-8"	1.000	Ea.	1.143	185.00	345.19
3'-0" x 6'-8"	1.000	Ea.	1.143	185.00	345.19
Mill finish, 2'-6" x 6'-8"	1.000	Ea.	1.067	217.00	390.68
2'-8" x 6'-8"	1.000	Ea.	1.143	217.00	394.94
3'-0" x 6'-8"	1.000	Ea.	1.143	236.00	424.23
Painted, 2'-6" x 6'-8"	1.000	Ea.	1.067	212.00	382.86
2'-8" x 6'-8"	1.000	Ea.	1.143	220.00	399.49
3'-0" x 6'-8"	1.000	Ea.	1.143	227.00	410.30
Wood, combination, storm & screen, crossbuck, 2'-6" x 6'-9"	1.000	Ea.	1.455	242.00	449.25
2'-8" x 6'-9"	1.000	Ea.	1.600	245.00	458.21
3'-0" x 6'-9"	1.000	Ea.	1.778	254.00	487.78
Full lite, 2'-6" x 6'-9"	1.000	Ea.	1.455	245.00	453.94
2'-8" x 6'-9"	1.000	Ea.	1.600	245.00	458.21
3'-0" x 6'-9"	1.000	Ea.	1.778	254.00	487.78
Windows, aluminum, combination storm & screen, basement, 1'-10" x 1'-0"	1.000	Ea.	.533	28.50	71.23
2'-9" x 1'-6"	1.000	Ea.	.533	31.00	75.14
3'-4" x 2'-0"	1.000	Ea.	.533	37.50	85.16
Double hung, anodized, 2'-0" x 3'-5"	1.000	Ea.	.533	74.00	141.81
2'-6" x 5'-0"	1.000	Ea.	.571	99.00	183.40
4'-0" x 6'-0"	1.000	Ea.	.640	210.00	358.41
Painted, 2'-0" x 3'-5"	1.000	Ea.	.533	88.00	163.49
2'-6" x 5'-0"	1.000	Ea.	.571	142.00	250.07
4'-0" x 6'-0"	1.000	Ea.	.640	254.00	426.65
Fixed window, anodized, 4'-6" x 4'-6"	1.000	Ea.	.640	113.00	207.99
5'-8" x 4'-6"	1.000	Ea.	.800	128.00	239.84
Painted, 4'-6" x 4'-6"	1.000	Ea.	.640	113.00	207.99
5'-8" x 4'-6"	1.000	Ea.	.800	128.00	239.84

Be sure to read pages 170-171 for proper use of this section.

Shutters/Blinds Systems

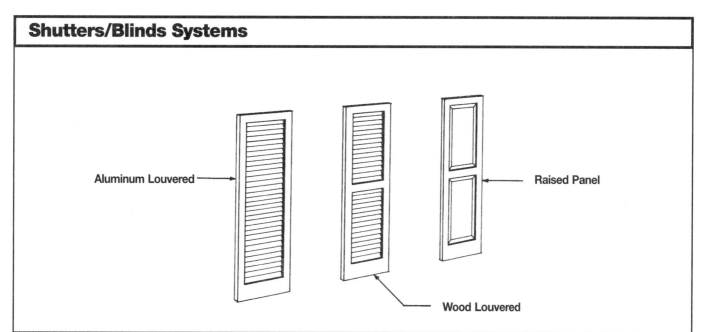

Aluminum Louvered

Raised Panel

Wood Louvered

SYSTEM DESCRIPTION	QUAN.	UNIT	LABOR-HOURS	COST PER PAIR	
				MAT.	TOTAL
Shutters, exterior blinds, aluminum, louvered, 1'-4" wide, 3"-0" long	1.000	Set	.800	43.50	107.98
4'-0" long	1.000	Set	.800	52.00	121.13
5'-4" long	1.000	Set	.800	68.50	146.72
6'-8" long	1.000	Set	.889	87.00	180.48
Wood, louvered, 1'-2" wide, 3'-3" long	1.000	Set	.800	82.50	168.47
4'-7" long	1.000	Set	.800	111.00	213.40
5'-3" long	1.000	Set	.800	126.00	236.71
1'-6" wide, 3'-3" long	1.000	Set	.800	87.50	176.22
4'-7" long	1.000	Set	.800	123.00	232.02
Polystyrene, solid raised panel, 3'-3" wide, 3'-0" long	1.000	Set	.800	172.00	307.94
3'-11" long	1.000	Set	.800	183.00	325.00
5'-3" long	1.000	Set	.800	240.00	413.43
6'-8" long	1.000	Set	.889	270.00	461.48
Polystyrene, louvered, 1'-2" wide, 3'-3" long	1.000	Set	.800	44.00	108.76
4'-7" long	1.000	Set	.800	54.50	125.11
5'-3" long	1.000	Set	.800	58.50	131.22
6'-8" long	1.000	Set	.889	95.50	192.92
Vinyl, louvered, 1'-2" wide, 4'-7" long	1.000	Set	.720	51.50	116.05
1'-4" x 6'-8" long	1.000	Set	.889	87.00	180.41

Be sure to read pages 170-171 for proper use of this section.

Insulation Systems

DESCRIPTION	QUAN.	UNIT	LABOR-HOURS	COST PER S.F.	
				MAT.	TOTAL
Poured insulation, cellulose fiber, R3.8 per inch (1" thick)	1.000	S.F.	.003	.04	.24
Fiberglass , R4.0 per inch (1" thick)	1.000	S.F.	.003	.03	.22
Mineral wool, R3.0 per inch (1" thick)	1.000	S.F.	.003	.03	.21
Polystyrene, R4.0 per inch (1" thick)	1.000	S.F.	.003	.20	.48
Vermiculite, R2.7 per inch (1" thick)	1.000	S.F.	.003	.15	.40
Perlite, R2.7 per inch (1" thick)	1.000	S.F.	.003	.15	.40
Reflective insulation, aluminum foil reinforced with scrim	1.000	S.F.	.004	.15	.45
Reinforced with woven polyolefin	1.000	S.F.	.004	.19	.51
With single bubble air space, R8.8	1.000	S.F.	.005	.30	.74
With double bubble air space, R9.8	1.000	S.F.	.005	.32	.77
Rigid insulation, fiberglass, unfaced,					
1-1/2" thick, R6.2	1.000	S.F.	.008	.45	1.10
2" thick, R8.3	1.000	S.F.	.008	.50	1.17
2-1/2" thick, R10.3	1.000	S.F.	.010	.63	1.49
3" thick, R12.4	1.000	S.F.	.010	.63	1.49
Foil faced, 1" thick, R4.3	1.000	S.F.	.008	.88	1.76
1-1/2" thick, R6.2	1.000	S.F.	.008	1.19	2.24
2" thick, R8.7	1.000	S.F.	.009	1.49	2.77
2-1/2" thick, R10.9	1.000	S.F.	.010	1.76	3.24
3" thick, R13.0	1.000	S.F.	.010	1.91	3.47
Foam glass, 1-1/2" thick R2.64	1.000	S.F.	.010	1.76	3.24
2" thick R5.26	1.000	S.F.	.011	3.07	5.32
Perlite, 1" thick R2.77	1.000	S.F.	.010	.29	.96
2" thick R5.55	1.000	S.F.	.011	.55	1.41
Polystyrene, extruded, blue, 2.2#/C.F., 3/4" thick R4	1.000	S.F.	.010	.37	1.09
1-1/2" thick R8.1	1.000	S.F.	.011	.74	1.70
2" thick R10.8	1.000	S.F.	.011	1.06	2.20
Molded bead board, white, 1" thick R3.85	1.000	S.F.	.010	.15	.74
1-1/2" thick, R5.6	1.000	S.F.	.011	.42	1.21
2" thick, R7.7	1.000	S.F.	.011	.57	1.44
Non-rigid insulation, batts					
Fiberglass, kraft faced, 3-1/2" thick, R11, 11" wide	1.000	S.F.	.005	.25	.64
15" wide	1.000	S.F.	.005	.25	.64
23" wide	1.000	S.F.	.005	.25	.64
6" thick, R19, 11" wide	1.000	S.F.	.006	.36	.86
15" wide	1.000	S.F.	.006	.36	.86
23" wide	1.000	S.F.	.006	.36	.86
9" thick, R30, 15" wide	1.000	S.F.	.006	.66	1.32
23" wide	1.000	S.F.	.006	.66	1.32
12" thick, R38, 15" wide	1.000	S.F.	.006	.84	1.60
23" wide	1.000	S.F.	.006	.84	1.60
Fiberglass, foil faced, 3-1/2" thick, R11, 15" wide	1.000	S.F.	.005	.37	.83
23" wide	1.000	S.F.	.005	.37	.83
6" thick, R19, 15" thick	1.000	S.F.	.005	.45	.95
23" wide	1.000	S.F.	.005	.45	.95
9" thick, R30, 15" wide	1.000	S.F.	.006	.78	1.51
23" wide	1.000	S.F.	.006	.78	1.51

Be sure to read pages 170-171 for proper use of this section.

Division 5
ROOFING

Gable End Roofing Systems

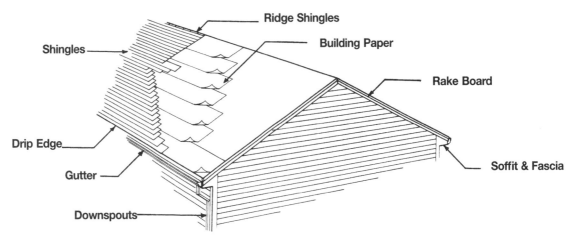

Ridge Shingles
Building Paper
Shingles
Rake Board
Drip Edge
Gutter
Soffit & Fascia
Downspouts

System Description	QUAN.	UNIT	LABOR-HOURS	MAT. COST	COST PER S.F.
ASPHALT, ROOF SHINGLES, CLASS A					
Shingles, inorganic class A, 210-235 lb./sq., 4/12 pitch	1.160	S.F.	.017	.37	
Drip edge, metal, 5" wide	.150	L.F.	.003	.04	
Building paper, #15 felt	1.300	S.F.	.002	.03	
Ridge shingles, asphalt	.042	L.F.	.001	.03	
Soffit & fascia, white painted aluminum, 1' overhang	.083	L.F.	.012	.20	
Rake trim, 1" x 6"	.040	L.F.	.002	.03	
Rake trim, prime and paint	.040	L.F.	.002	.01	
Gutter, seamless, aluminum painted	.083	L.F.	.006	.10	Contractor's Fee
Downspouts, aluminum painted	.035	L.F.	.002	.04	**3.62**
TOTAL			.047	.85	
WOOD, CEDAR SHINGLES NO. 1 PERFECTIONS, 18" LONG					
Shingles, wood, cedar, No. 1 perfections, 4/12 pitch	1.160	S.F.	.035	2.11	
Drip edge, metal, 5" wide	.150	L.F.	.003	.04	
Building paper, #15 felt	1.300	S.F.	.002	.03	
Ridge shingles, cedar	.042	L.F.	.001	.11	
Soffit & fascia, white painted aluminum, 1' overhang	.083	L.F.	.012	.20	
Rake trim, 1" x 6"	.040	L.F.	.002	.03	
Rake trim, prime and paint	.040	L.F.	.002	.01	
Gutter, seamless, aluminum, painted	.083	L.F.	.006	.10	Contractor's Fee
Downspouts, aluminum, painted	.035	L.F.	.002	.04	**7.39**
TOTAL			.065	2.67	

Be sure to read pages 170-171 for proper use of this section.

Hip Roof Roofing Systems

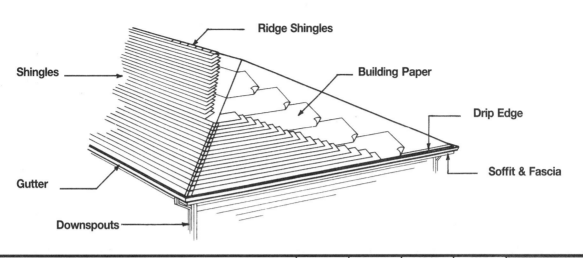

System Description	QUAN.	UNIT	LABOR-HOURS	MAT. COST	COST PER S.F.
ASPHALT, ROOF SHINGLES, CLASS A					
Shingles, inorganic, class A, 210-235 lb./sq. 4/12 pitch	1.570	S.F.	.023	.50	
Drip edge, metal, 5" wide	.122	L.F.	.002	.03	
Building paper, #15 asphalt felt	1.800	S.F.	.002	.04	
Ridge shingles, asphalt	.075	L.F.	.002	.06	
Soffit & fascia, white painted aluminum, 1' overhang	.120	L.F.	.017	.29	
Gutter, seamless, aluminum, painted	.120	L.F.	.008	.14	
Downspouts, aluminum, painted	.035	L.F.	.002	.04	**Contractor's Fee**
TOTAL			.056	1.10	**4.57**
WOOD, CEDAR SHINGLES, NO. 1 PERFECTIONS, 18" LONG					
Shingles, red cedar, No. 1 perfections, 5" exp., 4/12 pitch	1.570	S.F.	.047	2.82	
Drip edge, metal, 5" wide	.122	L.F.	.002	.03	
Building paper, #15 asphalt felt	1.800	S.F.	.002	.04	
Ridge shingles, wood, cedar	.075	L.F.	.002	.20	
Soffit & fascia, white painted aluminum, 1' overhang	.120	L.F.	.017	.29	
Gutter, seamless, aluminum, painted	.120	L.F.	.008	.14	
Downspouts, aluminum, painted	.035	L.F.	.002	.04	**Contractor's Fee**
TOTAL			.080	3.56	**9.66**

Be sure to read pages 170-171 for proper use of this section.

Gambrel Roofing Systems

System Description	QUAN.	UNIT	LABOR-HOURS	MAT. COST	COST PER S.F.
ASPHALT, ROOF SHINGLES, CLASS A					
Shingles, asphalt, inorganic, class A, 210-235 lb./sq.	1.450	S.F.	.022	.47	
Drip edge, metal, 5" wide	.146	L.F.	.003	.04	
Building paper, #15 asphalt felt	1.500	S.F.	.002	.04	
Ridge shingles, asphalt	.042	L.F.	.001	.03	
Soffit & fascia, painted aluminum, 1' overhang	.083	L.F.	.012	.20	
Rake trim, 1" x 6"	.063	L.F.	.003	.05	
Rake trim, prime and paint	.063	L.F.	.003	.02	
Gutter, seamless, alumunum, painted	.083	L.F.	.006	.10	
Downspouts, aluminum, painted	.042	L.F.	.002	.05	Contractor's Fee
TOTAL			.054	1.00	**4.14**
WOOD, CEDAR SHINGLES, NO. 1 PERFECTIONS, 18" LONG					
Shingles, wood, red cedar, No. 1 perfections, 5" exposure	1.450	S.F.	.044	2.64	
Drip edge, metal, 5" wide	.146	L.F.	.003	.04	
Building paper, #15 asphalt felt	1.500	S.F.	.002	.04	
Ridge shingles, wood	.042	L.F.	.001	.11	
Soffit & fascia, white painted aluminum, 1' overhang	.083	L.F.	.012	.20	
Rake trim, 1" x 6"	.063	L.F.	.003	.05	
Rake trim, prime and paint	.063	L.F.	.001	.01	
Gutter, seamless, aluminum, painted	.083	L.F.	.006	.10	
Downspouts, aluminum, painted	.042	L.F.	.002	.05	Contractor's Fee
TOTAL			.074	3.24	**8.75**

Be sure to read pages 170-171 for proper use of this section.

Mansard Roofing Systems

System Description	QUAN.	UNIT	LABOR-HOURS	MAT. COST	COST PER S.F.
ASPHALT, ROOF SHINGLES, CLASS A					
Shingles, standard inorganic class A 210-235 lb./sq.	2.210	S.F.	.032	.68	
Drip edge, metal, 5″ wide	.122	L.F.	.002	.03	
Building paper, #15 asphalt felt	2.300	S.F.	.003	.05	
Ridge shingles, asphalt	.090	L.F.	.002	.07	
Soffit & fascia, white painted aluminum, 1′ overhang	.122	L.F.	.018	.29	
Gutter, seamless, aluminum, painted	.122	L.F.	.008	.15	
Downspouts, aluminum, painted	.042	L.F.	.002	.05	Contractor's Fee
TOTAL			.067	1.32	**5.43**
WOOD, CEDAR SHINGLES, NO. 1 PERFECTIONS, 18″ LONG					
Shingles, wood, red cedar, No. 1 perfections, 5″ exposure	2.210	S.F.	.064	3.87	
Drip edge, metal, 5″ wide	.122	L.F.	.002	.03	
Building paper, #15 asphalt felt	2.300	S.F.	.003	.05	
Ridge shingles, wood	.090	L.F.	.003	.24	
Soffit & fascia, white painted aluminum, 1′ overhang	.122	L.F.	.018	.29	
Gutter, seamless, aluminum, painted	.122	L.F.	.008	.15	
Downspouts, aluminum, painted	.042	L.F.	.002	.05	Contractor's Fee
TOTAL			.100	4.68	**12.38**

Be sure to read pages 170-171 for proper use of this section.

Shed Roofing Systems

System Description	QUAN.	UNIT	LABOR-HOURS	MAT. COST	COST PER S.F.
ASPHALT, ROOF SHINGLES, CLASS A					
Shingles, inorganic class A 210-235 lb./sq. 4/12 pitch	1.230	S.F.	.019	.40	
Drip edge, metal, 5″ wide	.100	L.F.	.002	.02	
Building paper, #15 asphalt felt	1.300	S.F.	.002	.03	
Soffit & fascia, white painted aluminum, 1′ overhang	.080	L.F.	.012	.19	
Rake trim, 1″ x 6″	.043	L.F.	.002	.03	
Rake trim, prime and paint	.043	L.F.	.002	.01	
Gutter, seamless, aluminum, painted	.040	L.F.	.003	.05	
Downspouts, painted aluminum	.020	L.F.	.001	.02	Contractor's Fee **3.24**
TOTAL			.043	.75	
WOOD, CEDAR SHINGLES, NO. 1 PERFECTIONS, 18″ LONG					
Shingles, red cedar, No. 1 perfections, 5″ exp., 4/12 pitch	1.230	S.F.	.035	2.11	
Drip edge, metal, 5″ wide	.100	L.F.	.002	.02	
Building paper, #15 asphalt felt	1.300	S.F.	.002	.03	
Soffit & fascia, white painted aluminum, 1′ overhang	.080	L.F.	.012	.19	
Rake trim, 1″ x 6″	.043	L.F.	.002	.03	
Rake trim, prime and paint	.043	L.F.	.001	.01	
Gutter, seamless, aluminum, painted	.040	L.F.	.003	.05	
Downspouts, painted aluminum	.020	L.F.	.001	.02	Contractor's Fee **6.71**
TOTAL			.058	2.46	

Be sure to read pages 170-171 for proper use of this section.

Gable Dormer Roofing Systems

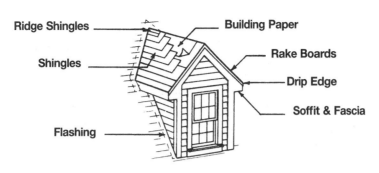

Ridge Shingles — Building Paper
Shingles — Rake Boards
— Drip Edge
— Soffit & Fascia
Flashing —

System Description	QUAN.	UNIT	LABOR-HOURS	MAT. COST	COST PER S.F.
ASPHALT, ROOF SHINGLES, CLASS A					
Shingles, standard inorganic class A 210-235 lb./sq	1.400	S.F.	.020	.43	
Drip edge, metal, 5" wide	.220	L.F.	.004	.05	
Building paper, #15 asphalt felt	1.500	S.F.	.002	.04	
Ridge shingles, asphalt	.280	L.F.	.007	.22	
Soffit & fascia, aluminum, vented	.220	L.F.	.032	.52	
Flashing, aluminum, mill finish, .013" thick	1.500	S.F.	.083	.59	Contractor's Fee
TOTAL			.148	1.85	**10.09**
WOOD, CEDAR, NO. 1 PERFECTIONS					
Shingles, red cedar, No.1 perfections, 18" long, 5" exp.	1.400	S.F.	.041	2.46	
Drip edge, metal, 5" wide	.220	L.F.	.004	.05	
Building paper, #15 asphalt felt	1.500	S.F.	.002	.04	
Ridge shingles, wood	.280	L.F.	.008	.74	
Soffit & fascia, aluminum, vented	.220	L.F.	.032	.52	
Flashing, aluminum, mill finish, .013" thick	1.500	S.F.	.083	.59	Contractor's Fee
TOTAL			.170	4.40	**15.22**
SLATE, BUCKINGHAM, BLACK					
Shingles, Buckingham, Virginia, black	1.400	S.F.	.064	8.33	
Drip edge, metal, 5" wide	.220	L.F.	.004	.05	
Building paper, #15 asphalt felt	1.500	S.F.	.002	.04	
Ridge shingles, slate	.280	L.F.	.011	2.62	
Soffit & fascia, aluminum, vented	.220	L.F.	.032	.52	
Flashing, copper, 16 oz.	1.500	S.F.	.104	3.80	Contractor's Fee
TOTAL			.217	15.36	**34.40**

Be sure to read pages 170-171 for proper use of this section.

Shed Dormer Roofing Systems

System Description	QUAN.	UNIT	LABOR-HOURS	MAT. COST	COST PER S.F.
ASPHALT, ROOF SHINGLES, CLASS A					
Shingles, standard inorganic class A 210-235 lb./sq.	1.100	S.F.	.016	.34	
Drip edge, aluminum, 5″ wide	.250	L.F.	.005	.06	
Building paper, #15 asphalt felt	1.200	S.F.	.002	.03	
Soffit & fascia, aluminum, vented, 1′ overhang	.250	L.F.	.036	.60	
Flashing, aluminum, mill finish, 0.013″ thick	.800	L.F.	.044	.31	
					Contractor's Fee 7.13
TOTAL			.103	1.34	
WOOD, CEDAR, NO. 1 PERFECTIONS, 18″ LONG					
Shingles, wood, red cedar, #1 perfections, 5″ exposure	1.100	S.F.	.032	1.94	
Drip edge, aluminum, 5″ wide	.250	L.F.	.005	.06	
Building paper, #15 asphalt felt	1.200	S.F.	.002	.03	
Soffit & fascia, aluminum, vented, 1′ overhang	.250	L.F.	.036	.60	
Flashing, aluminum, mill finish, 0.013″ thick	.800	L.F.	.044	.31	
					Contractor's Fee 10.47
TOTAL			.119	2.94	
SLATE, BUCKINGHAM, BLACK					
Shingles, slate, Buckingham, black	1.100	S.F.	.050	6.55	
Drip edge, aluminum, 5″ wide	.250	L.F.	.005	.06	
Building paper, #15 asphalt felt	1.200	S.F.	.002	.03	
Soffit & fascia, aluminum, vented, 1′ overhang	.250	L.F.	.036	.60	
Flashing, copper, 16 oz.	.800	L.F.	.056	2.02	
					Contractor's Fee 21.62
TOTAL			.149	9.26	

Be sure to read pages 170-171 for proper use of this section.

Built-up Roofing Systems

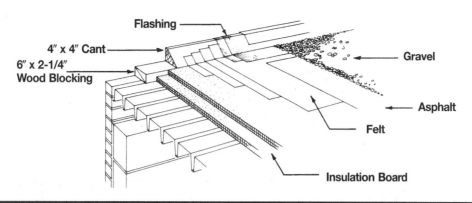

Flashing

4" x 4" Cant

6" x 2-1/4" Wood Blocking

Gravel

Asphalt

Felt

Insulation Board

System Description	QUAN.	UNIT	LABOR-HOURS	MAT. COST	COST PER S.F.
ASPHALT, ORGANIC, 4-PLY, INSULATED DECK					
Membrane, asphalt, 4-plies #15 felt, gravel surfacing	1.000	S.F.	.025	.60	
Insulation board, 2-layers of 1-1/16" glass fiber	2.000	S.F.	.016	1.68	
Wood blocking, 2" x 6"	.040	L.F.	.004	.07	
Treated 4" x 4" cant strip	.040	L.F.	.001	.06	
Flashing, aluminum, 0.040" thick	.050	S.F.	.003	.08	Contractor's Fee
TOTAL			.049	2.49	**6.40**
ASPHALT, INORGANIC, 3-PLY, INSULATED DECK					
Membrane, asphalt, 3-plies type IV glass felt, gravel surfacing	1.000	S.F.	.028	.59	
Insulation board, 2-layers of 1-1/16" glass fiber	2.000	S.F.	.016	1.68	
Wood blocking, 2" x 6"	.040	L.F.	.004	.07	
Treated 4" x 4" cant strip	.040	L.F.	.001	.06	
Flashing, aluminum, 0.040" thick	.050	S.F.	.003	.08	Contractor's Fee
TOTAL			.052	2.48	**6.51**
COAL TAR, ORGANIC, 4-PLY, INSULATED DECK					
Membrane, coal tar, 4-plies #15 felt, gravel surfacing	1.000	S.F.	.027	1.12	
Insulation board, 2-layers of 1-1/16" glass fiber	2.000	S.F.	.016	1.68	
Wood blocking, 2" x 6"	.040	L.F.	.004	.07	
Treated 4" x 4" cant strip	.040	L.F.	.001	.06	
Flashing, aluminum, 0.040" thick	.050	S.F.	.003	.08	Contractor's Fee
TOTAL			.051	3.01	**7.28**
COAL TAR, INORGANIC, 3-PLY, INSULATED DECK					
Membrane, coal tar, 3-plies type IV glass felt, gravel surfacing	1.000	S.F.	.029	.93	
Insulation board, 2-layers of 1-1/16" glass fiber	2.000	S.F.	.016	1.68	
Wood blocking, 2" x 6"	.040	L.F.	.004	.07	
Treated 4" x 4" cant strip	.040	L.F.	.001	.06	
Flashing, aluminum, 0.040" thick	.050	S.F.	.003	.08	Contractor's Fee
TOTAL			.053	2.82	**7.12**

Be sure to read pages 170-171 for proper use of this section.

Division 7
SPECIALTIES

Greenhouse Systems

SYSTEM DESCRIPTION	QUAN.	UNIT	LABOR-HOURS	COST EACH	
				MAT.	TOTAL
Economy, lean to, shell only, not including 2' stub wall, fndtn, flrs, heat					
4' x 16'	1.000	Ea.	26.212	2000.00	3225.00
4' x 24'	1.000	Ea.	30.259	2325.00	3750.00
6' x 10'	1.000	Ea.	16.552	1675.00	2450.00
6' x 16'	1.000	Ea.	23.034	2350.00	3425.00
6' x 24'	1.000	Ea.	29.793	3025.00	4425.00
8' x 10'	1.000	Ea.	22.069	2250.00	3275.00
8' x 16'	1.000	Ea.	38.400	3900.00	5700.00
8' x 24'	1.000	Ea.	49.655	5050.00	7375.00
Free standing, 8' x 8'	1.000	Ea.	17.356	2625.00	3435.00
8' x 16'	1.000	Ea.	30.211	4575.00	5975.00
8' x 24'	1.000	Ea.	39.051	5900.00	7725.00
10' x 10'	1.000	Ea.	18.824	3150.00	4030.00
10' x 16'	1.000	Ea.	24.095	4025.00	5150.00
10' x 24'	1.000	Ea.	31.624	5300.00	6775.00
14' x 10'	1.000	Ea.	20.741	3925.00	4890.00
14' x 16'	1.000	Ea.	24.889	4700.00	5850.00
14' x 24'	1.000	Ea.	33.349	6300.00	7850.00
Standard,lean to,shell only,not incl.2'stub wall, fndtn,flrs, heat 4'x10'	1.000	Ea.	28.235	2150.00	3475.00
4' x 16'	1.000	Ea.	39.341	3000.00	4850.00
4' x 24'	1.000	Ea.	45.412	3475.00	5600.00
6' x 10'	1.000	Ea.	24.827	2525.00	3675.00
6' x 16'	1.000	Ea.	34.538	3500.00	5125.00
6' x 24'	1.000	Ea.	44.689	4525.00	6625.00
8' x 10'	1.000	Ea.	33.103	3350.00	4900.00
8' x 16'	1.000	Ea.	57.600	5850.00	8550.00
8' x 24'	1.000	Ea.	74.482	7550.00	11025.00
Free standing, 8' x 8'	1.000	Ea.	26.034	3925.00	5150.00
8' x 16'	1.000	Ea.	45.316	6850.00	8975.00
8' x 24'	1.000	Ea.	58.577	8850.00	11575.00
10' x 10'	1.000	Ea.	28.236	4725.00	6050.00
10' x 16'	1.000	Ea.	36.142	6050.00	7750.00
10' x 24'	1.000	Ea.	47.436	7950.00	10175.00
14' x 10'	1.000	Ea.	31.112	5875.00	7325.00
14' x 16'	1.000	Ea.	37.334	7050.00	8800.00
14' x 24'	1.000	Ea.	50.030	9450.00	11775.00
Deluxe,lean to,shell only,not incl.2'stub wall, fndtn, flrs or heat, 4'x10'	1.000	Ea.	20.645	3750.00	4710.00
4' x 16'	1.000	Ea.	33.032	5975.00	7500.00
4' x 24'	1.000	Ea.	49.548	8975.00	11275.00
6' x 10'	1.000	Ea.	30.968	5600.00	7050.00
6' x 16'	1.000	Ea.	49.548	8975.00	11275.00
6' x 24'	1.000	Ea.	74.323	13500.00	16950.00
8' x 10'	1.000	Ea.	41.290	7475.00	9400.00
8' x 16'	1.000	Ea.	66.065	12000.00	15075.00
8' x 24'	1.000	Ea.	99.097	18000.00	22600.00
Freestanding, 8' x 8'	1.000	Ea.	18.618	5150.00	6020.00
8' x 16'	1.000	Ea.	37.236	10300.00	12050.00
8' x 24'	1.000	Ea.	55.855	15500.00	18100.00
10' x 10'	1.000	Ea.	29.091	8050.00	9400.00
10' x 16'	1.000	Ea.	46.546	12900.00	15075.00
10' x 24'	1.000	Ea.	69.818	19300.00	22575.00
14' x 10'	1.000	Ea.	40.727	11300.00	13200.00
14' x 16'	1.000	Ea.	65.164	18000.00	21050.00
14' x 24'	1.000	Ea.	97.746	27000.00	31575.00

Be sure to read pages 170-171 for proper use of this section.

Wood Deck Systems

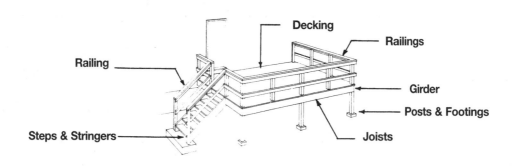

Decking
Railings
Railing
Girder
Posts & Footings
Steps & Stringers
Joists

System Description	QUAN.	UNIT	LABOR-HOURS	MAT. COST	COST PER S.F.
8′ X 12′ DECK, PRESSURE TREATED LUMBER, JOISTS 16″ O.C.					
Decking, 2″ x 6″ lumber	2.080	L.F.	.027	1.29	
Lumber preservative	2.080	L.F.		.27	
Joists, 2″ x 8″, 16″ O.C.	1.000	L.F.	.015	.95	
Lumber preservative	1.000	L.F.		.17	
Girder, 2″ x 10″	.125	L.F.	.002	.17	
Lumber preservative	.125	L.F.		.03	
Hand excavation for footings	.250	L.F.	.006		
Concrete footings	.250	L.F.	.006	.22	
4″ x 4″ Posts	.250	L.F.	.010	.36	
Lumber preservative	.250	L.F.		.04	
Stairs, 2″ x 10″ stringers, 2″ x 10″ steps	1.000	Set	.020	3.15	
Railings, 2″ x 4″	1.000	L.F.	.026	.39	Contractor's Fee
Lumber preservative	1.000	L.F.		.09	
TOTAL			.112	7.13	**16.59**
12′ X 16′ DECK, PRESSURE TREATED LUMBER, JOISTS 24″ O.C.					
Decking, 2″ x 6″	2.080	L.F.	.027	1.29	
Lumber preservative	2.080	L.F.		.27	
Joists, 2″ x 10″, 24″ O.C.	.800	L.F.	.014	1.10	
Lumber preservative	.800	L.F.		.17	
Girder, 2″ x 10″	.083	L.F.	.001	.11	
Lumber preservative	.083	L.F.		.02	
Hand excavation for footings	.122	L.F.	.006		
Concrete footings	.122	L.F.	.006	.22	
4″ x 4″ Posts	.122	L.F.	.005	.18	
Lumber preservative	.122	L.F.		.02	
Stairs, 2″ x 10″ stringers, 2″ x 10″ steps	1.000	Set	.012	1.89	
Railings, 2″ x 4″	.670	L.F.	.017	.26	Contractor's Fee
Lumber preservative	.670	L.F.		.06	
TOTAL			.088	5.59	**13.03**
12′ X 24′ DECK, REDWOOD OR CEDAR, JOISTS 16″ O.C.					
Decking, 2″ x 6″ redwood	2.080	L.F.	.027	4.64	
Joists, 2″ x 10″, 16″ O.C.	1.000	L.F.	.018	3.85	
Girder, 2″ x 10″	.083	L.F.	.001	.32	
Hand excavation for footings	.111	L.F.	.006		
Concrete footings	.111	L.F.	.006	.22	
Lumber preservative	.111	L.F.		.02	
Post, 4″ x 4″, including concrete footing	.111	L.F.	.009	.68	
Stairs, 2″ x 10″ stringers, 2″ x 10″ steps	1.000	Set	.012	1.61	Contractor's Fee
Railings, 2″ x 4″	.540	L.F.	.005	.80	
TOTAL			.084	12.14	**22.92**

Be sure to read pages 170-171 for proper use of this section.

Adjusting Project Costs To Your Location

Adjusting Project Costs to Your Location

Costs shown in *Exterior Home Improvement Costs* are based on National Averages for materials and installation. To adjust these costs to a specific location, simply multiply the base cost by the factor for that city. The data is arranged alphabetically by state and postal zip code numbers. For a city not listed, use the factor for a nearby city with similar economic characteristics.

STATE	CITY	Residential
ALABAMA		
350-352	Birmingham	.85
354	Tuscaloosa	.80
355	Jasper	.76
356	Decatur	.79
357-358	Huntsville	.81
359	Gadsden	.80
360-361	Montgomery	.82
362	Anniston	.73
363	Dothan	.79
364	Evergreen	.79
365-366	Mobile	.81
367	Selma	.79
368	Phenix City	.82
369	Butler	.79
ALASKA		
995-996	Anchorage	1.25
997	Fairbanks	1.25
998	Juneau	1.24
999	Ketchikan	1.30
ARIZONA		
850,853	Phoenix	.92
852	Mesa/Tempe	.87
855	Globe	.88
856-857	Tucson	.90
859	Show Low	.89
860	Flagstaff	.92
863	Prescott	.90
864	Kingman	.89
865	Chambers	.88
ARKANSAS		
716	Pine Bluff	.80
717	Camden	.70
718	Texarkana	.74
719	Hot Springs	.69
720-722	Little Rock	.81
723	West Memphis	.79
724	Jonesboro	.79
725	Batesville	.75
726	Harrison	.76
727	Fayetteville	.69
728	Russellville	.77
729	Fort Smith	.83
CALIFORNIA		
900-902	Los Angeles	1.08
903-905	Inglewood	1.06
906-908	Long Beach	1.07
910-912	Pasadena	1.07
913-916	Van Nuys	1.09
917-918	Alhambra	1.08
919-921	San Diego	1.10
922	Palm Springs	1.09
923-924	San Bernardino	1.08
925	Riverside	1.11
926-927	Santa Ana	1.09
928	Anaheim	1.10
930	Oxnard	1.13
931	Santa Barbara	1.11
932-933	Bakersfield	1.11

STATE	CITY	Residential
934	San Luis Obispo	1.14
935	Mojave	1.09
936-938	Fresno	1.12
939	Salinas	1.12
940-941	San Francisco	1.21
942,956-958	Sacramento	1.11
943	Palo Alto	1.15
944	San Mateo	1.16
945	Vallejo	1.11
946	Oakland	1.16
947	Berkeley	1.16
948	Richmond	1.14
949	San Rafael	1.26
950	Santa Cruz	1.16
951	San Jose	1.22
952	Stockton	1.13
953	Modesto	1.13
954	Santa Rosa	1.14
955	Eureka	1.10
959	Marysville	1.10
960	Redding	1.11
961	Susanville	1.11
COLORADO		
800-802	Denver	.99
803	Boulder	.88
804	Golden	.97
805	Fort Collins	.98
806	Greeley	.90
807	Fort Morgan	.97
808-809	Colorado Springs	.94
810	Pueblo	.94
811	Alamosa	.89
812	Salida	.89
813	Durango	.88
814	Montrose	.86
815	Grand Junction	.90
816	Glenwood Springs	.95
CONNECTICUT		
060	New Britain	1.04
061	Hartford	1.04
062	Willimantic	1.03
063	New London	1.05
064	Meriden	1.03
065	New Haven	1.04
066	Bridgeport	1.02
067	Waterbury	1.05
068	Norwalk	1.01
069	Stamford	1.04
D.C.		
200-205	Washington	.93
DELAWARE		
197	Newark	1.00
198	Wilmington	1.00
199	Dover	1.00
FLORIDA		
320,322	Jacksonville	.83
321	Daytona Beach	.87
323	Tallahassee	.75

Adjusting Project Costs to Your Location

STATE	CITY	Residential	STATE	CITY	Residential
324	Panama City	.70	463-464	Gary	1.04
325	Pensacola	.84	465-466	South Bend	.94
326,344	Gainesville	.84	467-468	Fort Wayne	.92
327-328,347	Orlando	.86	469	Kokomo	.93
329	Melbourne	.90	470	Lawrenceburg	.93
330-332,340	Miami	.83	471	New Albany	.93
333	Fort Lauderdale	.83	472	Columbus	.96
334,349	West Palm Beach	.86	473	Muncie	.94
335-336,346	Tampa	.80	474	Bloomington	.96
337	St. Petersburg	.81	475	Washington	.93
338	Lakeland	.79	476-477	Evansville	.95
339,341	Fort Myers	.79	478	Terre Haute	.96
342	Sarasota	.78	479	Lafayette	.92
GEORGIA			**IOWA**		
300-303,399	Atlanta	.85	500-503,509	Des Moines	.97
304	Statesboro	.72	504	Mason City	.86
305	Gainesville	.76	505	Fort Dodge	.84
306	Athens	.77	506-507	Waterloo	.88
307	Dalton	.68	508	Creston	.89
308-309	Augusta	.76	510-511	Sioux City	.95
310-312	Macon	.81	512	Sibley	.80
313-314	Savannah	.80	513	Spencer	.80
315	Waycross	.74	514	Carroll	.84
316	Valdosta	.76	515	Council Bluffs	.96
317	Albany	.77	516	Shenandoah	.82
318-319	Columbus	.78	520	Dubuque	.99
			521	Decorah	.88
HAWAII			522-524	Cedar Rapids	1.01
967	Hilo	1.27	525	Ottumwa	.94
968	Honolulu	1.27	526	Burlington	.92
			527-528	Davenport	.98
STATES & POSS.					
969	Guam	1.37	**KANSAS**		
			660-662	Kansas City	.96
IDAHO			664-666	Topeka	.86
832	Pocatello	.94	667	Fort Scott	.85
833	Twin Falls	.79	668	Emporia	.81
834	Idaho Falls	.83	669	Belleville	.87
835	Lewiston	1.09	670-672	Wichita	.89
836-837	Boise	.94	673	Independence	.82
838	Coeur d'Alene	.95	674	Salina	.85
			675	Hutchinson	.79
ILLINOIS			676	Hays	.84
600-603	North Suburban	1.11	677	Colby	.85
604	Joliet	1.11	678	Dodge City	.84
605	South Suburban	1.10	679	Liberal	.78
606	Chicago	1.13			
609	Kankakee	1.00	**KENTUCKY**		
610-611	Rockford	1.05	400-402	Louisville	.95
612	Rock Island	1.06	403-405	Lexington	.87
613	La Salle	1.06	406	Frankfort	.92
614	Galesburg	1.09	407-409	Corbin	.78
615-616	Peoria	1.09	410	Covington	.98
617	Bloomington	1.05	411-412	Ashland	.96
618-619	Champaign	1.04	413-414	Campton	.77
620-622	East St. Louis	1.00	415-416	Pikeville	.82
623	Quincy	.99	417-418	Hazard	.76
624	Effingham	1.02	420	Paducah	.97
625	Decatur	1.01	421-422	Bowling Green	.96
626-627	Springfield	1.01	423	Owensboro	.91
628	Centralia	.99	424	Henderson	.94
629	Carbondale	.97	425-426	Somerset	.75
			427	Elizabethtown	.94
INDIANA					
460	Anderson	.95	**LOUISIANA**		
461-462	Indianapolis	.98	700-701	New Orleans	.86

225

Adjusting Project Costs to Your Location

STATE	CITY	Residential
703	Thibodaux	.85
704	Hammond	.84
705	Lafayette	.84
706	Lake Charles	.83
707-708	Baton Rouge	.82
710-711	Shreveport	.81
712	Monroe	.79
713-714	Alexandria	.78
MAINE		
039	Kittery	.86
040-041	Portland	.91
042	Lewiston	.92
043	Augusta	.88
044	Bangor	.93
045	Bath	.89
046	Machias	.87
047	Houlton	.89
048	Rockland	.86
049	Waterville	.86
MARYLAND		
206	Waldorf	.87
207-208	College Park	.90
209	Silver Spring	.89
210-212	Baltimore	.91
214	Annapolis	.89
215	Cumberland	.87
216	Easton	.73
217	Hagerstown	.90
218	Salisbury	.76
219	Elkton	.82
MASSACHUSETTS		
010-011	Springfield	1.04
012	Pittsfield	.99
013	Greenfield	1.02
014	Fitchburg	1.08
015-016	Worcester	1.10
017	Framingham	1.06
018	Lowell	1.08
019	Lawrence	1.09
020-022, 024	Boston	1.14
023	Brockton	1.06
025	Buzzards Bay	1.02
026	Hyannis	1.04
027	New Bedford	1.06
MICHIGAN		
480,483	Royal Oak	1.03
481	Ann Arbor	1.04
482	Detroit	1.07
484-485	Flint	.99
486	Saginaw	.97
487	Bay City	.96
488-489	Lansing	1.01
490	Battle Creek	1.01
491	Kalamazoo	1.00
492	Jackson	.99
493,495	Grand Rapids	.88
494	Muskegon	.95
496	Traverse City	.87
497	Gaylord	.87
498-499	Iron Mountain	.98
MINNESOTA		
550-551	Saint Paul	1.09

STATE	CITY	Residential
553-555	Minneapolis	1.11
556-558	Duluth	1.04
559	Rochester	1.03
560	Mankato	1.00
561	Windom	.90
562	Willmar	.93
563	St. Cloud	1.11
564	Brainerd	1.06
565	Detroit Lakes	.88
566	Bemidji	.91
567	Thief River Falls	.87
MISSISSIPPI		
386	Clarksdale	.70
387	Greenville	.80
388	Tupelo	.71
389	Greenwood	.72
390-392	Jackson	.80
393	Meridian	.76
394	Laurel	.72
395	Biloxi	.84
396	McComb	.72
397	Columbus	.70
MISSOURI		
630-631	St. Louis	1.00
633	Bowling Green	.92
634	Hannibal	.99
635	Kirksville	.86
636	Flat River	.94
637	Cape Girardeau	.93
638	Sikeston	.90
639	Poplar Bluff	.90
640-641	Kansas City	1.04
644-645	St. Joseph	.90
646	Chillicothe	.82
647	Harrisonville	.98
648	Joplin	.84
650-651	Jefferson City	.98
652	Columbia	.99
653	Sedalia	.99
654-655	Rolla	.95
656-658	Springfield	.86
MONTANA		
590-591	Billings	.93
592	Wolf Point	.92
593	Miles City	.91
594	Great Falls	.92
595	Havre	.90
596	Helena	.91
597	Butte	.90
598	Missoula	.89
599	Kalispell	.88
NEBRASKA		
680-681	Omaha	.92
683-685	Lincoln	.88
686	Columbus	.74
687	Norfolk	.84
688	Grand Island	.88
689	Hastings	.82
690	Mccook	.79
691	North Platte	.86
692	Valentine	.77
693	Alliance	.75

Adjusting Project Costs to Your Location

STATE	CITY	Residential
NEVADA		
889-891	Las Vegas	1.05
893	Ely	.93
894-895	Reno	.95
897	Carson City	.96
898	Elko	.90
NEW HAMPSHIRE		
030	Nashua	.94
031	Manchester	.94
032-033	Concord	.93
034	Keene	.78
035	Littleton	.82
036	Charleston	.77
037	Claremont	.76
038	Portsmouth	.93
NEW JERSEY		
070-071	Newark	1.14
072	Elizabeth	1.09
073	Jersey City	1.11
074-075	Paterson	1.12
076	Hackensack	1.10
077	Long Branch	1.10
078	Dover	1.11
079	Summit	1.08
080,083	Vineland	1.11
081	Camden	1.11
082,084	Atlantic City	1.11
085-086	Trenton	1.12
087	Point Pleasant	1.10
088-089	New Brunswick	1.12
NEW MEXICO		
870-872	Albuquerque	.88
873	Gallup	.88
874	Farmington	.88
875	Santa Fe	.88
877	Las Vegas	.88
878	Socorro	.87
879	Truth/Consequences	.87
880	Las Cruces	.84
881	Clovis	.89
882	Roswell	.90
883	Carrizozo	.91
884	Tucumcari	.90
NEW YORK		
100-102	New York	1.35
103	Staten Island	1.31
104	Bronx	1.30
105	Mount Vernon	1.20
106	White Plains	1.19
107	Yonkers	1.22
108	New Rochelle	1.20
109	Suffern	1.14
110	Queens	1.30
111	Long Island City	1.31
112	Brooklyn	1.31
113	Flushing	1.32
114	Jamaica	1.30
115,117,118	Hicksville	1.26
116	Far Rockaway	1.32
119	Riverhead	1.27
120-122	Albany	.97
123	Schenectady	.98
124	Kingston	1.11

STATE	CITY	Residential
125-126	Poughkeepsie	1.13
127	Monticello	1.09
128	Glens Falls	.95
129	Plattsburgh	.95
130-132	Syracuse	.99
133-135	Utica	.91
136	Watertown	.92
137-139	Binghamton	.94
140-142	Buffalo	1.05
143	Niagara Falls	1.07
144-146	Rochester	.99
147	Jamestown	.98
148-149	Elmira	.95
NORTH CAROLINA		
270,272-274	Greensboro	.75
271	Winston-Salem	.74
275-276	Raleigh	.76
277	Durham	.75
278	Rocky Mount	.68
279	Elizabeth City	.70
280	Gastonia	.74
281-282	Charlotte	.74
283	Fayetteville	.75
284	Wilmington	.73
285	Kinston	.67
286	Hickory	.66
287-288	Asheville	.73
289	Murphy	.66
NORTH DAKOTA		
580-581	Fargo	.79
582	Grand Forks	.77
583	Devils Lake	.76
584	Jamestown	.76
585	Bismarck	.80
586	Dickinson	.81
587	Minot	.82
588	Williston	.76
OHIO		
430-432	Columbus	.98
433	Marion	.92
434-436	Toledo	1.02
437-438	Zanesville	.93
439	Steubenville	.97
440	Lorain	1.05
441	Cleveland	1.09
442-443	Akron	1.02
444-445	Youngstown	1.01
446-447	Canton	.97
448-449	Mansfield	.96
450	Hamilton	1.00
451-452	Cincinnati	1.00
453-454	Dayton	.94
455	Springfield	.95
456	Chillicothe	1.02
457	Athens	.92
458	Lima	.96
OKLAHOMA		
730-731	Oklahoma City	.82
734	Ardmore	.83
735	Lawton	.84
736	Clinton	.80
737	Enid	.83
738	Woodward	.82

Adjusting Project Costs to Your Location

STATE	CITY	Residential
739	Guymon	.69
740-741	Tulsa	.84
743	Miami	.86
744	Muskogee	.75
745	Mcalester	.76
746	Ponca City	.82
747	Durant	.79
748	Shawnee	.79
749	Poteau	.85
OREGON		
970-972	Portland	1.08
973	Salem	1.06
974	Eugene	1.05
975	Medford	1.05
976	Klamath Falls	1.05
977	Bend	1.06
978	Pendleton	1.03
979	Vale	.98
PENNSYLVANIA		
150-152	Pittsburgh	1.04
153	Washington	1.02
154	Uniontown	1.01
155	Bedford	1.03
156	Greensburg	1.02
157	Indiana	1.05
158	Dubois	1.04
159	Johnstown	1.04
160	Butler	1.01
161	New Castle	1.01
162	Kittanning	1.02
163	Oil City	.91
164-165	Erie	.98
166	Altoona	1.04
167	Bradford	.99
168	State College	.96
169	Wellsboro	.93
170-171	Harrisburg	.98
172	Chambersburg	.96
173-174	York	.97
175-176	Lancaster	.95
177	Williamsport	.91
178	Sunbury	.95
179	Pottsville	.95
180	Lehigh Valley	1.04
181	Allentown	1.01
182	Hazleton	.97
183	Stroudsburg	1.01
184-185	Scranton	.95
186-187	Wilkes-Barre	.93
188	Montrose	.93
189	Doylestown	.94
190-191	Philadelphia	1.13
193	Westchester	1.08
194	Norristown	1.09
195-196	Reading	.97
PUERTO RICO		
009	San Juan	.86
RHODE ISLAND		
028	Newport	1.02
029	Providence	1.02
SOUTH CAROLINA		
290-292	Columbia	.72

STATE	CITY	Residential
293	Spartanburg	.71
294	Charleston	.73
295	Florence	.71
296	Greenville	.70
297	Rock Hill	.64
298	Aiken	.80
299	Beaufort	.68
SOUTH DAKOTA		
570-571	Sioux Falls	.88
572	Watertown	.84
573	Mitchell	.83
574	Aberdeen	.84
575	Pierre	.84
576	Mobridge	.84
577	Rapid City	.85
TENNESSEE		
370-372	Nashville	.86
373-374	Chattanooga	.82
375,380-381	Memphis	.84
376	Johnson City	.80
377-379	Knoxville	.80
382	Mckenzie	.69
383	Jackson	.68
384	Columbia	.76
385	Cookeville	.68
TEXAS		
750	Mckinney	.88
751	Waxahackie	.82
752-753	Dallas	.89
754	Greenville	.78
755	Texarkana	.87
756	Longview	.84
757	Tyler	.91
758	Palestine	.72
759	Lufkin	.76
760-761	Fort Worth	.83
762	Denton	.87
763	Wichita Falls	.80
764	Eastland	.73
765	Temple	.77
766-767	Waco	.81
768	Brownwood	.72
769	San Angelo	.79
770-772	Houston	.87
773	Huntsville	.73
774	Wharton	.75
775	Galveston	.86
776-777	Beaumont	.82
778	Bryan	.81
779	Victoria	.78
780	Laredo	.76
781-782	San Antonio	.82
783-784	Corpus Christi	.80
785	Mc Allen	.78
786-787	Austin	.78
788	Del Rio	.68
789	Giddings	.72
790-791	Amarillo	.81
792	Childress	.75
793-794	Lubbock	.78
795-796	Abilene	.79
797	Midland	.78
798-799,885	El Paso	.79

Adjusting Project Costs to Your Location

STATE	CITY	Residential
UTAH		
840-841	Salt Lake City	.90
842,844	Ogden	.90
843	Logan	.91
845	Price	.81
846-847	Provo	.90
VERMONT		
050	White River Jct.	.73
051	Bellows Falls	.73
052	Bennington	.72
053	Brattleboro	.74
054	Burlington	.85
056	Montpelier	.84
057	Rutland	.87
058	St. Johnsbury	.74
059	Guildhall	.73
VIRGINIA		
220-221	Fairfax	.89
222	Arlington	.89
223	Alexandria	.89
224-225	Fredericksburg	.83
226	Winchester	.78
227	Culpeper	.78
228	Harrisonburg	.75
229	Charlottesville	.83
230-232	Richmond	.86
233-235	Norfolk	.82
236	Newport News	.82
237	Portsmouth	.81
238	Petersburg	.86
239	Farmville	.74
240-241	Roanoke	.76
242	Bristol	.79
243	Pulaski	.73
244	Staunton	.76
245	Lynchburg	.80
246	Grundy	.72
WASHINGTON		
980-981,987	Seattle	1.00
982	Everett	.97
983-984	Tacoma	1.05
985	Olympia	1.05
986	Vancouver	1.11
988	Wenatchee	.94
989	Yakima	1.02
990-992	Spokane	.99
993	Richland	1.00
994	Clarkston	.99
WEST VIRGINIA		
247-248	Bluefield	.89
249	Lewisburg	.90
250-253	Charleston	.93
254	Martinsburg	.75
255-257	Huntington	.93
258-259	Beckley	.90
260	Wheeling	.93
261	Parkersburg	.92
262	Buckhannon	.97
263-264	Clarksburg	.97
265	Morgantown	.97
266	Gassaway	.93
267	Romney	.90
268	Petersburg	.95

STATE	CITY	Residential
WISCONSIN		
530,532	Milwaukee	1.02
531	Kenosha	1.02
534	Racine	1.06
535	Beloit	1.00
537	Madison	1.00
538	Lancaster	.91
539	Portage	.98
540	New Richmond	1.03
541-543	Green Bay	1.00
544	Wausau	.98
545	Rhinelander	.98
546	La Crosse	.98
547	Eau Claire	1.04
548	Superior	1.03
549	Oshkosh	.97
WYOMING		
820	Cheyenne	.86
821	Yellowstone Nat. Pk.	.80
822	Wheatland	.83
823	Rawlins	.81
824	Worland	.78
825	Riverton	.81
826	Casper	.86
827	Newcastle	.79
828	Sheridan	.83
829-831	Rock Springs	.83
CANADIAN FACTORS (reflect Canadian currency)		
ALBERTA		
	Calgary	.99
	Edmonton	.99
BRITISH COLUMBIA		
	Vancouver	1.05
	Victoria	1.04
MANITOBA		
	Winnipeg	.97
NEW BRUNSWICK		
	Moncton	.92
	Saint John	.96
NEWFOUNDLAND		
	St. John's	.94
NOVA SCOTIA		
	Halifax	.96
ONTARIO		
	Hamilton	1.12
	Kitchener	1.05
	London	1.08
	Oshawa	1.09
	Ottawa	1.09
	Sudbury	1.04
	Thunder Bay	1.05
	Toronto	1.12
	Windsor	1.06
PRINCE EDWARD ISLAND		
	Charlottetown	.92

Adjusting Project Costs to Your Location

STATE	CITY	Residential
QUEBEC		
	Chicoutimi	1.02
	Montreal	1.08
	Quebec	1.10
SASKATCHEWAN		
	Regina	.92
	Saskatoon	.91

GLOSSARY

A

Access door or panel

A means of access for the inspection, repair, or service of concealed systems, such as air conditioning equipment.

Anchor bolt, foundation bolt, hold-down bolt

A threaded bolt, usually embedded in a foundation, for securing a sill, framework, or machinery.

Apron

(1) A piece of finished trim placed under a window stool. (2) A slab of concrete extending beyond the entrance to a building, particularly at an entrance for vehicular traffic.

Architectural millwork, custom millwork

Millwork manufactured to meet specifications of a particular job, as distinguished from stock millwork.

Asphalt shingles, composition shingles, strip slates

Roofing felt, saturated with asphalt and coated on the weather side with a harder asphalt and aggregate particles, which has been cut into shingles for application to a sloped roof.

B

Backer

Three studs nailed together in a U-shape, to which a partition is attached.

Baluster, banister

One of a series of short, vertical supporting elements for a handrail or a coping.

Base flashing

In roofing, the flashing supplied by the upturned edges of a watertight membrane.

Batt insulation

Thermal- or sound-insulating material, such as fiberglass or expanded shale, which has been fashioned into a flexible, blanket-like form. It often has a vapor barrier on one side. Batt insulation is manufactured in dimensions which facilitate its installation between the studs or joists of a frame construction.

Bead

Any molding, stop, or caulking used around a glass or panel to hold it in position.

Beam

A large horizontal structure of wood or steel.

Bearing

(1) The section of a structural member, such as a beam or truss, that rests on the supports. (2) Descriptive of any wall that provides support to the floor and/or roof of a building.

Bearing wall

Any wall that supports a vertical load as well as its own weight.

Bed

(1) The mortar into which masonry units are set. (2) Sand or other aggregate on which pipe or conduit is laid in a trench.

B-labeled door

A door carrying a certification from Underwriters' Laboratories that it is of a construction that will pass the standard fire door test for the length of time required for a Class B opening.

Blocking

(1) Small pieces of wood used to secure, join, or reinforce members, or to fill spaces between members. (2) Small wood blocks used for shimming.

Blueprint

Negative image reproduction having white lines on a blue background and made either from an original or from a positive intermediate print.

Board and batten

A method of siding in which the joints between vertically placed boards or plywood are covered by narrow strips of wood.

Board foot

The basic unit of measure for lumber. One board foot is equal to a 1″ board 1′ in width and 1′ in length.

Bonding agent

A substance applied to a suitable substrate to create a bond between it and a succeeding layer.

Box nail

Nail similar to a common nail, but with a smaller diameter shank.

Box sill

A common method of frame construction using a header nailed across the ends of floor joists where they rest on the sill.

Brace

(1) A diagonal tie that interconnects scaffold members. (2) A temporary support for aligning vertical concrete formwork. (3) A horizontal or inclined member used to hold sheeting in place. (4) A hand tool with a handle, crank, and chuck used for turning a bit or auger.

Brick

A solid masonry unit of clay or shale, formed into a rectangular prism while plastic, and then burned or fired in a kiln.

Building paper

A heavy, asphalt-impregnated paper used as a lining and/or vapor barrier between sheathing and an outside wall covering, or as a lining between rough and finish flooring.

Butt hinge

A common form of hinge consisting of two plates, each with one meshing knuckle edge and connected by means of a removable or fixed pin through the knuckles.

Butt joint

(1) A square joint between two members at right angles to each other. The contact surface of the outstanding member is cut square and is flush to the surface of the other member. (2) A joint in which the ends of two members butt each other so only tensile or compressive loads are transferred.

C

Cantilever

Any part of a structure that projects beyond its main support and is balanced on it.

Casement window

A window assembly having at least one casement or vertically hinged sash.

Casing

A piece of wood or metal trim that finishes off the frame of a door or window.

Caulk

(1) To fill a joint, crack, or opening with a sealer material. (2) The filling of joints in bell-and-spigot pipe with lead and oakum.

Cement

Any chemical binder that makes bodies adhere to it or to each other, such as glue, paste, or Portland cement.

Class A, B, C, D, E,

Fire-resistance ratings applied to building components such as doors or windows. The term "class" also refers to the opening into which the door or window will be fitted.

Common brick

Brick not selected for color or texture, and thus useful as filler or backing. Though usually not less durable or of lower quality than face brick, common brick typically costs less. Greater dimensional variations are also permitted.

Common nail

Nail used in framing and rough carpentry having a flat head about twice the diameter of its shank.

Concrete

A composite material that consists essentially of a binding medium within which are embedded particles or fragments of aggregate.

Cornice

The horizontal projection of a roof overhang at the eaves, consisting of lookout, soffit, and fascia.

Crown

The high point of a piece of lumber with a curve in it.

D

d

Abbreviation for penny. Refers only to nail size.

Dampproofing

An application of a water-resisting treatment or material to the surface of a concrete or masonry wall to prevent passage or absorption of water or moisture.

Deck

(1) An uncovered wood platform usually attached to or on the roof of a structure. (2) The flooring of a building. (3) The structural assembly to which a roof covering is applied.

Dressed lumber

Lumber that has been processed through a planing machine for the purpose of attaining a smooth surface and uniformity of size on at least one side or edge.

Drip edge

The edge of a roof that drips into a gutter or into the open.

Drywall

The term commonly applied to interior finish construction using preformed sheets, such as gypsum wallboard, as opposed to using plaster.

E

Eave

The part of a roof that projects beyond its supporting walls.

Estimate

The anticipated cost of materials, labor, services, or any combination of these for a proposed construction project.

F

Fascia, facia

(1) A board used on the outside vertical face of a cornice. (2) The board connecting the top of the siding with the bottom of a soffit. (3) A board nailed across the ends of the rafters at the eaves. (4) The edge beam of a bridge. (5) A flat member or band at the surface of a building.

Finish flooring

The material used to make the wearing surface of a floor, such as hardwood, tile, or terrazzo.

Flashing

A thin, impervious sheet of material placed in construction to prevent water penetration or direct the flow of water. Flashing is used especially at roof hips and valleys, roof penetrations, joints between a roof and a vertical wall, and in masonry walls to direct the flow of water and moisture.

Foundation

The material or materials through which the load of a structure is transmitted to the earth.

Frame

The wood skeleton of a building. Also called "framing."

Framing

(1) Structural timbers assembled into a given construction system. (2) Any construction work involving and incorporating a frame, as around a window or door opening. (3) The unfinished structure, or underlying rough timbers of a building, including walls, roofs, and floors.

Furring

(1) Strips of wood or metal fastened to a wall or other surface to even it, to form an air space, to give appearance of greater thickness, or for the application of an interior finish such as plaster. (2) Lumber 1″ in thickness (nominal) and less than 4″ in width, frequently the product of resawing a wider piece.

G

Gable

The portion of the end of a building that extends from the eaves upward to the peak or ridge of the roof.

Gable roof

A roof shape characterized by two sections of roof of constant slope that meet at a ridge; peaked roof.

Gambrel roof

A roof whose slope on each side is interrupted by an obtuse angle that forms two pitches on each side, the lower slope being steeper than the upper.

Girder

A large principal beam of steel, reinforced concrete, wood, or combination of these, used to support other structural members at isolated points along its length.

Glue laminated, glu-lam

The result of a process in which individual pieces of lumber or veneer are bonded together with adhesives to make a single piece in which the grain of all the constituent pieces is parallel.

Grade

(1) A designation of quality, especially of lumber and plywood. (2) Ground level. (3) The slope of the ground on a building site.

Grain

The direction of fibers in wood.

Ground

(1) A strip of wood that is fixed in a wall of concrete or masonry to provide a place for attaching wood trim or furring strips. (2) A screed, strip of wood, or bead of metal fastened around an opening in a wall and acting as a thickness guide for plastering or as a fastener for trim.

Grout

(1) An hydrous mortar whose consistency allows it to be placed or pumped into small joints or cavities, as between pieces of ceramic clay, slate, and floor tile. (2) Various mortar mixes used in foundation work to fill voids in soils, usually through successive injections through drilled holes.

Gutter

(1) A shallow channel of wood, metal, or PVC positioned just below and following along the eaves of a building for the purpose of collecting and diverting water from a roof. (2) In electrical wiring, the rectangular space allowed around the interior of an electrical panel for the installation of feeder and branch wiring conductors.

Gypsum

A naturally occurring, soft, whitish mineral (hydrous calcium sulfate) that, after processing, is used as a retarding agent in Portland cement and as the primary ingredient in plaster, gypsum board, and related products.

H

Header

(1) A rectangular masonry unit laid across the thickness of a wall, so as to expose its end(s). (2) A lintel. (3) A member extending horizontally between two joists to support tailpieces. (4) In piping, a chamber, pipe, or conduit having several openings through which it collects or distributes material from other pipes or conduits. (5) The wood surrounding an area of asphaltic concrete paving.

Hip

(1) The exterior inclining angle created by the junction of the sides of adjacent sloping roofs, excluding the ridge angle. (2) The rafter at this angle. (3) In a truss, the joint at which the upper chord meets an inclined end post.

Hip roof

A roof shape characterized by four or more sections of constant slope, all of which run from a uniform eave height to the ridge.

J

Jamb

An exposed upright member on each side of a window frame, door frame, or door lining. In a window, these jambs outside the frame are called "reveals."

Joint

(1) The point, area, position, or condition at which two or more things are jointed. (2) The space, however small, where two surfaces meet. (3) The mortar-filled space between adjacent masonry units. (4) The place where seperate but adjacent timbers are

connected, as by nails or screws, or by mortises and tenons, glue, etc.

Joist

A piece of lumber 2″ or 4″ thick and 6″ or more wide, used horizontally as a support for a ceiling or floor. Also, such a support made from steel, aluminum, or other material.

K

Keyway

A recess or groove in one lift or placement of concrete that is filled with concrete of the next lift, giving shear strength to the joint. Also called a "key."

Kneewall

A short wall under a slope, usually in attic space.

L

Labor-hour

A unit describing the work performed by one person in one hour.

Lally column

A trade name for a pipe column 3″ to 6″ in diameter, sometimes filled with concrete.

Layout

A design scheme or plan showing the proposed arrangement of objects and spaces within and outside a structure.

Level

(1) A term used to describe any horizontal surface that has all points at the same elevation and thus does not tilt or slope. (2) In surveying, an instrument that measures heights from an established reference. (3) A spirit level, consisting of small tubes of liquid with bubbles in each. The small tubes are positioned in a length of wood or metal that is hand held and, by observing the position of the bubbles, used to find and check level surfaces.

Lintel

A horizontal supporting member, installed above an opening such as a window or a door, that serves to carry the weight of the wall above it.

M

Main beam

A structural beam that transmits its load directly to columns, rather than to another beam.

Mansard roof

A type of roof with two slopes on each of four sides, the lower slope much steeper than the upper and ending at a constant eave height.

Millwork

All the building products made of wood that are produced in a planing mill such as moldings, door and window frames, doors, windows, blinds, and stairs. Millwork does not include flooring, ceilings, and siding.

Mortar

A plastic mixture used in masonry construction that can be troweled and hardens in place.

N

Nominal dimension

The size designation for most lumber, plywood, and other panel products.

Nominal size

The rounded-off, simplified dimensional name given to lumber.

Nonbearing wall

A dividing wall that supports none of the structure above it.

Nosing

The rounded front edge of a stair tread that extends over the riser.

O

On center (O.C.)

Layout spacing designation that refers to distance from the center of one framing member to another.

Outlet

The point in an electrical wiring circuit at which the current is supplied to an appliance or device.

P

Paint

(1) A mixture of a solid pigment in a liquid vehicle that dries to a protective and decorative coating. (2) The resultant dry coating.

Panel door

A door constructed with panels, usually shaped to a pattern, installed between the stiles and rails that form the outside frame of the door.

Paneling

The material used to cover an interior wall. Paneling may be made from a 4/4 sheet milled to a pattern and may be either hardwood or softwood plywood, often prefinished or overlaid with a decorative finish, or hardboard, also usually prefinished.

Particle board

A generic term used to describe panel products made from discrete particles of wood or other ligno-cellulosic material rather than from fibers. The wood particles are mixed with resins and formed into a solid board under heat and pressure.

Partition

An interior wall that divides a building into rooms or areas, usually nonload-bearing.

Pitch

The angle or inclination of a roof, which varies according to the climate and roofing materials used.

Plaster

A cementitious material or combination of cementitious material and aggregate that, when mixed with a suitable amount of water, forms a plastic mass or paste.

Plumb

Straight up and down, perfectly vertical.

Plumbing system

The water supply and distribution pipes; plumbing fixtures and traps; soil, waste, and vent pipes; building drains and sewers; and respective devices and appurtenances within a building.

Plywood

A flat panel made up of a number of thin sheets, or veneers, of wood, in which the grain direction of each ply, or layer, is at right angles to the one adjacent to it. The veneer sheets are united under pressure by a bonding agent.

Polystyrene foam

A low-cost, foamed plastic weighing about 1 lb. per cu. ft., with good insulating properties; it is resistant to grease.

Polyurethane

Reaction product of an isocyanate with any of a wide variety of other compounds containing an active hydrogen group. Polyurethane is used to formulate tough, abrasion-resistant coatings.

Polyvinyl chloride (PVC)

A synthetic resin prepared by the polymerization of vinyl chloride, used in the manufacture of nonmetallic waterstops for concrete, floor coverings, pipe and fittings.

Post-and-beam framing

Framing in which the horizontal members are supported by a distinct column, as opposed to a wall.

Prehung door

A packaged unit consisting of a finished door on a frame with all necessary hardware and trim.

Purlin

In roofs, horizontal member supporting the common rafters.

Q

Quotation

A price quoted by a contractor, subcontractor, material supplier, or vendor to furnish materials, labor, or both.

R

Rail

A horizontal member supported by vertical posts.

Resilient flooring

A durable floor covering that has the ability to resume its original shape, such as linoleum.

Retaining wall

(1) A structure used to sustain the pressure of the earth behind it. (2) Any wall subjected to lateral pressure other than wind pressure.

Rib

One of a number of parallel structural members backing sheathing.

Rise and run

The angle of inclination or slope of a member or structure, expressed as the ratio of the vertical rise to the horizontal run.

Riser

A vertical member between two stair treads.

Rough flooring

Any materials used to construct an unfinished floor.

Rough opening (R.O.)

Any opening formed by the framing members to accommodate doors or windows.

Run

(1) In a roof with a ridge, the horizontal distance between the edge of the rafter plate (building line) and the center line of the ridgeboard. (2) In a stairway, the horizontal distance between the top and bottom risers plus the width of one tread.

"R" value

A measure of a material's resistance to heat flow given a thickness of material. The term is the reciprocal of the "U" value. The higher the "R" value, the more effective the particular insulation.

S

Sheathing

(1) The material forming the contact face of forms. Also called "lagging" or "sheeting."(2) Plywood, waferboard, oriented strand board, or lumber used to close up side walls, floors, or roofs preparatory to the installation of finish materials on the surface.

Shed dormer

A dormer window having vertical framing projecting from a sloping roof, and an eave line parallel to the eave line of the principal roof.

Shim

A thin piece of material, often tapered (such as a wood shingle) inserted between building materials for the purpose of straightening or making their surfaces flush at a joint.

Shingle

A roof-covering unit made of asphalt, wood, slate, cement, or other material cut into stock sizes and applied on sloping roofs in an overlapping pattern.

Shoe

Any piece of timber, metal, or stone receiving the lower end of virtually any member.

Sill

(1) The horizontal member forming the bottom of a window or exterior door frame. (2) As applied to general construction, the lowest member of the frame of the structure, resting on the foundation and supporting the frame.

Sized lumber

Lumber uniformly manufactured to net surfaced sizes. Sized lumber may be rough, surfaced, or partly surfaced on one or more faces.

Slab

A flat, horizontal (or nearly so) molded layer of plain or reinforced concrete, usually of uniform but sometimes of variable thickness, either on the ground or supported by beams, columns, walls, or other framework.

Sleeper

Lumber laid on a concrete floor as a nailing base for wood flooring.

Slope

The pitch of a roof, expressed as inches of rise per 12" of run.

Soffit

The underside of a projection, such as a cornice.

Span

(1) The distance between supports of a member. (2) The measure of distance between two supporting members.

Spread footing

A generally rectangular prism of concrete, larger in lateral dimensions than the column or wall it supports, that distributes the load of a column or wall to the subgrade.

Stair

(1) A single step. (2) A series of steps or flights of steps connected by landings, used for passage from one level to another.

Stepped footing

A wall footing with horizontal steps to accommodate a sloping grade or bearing stratum.

Stock lumber

Lumber cut to standard sizes and readily available from suppliers.

Strapping

(1) Flexible metal bands used to bind units for ease of handling and storage. (2) Another name for *furring*.

Stringer

(1) A secondary flexural member parallel to the longitudinal axis of a bridge or other structure. (2) A horizontal timber used to support joists or other cross members.

Structural lumber

Any lumber with nominal dimensions of 2″ or more in thickness and 4″ or more in width that is intended for use where working stresses are required.

Stud

(1) A vertical member of appropriate size (2 x 4 to 4 x 10 in. or 50 x 100 to 100 x 250 mm) and spacing (16″ to 30″ or 400 to 750 mm) to support sheathing or concrete forms. (2) A framing member, usually cut to a precise length at the mill, designed to be used in framing building walls with little or no trimming before it is set in place.

Subfloor sheathing

The rough floor, usually plywood, laid across floor joists and under finish flooring.

T

Thermal barrier

An element of low conductivity placed between two conductive materials to limit heat flow, for use in metal windows or curtain walls that are to be used in cold climates.

Threshold

A shaped strip on the floor between the jambs of a door, used to separate different types of flooring or to provide weather protection at an exterior door.

Toe

(1) Any projection from the base of a construction or object to give it increased bearing and stability. (2) That part of the base of a retaining wall that projects beyond the face, away from the retained material.

Toenailing

To drive a nail at an angle to join two pieces of wood.

Tongue and groove

Lumber machined to have a groove on one side and a protruding tongue on the other so that pieces will fit snugly together, with the tongue of one fitting into the groove of the other.

Top plate

A member on top of a stud wall on which joists rest to support an additional floor or form a ceiling.

Truss

A structural component composed of a combination of members, usually in a triangular arrangement, to form a rigid framework; often used to support a roof.

V

Vapor barrier

Material used to prevent the passage of vapor or moisture into a structure or another material, thus preventing condensation within them.

Veneer

(1) A masonry facing that is attached to the backup but not so bonded as to act with it under load. (2) Wood peeled, sawn, or sliced into sheets of a given constant thickness and combined with glue to produce plywood.

Ventilation

A natural or mechanical process by which air is introduced to or removed from a space, with or without heating, cooling, or purification treatment.

W

Waterstop

A thin sheet of metal, rubber, plastic, or other material inserted across a joint to obstruct the seeping of water through the joint.

Wood preservative

Any chemical preservative for wood, applied by washing on or pressure-impregnating. Products used include creosote, sodium fluoride, copper sulfate, and tar or pitch.

Z

Z-bar

A Z-shaped member that is used as a main runner in some types of acoustical ceiling.

Zoning permit

A permit issued by appropriate government officials authorizing land to be used for a specific purpose.

ABBREVIATIONS

A	Area	kW	Kilowatt
Ab	Above	L or Ldr	Leader
Abs	Absolute	L or Lth	Length
AFF	Above finished floor	Lav	Lavatory
Al	Aluminum	Lb	Pound
Avg	Average	LF	Linear feet
BCF	Backfill	LH	Labor-Hours
BOCA	Building Officials and Code Administrators	Mat	Material
		Max	Maximum
Br	Branch	Mfr	Manufacturer
BTU	British thermal unit	Min	Minimum (or minute)
C	Centigrade	MS	Milled steel
°C	Degrees centigrade	NTS	Not to scale
C to C	Center to Center	Oz	Ounce
CA	Compressed air	P&T	Pressure and temperature
CF	Cubic Feet	Pb	Lead
Cfm	Cubic feet per minute	PG	Pressure gauge
CI	Cubic Inches	pH	Hydrogen concentration
Circ.	Circulator/Circulation	PIV	Post indicator valve
CL el	Centerline elevation	PO	Plugged outlet
Clg	Ceiling	Ppm	Parts per million
CTE	Connection to existing	Press	Pressure
Cu	Copper	PSI	Pounds per square inch
CW	Cold water	PVC	Polyvinylchloride
CY	Cubic yard	Qt	Quart
D	Drain	Qty	Quantity
Deg or °	Degrees	R	Hydraulic radius
Dn	Down	Rad	Radius
Dp	Deep	RCP	Reinforced concrete pipe
Dwg	Drawing	RD	Rate of demand (or roof drain)
Elev.	Elevation		
Exc	Excavation	Red	Reducer
F	Fahrenheit	RT	Running trap
°F	Degrees fahrenheit	RV	Relief valve
FAI	Fresh air intake	S	Soil
FF	Finish floor	S&W	Soil and waste
FG	Finish grade	SA	Shock absorber
Fig	Figure	San	Sanitary
Fixt	Fixture	Sb	Antimony
Flr	Floor	Sc	Sillcock
Ga	Gauge	Sec	Second
Gal	Gallon (231 CI)	SF	Square foot
Galv.	Galvanized	Shwr	Shower
Gas	Gallons	SI	Square inches
H	Hydrogen or handicapped	Spec	Specification
HClg	Hung ceiling	Std	Standard
Hd	Head	T	Temperature (or time)
HP	Horsepower	Therm	Thermometer
Hr	Hour	V	Vent
HT	House trap	Vac	Vacuum
Htr	Heater	Vel	Velocity
HW	Hot water	Vol	Volume
In	Inch	Wgt	Weight
Jt	Joint		

INDEX

REFERENCE BOOKS

Interior Home Improvement Costs 8th Edition

Estimates for today's most-wanted interior projects. Over 66 remodeling jobs, with new coverage of a home entertainment center, home offices, in-law apartment conversions, and remodeling for handicapped residents. Includes:

- Kitchen/Bath Remodeling
- Attic, Basement, and Garage Conversions
- Painting, Wallcovering, and Flooring
- Fireplaces
- Plumbing and Electrical
- Home Offices
- Stairs and Doors
- Walls and Ceilings

$19.95 per copy
Over 250 pages, illustrated, softcover
Catalog No. 67308D   ISBN 0-87629-656-8

Means Illustrated Construction Dictionary 3rd Edition

Your comprehensive guide to understanding the words, terms, phrases, concepts, slang, abbreviations, symbols, and acronyms of today's construction. Includes 19,000 construction words, terms, phrases, symbols, weights, measures, and equivalents.

Long regarded as the Industry's finest, the *Means Illustrated Construction Dictionary* is now even better. With the addition of over 1000 new terms and hundreds of new illustrations, it is the clear choice for the most comprehensive and current information.

The companion CD-ROM that comes with this new edition adds many extra features: larger graphics, expanded definitions, and links to both CSI MasterFormat numbers and production information.

$99.95 per copy
Over 800 pages, illustrated, hardcover
Catalog No. 67292A   ISBN 0-87629-538-3

Residential & Light Commercial Construction Standards

A Unique Collection of Industry Standards That Define Quality in Construction—For Contractors & Subcontractors, Owners, Developers, Architects & Engineers, Attorneys & Insurance Personnel

Compiled from the nation's major building codes, and from scores of publications and reports from professional institutes and other authorities, this one-of-a-kind resource enables you to:

- Set a standard for subcontractors and employees
- Protect yourself against defect claims
- Resolve disputes
- Overview installation methods
- Answer client questions

$59.95 per copy
Over 500 pages, illustrated, softcover
Catalog No. 67322   ISBN 0-87629-499-9

Builder's Essentials
Plan Reading & Material Takeoff

A complete course in reading and interpreting building plans—and performing quantity takeoffs to professional standards.

Organized by CSI division, this book shows and explains, in clear language and with over 160 illustrations, typical working drawings encountered by contractors in residential and light commercial construction. The author describes not only how all common features are represented, but how to translate that information into a materials list. Each chapter uses plans, details and tables, and a summary checklist.

$35.95 per copy
Over 420 pages, illustrated, softcover
Catalog No. 67307   ISBN 0-87629-348-8

Contractor's Pricing Guide:
Residential Detailed Costs 2002

Every aspect of residential construction, from overhead costs to residential lighting and wiring, is in here. All the detail you need to accurately estimate the costs of your work with or without markups— labor-hours required, typical crews and equipment are included as well. When you need a detailed estimate, this publication has all the costs to help you come up with a complete, on-the-money price you can rely on to win profitable work.

$36.95 per copy
Over 300 pages, with charts and tables, 8-1/2 x 11
Catalog No. 60332   ISBN 0-87629-648-7

Contractor's Pricing Guide:
Residential Repair & Remodeling Costs 2002

This book provides total unit price costs for every aspect of the most common repair & remodeling projects. Organized in the order of construction by component and activity, it includes demolition and installation, cleaning, painting, and more.

With simplified estimating methods; clear, concise descriptions; and technical specifications for each component, the book is a valuable tool for contractors who want to speed up their estimating time, while making sure their costs are on target.

$36.95 per copy
Over 200 pages, illustrated, 8-1/2 x 11
Catalog No. 60342   ISBN 0-87629-655-X

MATERIAL SHOPPING LIST

Use this form to develop a shopping list of materials needed to complete your home improvement project. Materials have been grouped under headings that will make it easier for you to find items at your lumberyard or home center. Use the blank lines under each category for other items you may select.

The shaded section of the form is provided to help you track multiple prices, if you do comparison shopping. Simply enter the name of each store at the top of the column, and put the prices obtained from each one in the spaces provided. Circle the best price for each item.

This form can become a Project Estimate by performing the following steps:

1. Enter the quantity and the cost of each item in the appropriate spaces in the **Project Estimate** section. Multiply **Quantity** times **Cost** to determine the **Total Cost** for each item.

2. Add all the numbers in the Total Cost column *on each page* to determine **Page Totals**. Enter each Page Total in the space provided at the bottom of the page.

3. On the last page, add all Page Totals to determine the **Project Subtotal**.

4. Multiply the Project Subtotal by your local sales tax (if necessary) to determine the **Project Total**. (For example, if the sales tax is 5%, multiply the Project Subtotal by 1.05)

ITEM DESCRIPTION	Unit	PRICE COMPARISON			PROJECT ESTIMATE		
		Store 1	Store 2	Store 3	Quantity ×	Cost Used =	Total Cost
Building Materials							
Concrete Field Mix							
Round Fiber Tube							
Roof Shingles							
Asphalt							
Wood							
Roll Roofing							
Roofing Cement							
Cement Board							
Roof Accessories							
Flashing							
Drip Edge							
Soffit Vent							
Ridge Vent							
Gutter							
Downspout							
Elbows							
Siding							
Vinyl							
Cedar Bevel							

PAGE TOTAL _____

ITEM DESCRIPTION	Unit	PRICE COMPARISON			PROJECT ESTIMATE		
		Store 1	Store 2	Store 3	Quantity ×	Cost Used =	Total Cost
1/2″ × 6″							
1/2″ × 8″							
Cedar Shingles							
Rough-sawn Cedar Boards, Random Width							
Rough-sawn Spruce, 1 × 8							
Exterior Mouldings							
Crown Moulding							
1 × 3							
Door Casings							
Window Casings							
Exterior Deck/Stair Parts							
Treads							
Railing							
Balusters							
Circular Stair – Metal							
Insulation							
Fiberglass Batt, Kraft Faced, 3-1/2″							
Fiberglass Batt, Kraft Faced, 6″							
Fiberglass Batt, Foil Faced, 6″							
Fiberglass Batt, Unfaced, 6″							
Fiberglass Batt, Unfaced, 9″							
Rigid Molded Beadboard, 1/2″							
Polyethylene Vapor Barrier							
Asphalt Dampproofing							
Asphalt Felt Building Paper							
Air Infiltration Barrier (Housewrap)							
Metal Bulkhead							
Caulking Compound and Sealants							
Masonry Materials							
Cement							
Lime							
Mortar							
Gypsum Board, 1/2″							
Lumber							
Framing Lumber							
1 × 4							
2 × 3							
2 × 4							
2 × 6							

PAGE TOTAL _____

ITEM DESCRIPTION	Unit	PRICE COMPARISON			PROJECT ESTIMATE		
		Store 1	Store 2	Store 3	Quantity ×	Cost Used =	Total Cost
2 × 8							
2 × 10							
2 × 12							
4 × 4							
Pressure Treated							
1 × 4							
2 × 4							
2 × 6							
2 × 8							
2 × 10							
2 × 12							
4 × 4							
Redwood							
2 × 4							
2 × 6							
2 × 8							
Plywood Sheathing							
1/2″ CDX							
5/8″ CDX							
3/4″ CDX							
Plywood Underlayment, 5/8″							
Finish Plywood							
1/2″ AA							
3/4″ AA							
3/4″ MDO							
Exterior Trim, Pine							
1 × 4							
1 × 6							
1 × 8							
1 × 10							
3″ Finials							
Doors & Windows							
Doors							
Sliding Door							
Vinyl Clad							
Wood							
French Door							
Prehung Entrance							
Combination Screen/Storm							
Garage—Overhead							

PAGE TOTAL _____

257

ITEM DESCRIPTION	Unit	PRICE COMPARISON			PROJECT ESTIMATE		
		Store 1	Store 2	Store 3	Quantity ×	Cost Used =	Total Cost
Door Grilles							
Windows							
Double Hung							
Wood							
Vinyl Clad							
Sliding							
Vinyl Replacement							
Storm—Aluminum							
Skylights							
Ventilating Bubble							
Fixed							
Operable							
Hardware and Fasteners							
Nails							
Screws							
Hinges							
Joist Hangers							
Timber Connectors							
Door Hardware							
Entrance Lock, Cylinder, Grip Handle							
Lockset							
Post Base, 4 × 4							
Post Cap, 4 × 4							
Plumbing							
Rough Plumbing							
Supply Pipe							
Drain Pipe							
Vent Pipe							
Traps							
Exterior Hose Bibb							
Paint							
Paint							
Stain							
Wood Preservative							
Drop Cloth							

PAGE TOTAL _____

ITEM DESCRIPTION	Unit	PRICE COMPARISON			PROJECT ESTIMATE			
		Store 1	Store 2	Store 3	Quantity	×	Cost Used	= Total Cost
Brushes								
Sandpaper								
Electrical								
Switches								
Light Switch								
Receptacles								
Duplex								
GFI								
GFI—Exterior								
Boxes								
Conduit								
Wire								
Boxes								
Fittings								
Lighting and Ceiling Fans								
Light Fixtures								
Exterior Entrance								
Post								
Spotlight								
Lamps								
Landscaping								
Landscape Timbers								
Redwood Edging								
Segmental Retaining Wall Stones								
Plastic Drainage Pipe								
Brick Pavers								
Edging Brick								
Concrete, Ready-Mix								
Sand								
Topsoil								
Mulch								
Fertilizer								
Lime								
Grass Seed								
Shrubs								

PAGE TOTAL _____

ITEM DESCRIPTION	Unit	PRICE COMPARISON			PROJECT ESTIMATE		
		Store 1	Store 2	Store 3	Quantity ×	Cost Used =	Total Cost
Bituminous Asphalt							
Pea Gravel							
Tools/Equipment							
Eye Protection							
Dust Mask							
Fire Extinguisher							
Hard Hat							
Work Boots							
Gloves							
Rubbish Barrels							
Rental Equipment							
Saw Blades							
Flat Bar, Cat's Paw							
Drill Bits							
Extension Cords							
Drop Light							
Saw Horses							

PAGE TOTAL _____

TOTAL FROM PREVIOUS PAGES _____

PROJECT SUBTOTAL _____

TAX @ _____ %

PROJECT TOTAL _____

NOTES

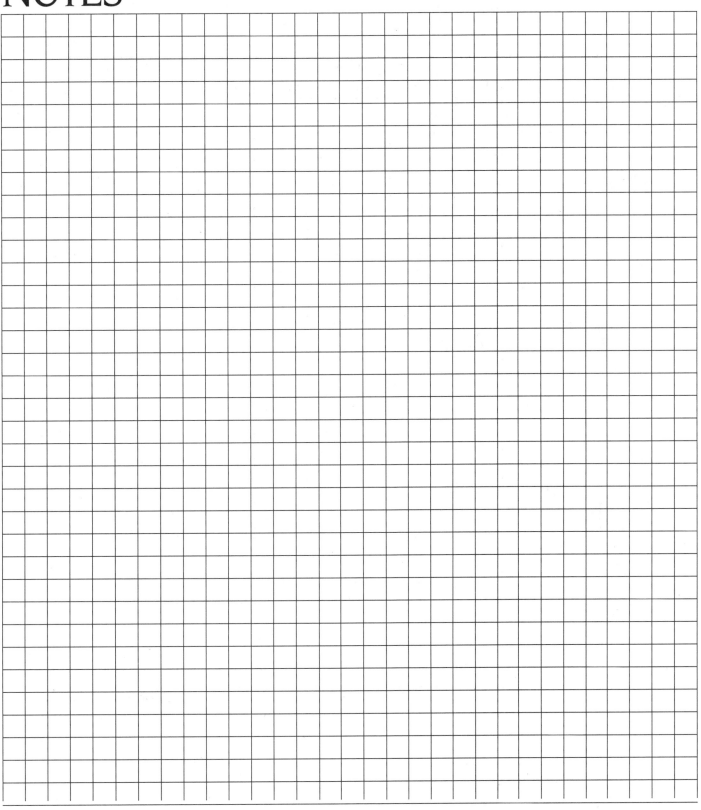

NOTES

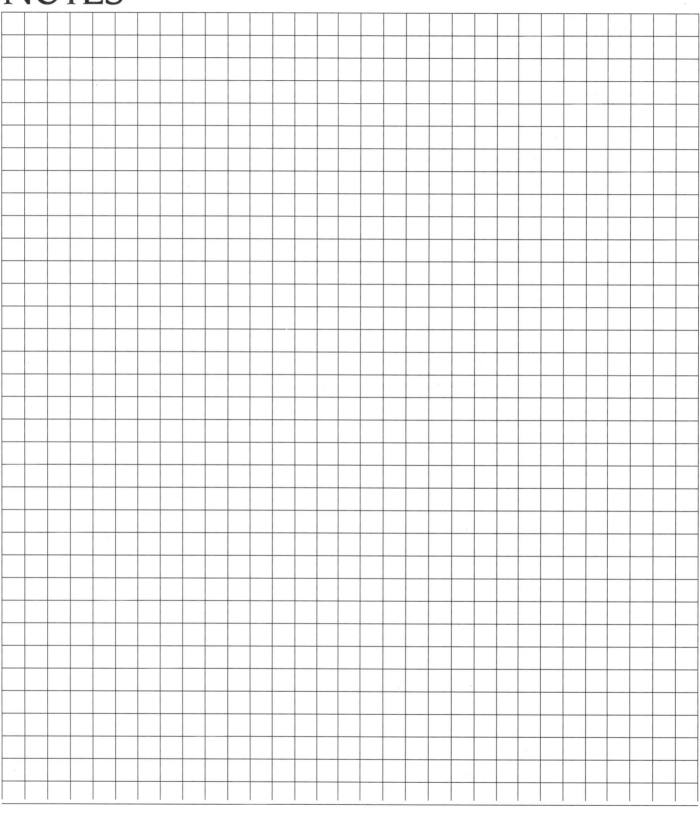

NOTES

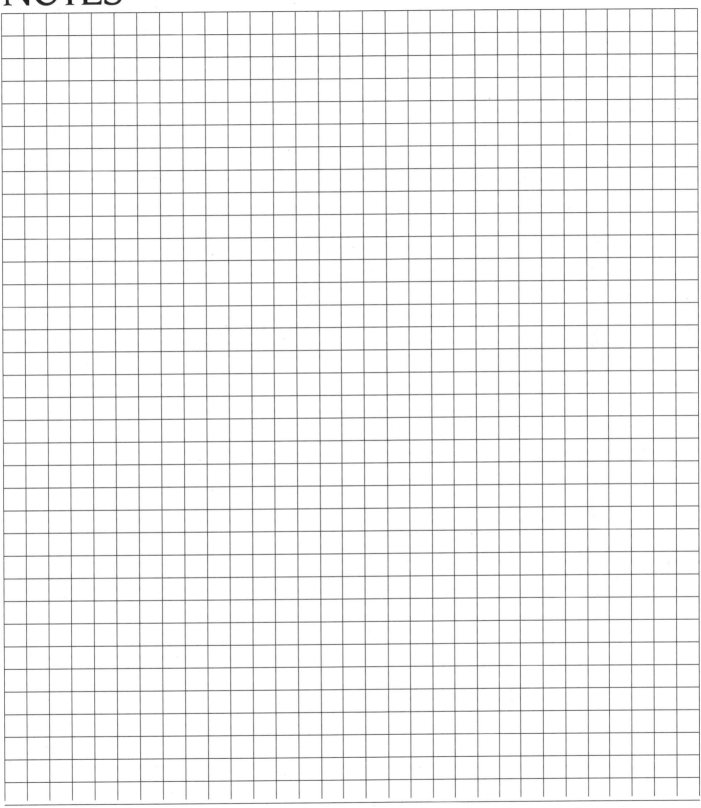

NOTES

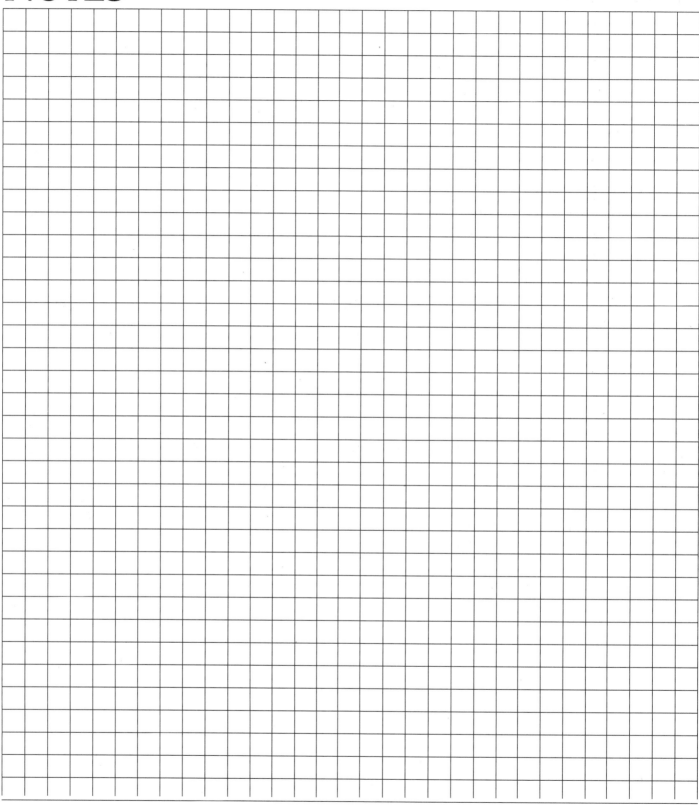

NOTES

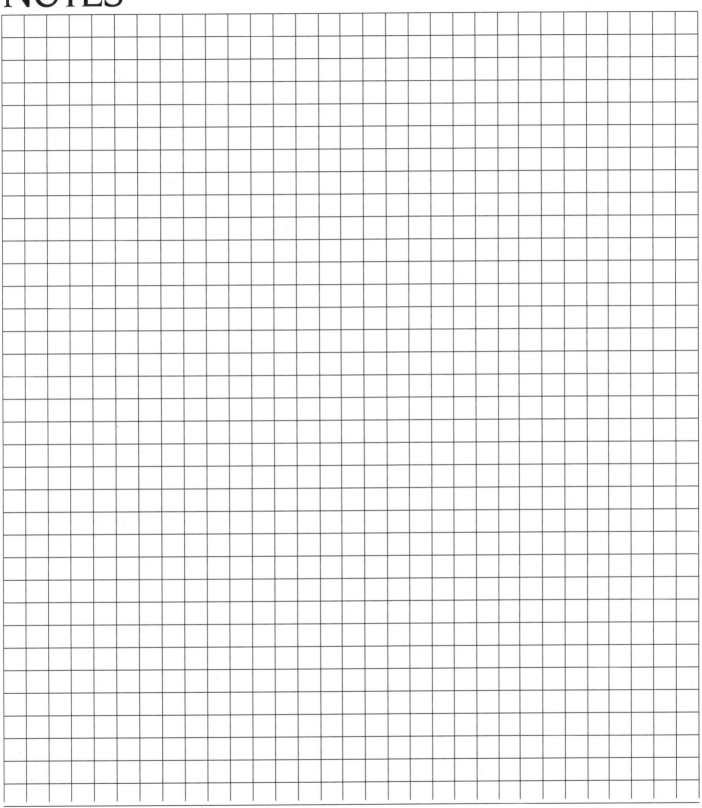

NOTES

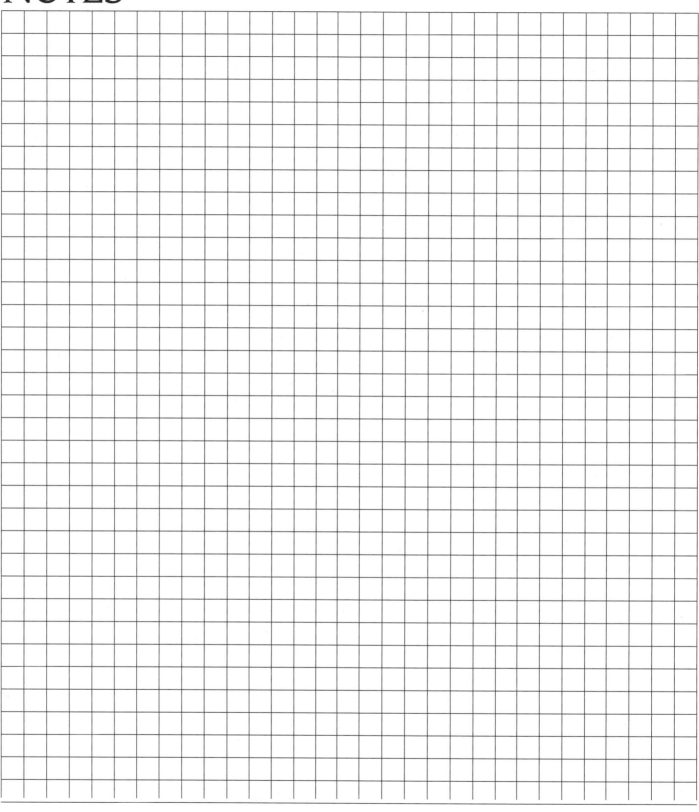

NOTES

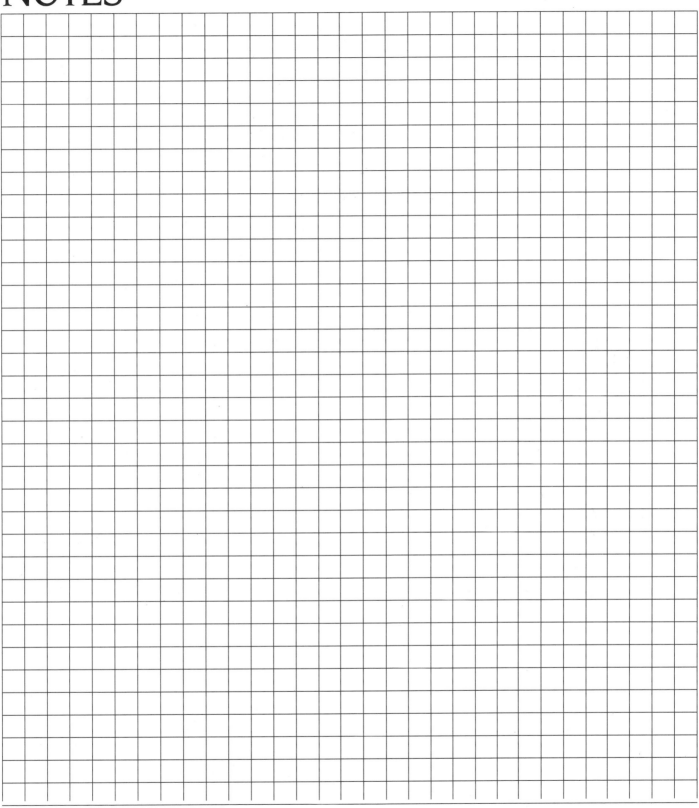

NOTES

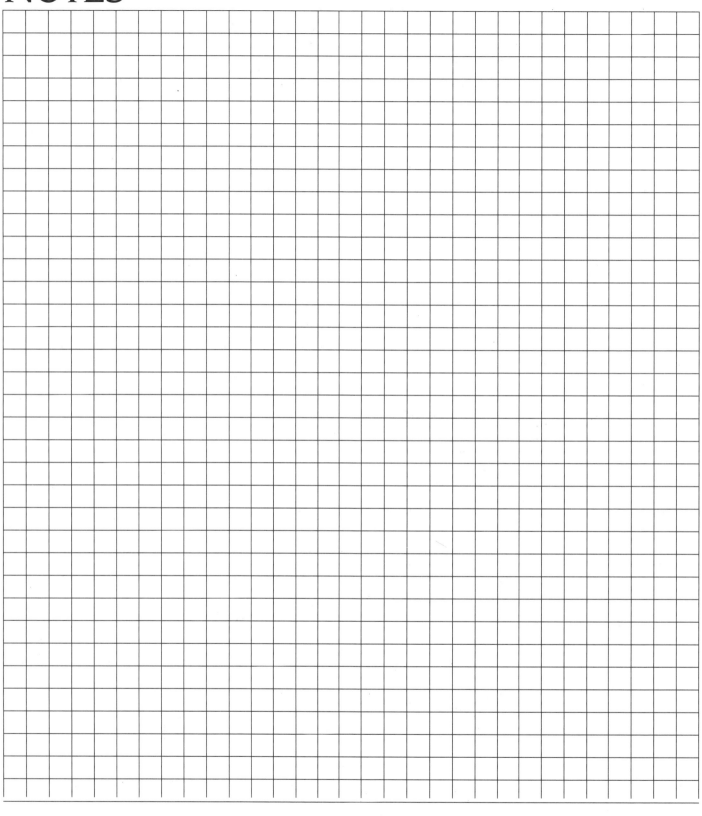

NOTES

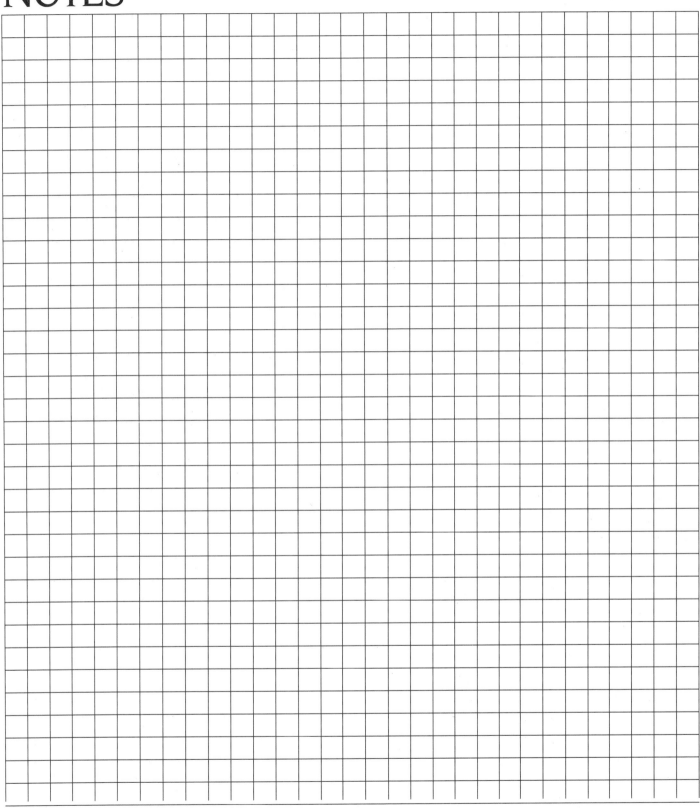

NOTES

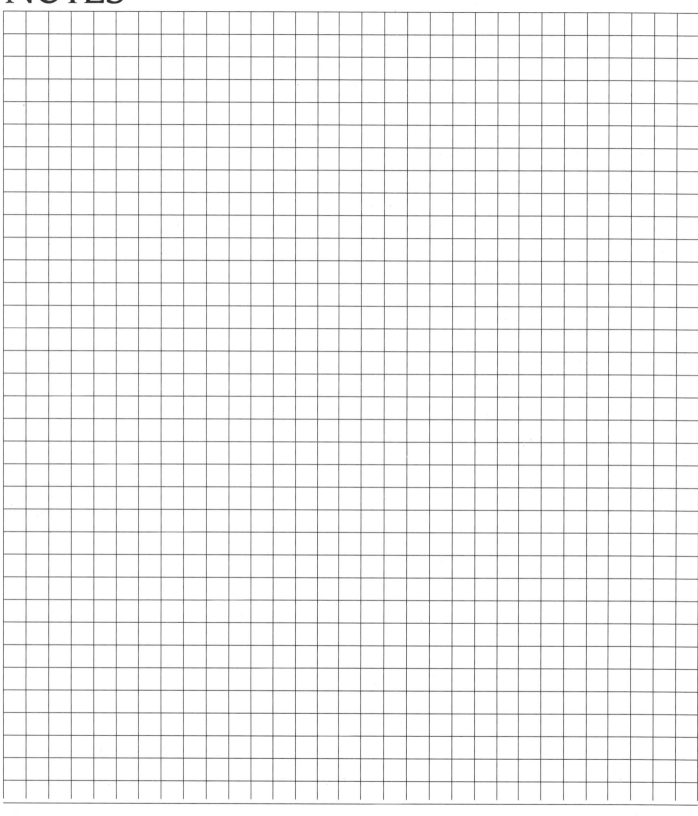